Exploring Geographic Information Systems

Exploring Geographic Information Systems

Nicholas Chrisman

University of Washington

John Wiley & Sons, Inc.

New York Chichester Brisbane Toronto Singapore Weinheim

ACQUISITIONS EDITOR Nanette Kauffman
MARKETING MANAGER Cathy Faduska
SENIOR PRODUCTION EDITOR Jeanie Berke
DESIGNER Harry Nolan
MANUFACTURING MANAGER Mark Cirillo
ILLUSTRATION EDITOR Edward Starr
COVER ILLUSTRATION David B. Bramwel

This book was set in 10/12 New Caledonia by HRS Electronic Text Management and printed and bound by RR Donnelly & Sons. The cover was printed by Phoenix Color Corp.

Library of Congress Cataloging in Publication Data:
Chrisman, Nicholas R.
Exploring geographic information systems / Nicholas Chrisman.
p. cm.
Includes bibliographical references and index.
ISBN 0-471-10842-1 (pbk. : alk. paper)
1. Geographic information systems. I. Title.
G70.212.C48 1996
025.06'91--dc21 96-46941
CIP

Printed in the United States of America
10 9 8 7 6 5 4 3 2 1

PREFACE

CALL FOR PARTICIPATION IN AN EXPLORATION

Geographic information systems have matured to serve important roles for many academic disciplines, government organizations, and commercial enterprises. The acronym GIS has become a buzzword for a technology. The technical aspects of GIS can seem overwhelming, with thousands of possible commands and even more sources of errors. Some seek understanding through learning a set of particular steps to follow. However, at the core of each system lies the geographic information; the meaning of the information provides the surest guide to understanding.

The choice of "exploring" in the title signals the nature of this book. Many textbooks (especially those on technical subjects) are written in a remote and impersonal voice that cloaks the author in anonymity and magisterial distance. I have tried to avoid that kind of writing in this book because the field of GIS is still too new to know its bounds and to have an orthodoxy accepted by all. Some of the presentation comes from my personal exploration of geographic information, but this is not a travelogue that follows the author on his personal path. The discoveries on my particular path hardly followed a logical or planned order. Instead, this book challenges the reader to participate in the exploration.

In an early period of GIS development, the aura of the computer made any automated analysis seem objective. The glow of new technology sometimes overwhelmed careful understanding of the problem. Some dubious assumptions slipped into applications without being questioned. This book challenges the reader to probe deeper than the technological slickness. In the end, you must be prepared to defend the choices you make. A good defense will come from understanding the information as much as the tool you apply.

The examples mentioned in the following chapters are intended to be representative, but readers should use these to embark on explorations of their own. You should go seek out applications that interest you, that deal with your locality, or that connect to your disciplinary background. To aid in this exploration, to provide access to color illustrations not possible within this book's budget, and to ensure that the material remains current in this volatile field, this book is accompanied by resources made available over the World Wide Web. The web site for this book `http://www.wiley.com/college/chrisman` will provide current resources for continued exploration.

ORGANIZATION OF THE BOOK

The organizing framework adopted for *Exploring Geographic Information Systems* simultaneously views the subject as a technical problem, as an empowering application, as a scientific endeavor, as an academic pursuit, and as a social necessity. Geographic information systems can be approached from each of these perspectives, but no single perspective is adequate to build the whole structure.

The sequence of the book is designed on the basis of 12 years of teaching a course that introduces geographic information systems, first at the University of Wisconsin–Madison, then at the University of Washington. I have always tried to focus this course on critical thinking, meaning the ability to make judgments concerning the fitness of tools and information sources. The critical thinking process works best when confronting real dilemmas, not simplified examples or abstract theories. This book tries to introduce a few selected examples, but there is not enough space to include as many as are needed to do justice to the diversity of applications.

There are many ways to address this diversity. Many of the discussions of GIS use the sequence input–processing–output to describe the operation of an information system. This order may follow the flow of work, but it may also miss some critical interconnections. *Exploring Geographic Information Systems* adopts a nested scheme for presenting the issues related to geographic information (Figure P-1). This book will progress from the inner rings that deal with the fundamentals of measurement and representation outward, but the progress will not be single-minded. Each ring incorporates the issues of those inside, building to a more complex structure. In presenting each ring, the interactions will be apparent because the examples of applications cannot be limited to the specific level. All geographic information is created inside a specific human context (a client, an institution, a marketplace). The demands and expectations of the human context become intermingled with what seem to be technical decisions.

The six rings of Figure P-1 will be taken in pairs to organize the three Parts of this book. At the innermost (Part1), geographic information consists of measurements organized in a system of representation. Part 1 begins with an introduction to geographic information and a brief review of the conventional approach to measurement.

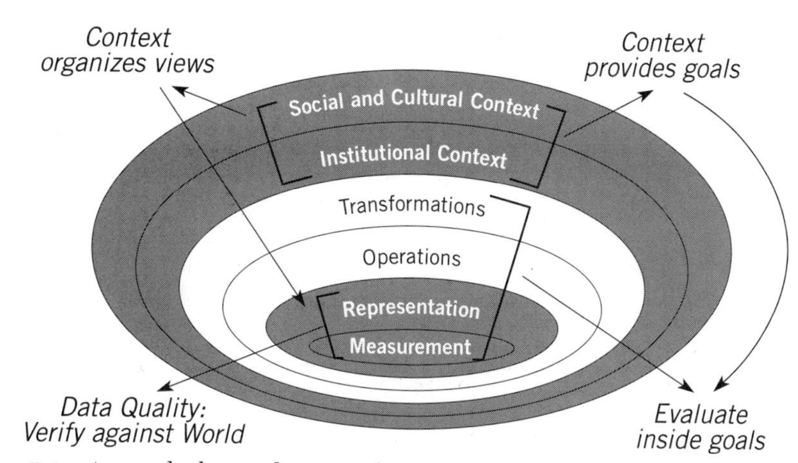

Figure P-1: A nested scheme of rings used to structure this book.

Chapter 2 presents a system of measurement frameworks with some examples. Chapter 3 converts the measurement frameworks to systems of representation, with substantial treatment of the conversion from paper maps to digital form.

The next ring concerns operations that discover latent relationships or create new information. These are the analytical core of a GIS, and occupy Part 2 of this book. Chapters 4 through 8 cover operations on geographic information roughly in order of increasing complexity. Each chapter will present some applications examples that demonstrate the operations and will discuss the data quality connections. Because the measurements available are not always organized in the desired form, a transformation may be required to convert a representation into another structure. Of course, this process often occurs at a price in resolution, accuracy, or some other element of data quality. Chapter 9 presents transformations as the culmination of the operations discussed in the earlier chapters.

The outer pair of rings may be missing from most of the technical literature on GIS. Yet, it should be apparent that an information system is developed and maintained for specific purposes and often at considerable expense. Part 3 considers the broader context of geographic information. Transformations and operations must be evaluated, not just in technical terms, but in terms of their fitness for the goals of a project. Chapter 10 considers the measures of evaluation and then the procedures followed to implement a GIS.

The nature of the institution charged with operating a GIS can also be critical, particularly because each institution operates inside a social context that may often be taken for granted. More often the literature on GIS takes an abstract view and discusses the technical implementation as if it were mathematical truth that would apply in all possible contexts. While there are some inescapable elements of the inner rings, this book will try to demonstrate that the goals and attitudes behind the information system should be openly examined. Chapter 11 discusses the institutional and cultural context that shapes the use of geographic information.

The exploration process cannot be contained by this or any book. Additional resources on the World Wide Web connect the topics of this book to the worldwide practice of GIS. I hope these chapters launch many explorations.

Nicholas Chrisman
Bainbridge Island, Washington

Acknowledgments

This book has been an exploration in itself. Colleagues and students who have worked with me at the Harvard Laboratory for Computer Graphics, the University of Wisconsin–Madison, and the University of Washington may recognize glimmers of past conversations and projects that helped me discover this way of presenting geographic information systems. The writing process actually began in 1991, but progress was extremely slow until I took the academic year 1993–1994 as a sabbatical leave on partial pay from the University of Washington. This support from my department and from the university was critical.

During the writing process, I had substantial assistance in the form of comments on the draft chapters. Graduate students, including Charles Hendrikson, Ilya Zaslavsky, and Daniel Karnes, helped keep the draft on track, but Francis Harvey provided the most thorough commentary across multiple drafts. The first draft was sprung on 60 unsuspecting students in Geography 460 in Fall Quarter 1994 while the review process at John Wiley & Sons began. A number of students provided helpful comments, particularly Steven Gott, a mathematics major, who saved me from a number of potential mistakes. In the Fall Quarter of 1995, another 60 students used an improved but still incomplete version. In 1996, another 106 students have helped in final proofreading.

In the first review process, David Cowen, Michael Demers, Richard Scott, and one anonymous reviewer provided insight on the manuscript's structure. In addition, Richard Scott made detailed comments on writing style. The organization was changed; chapters split and moved into new places. In April 1995, the revised draft went out for a second review. Gregory Elmes, Aaron Moody, Daniel Sui, and another anonymous reviewer returned helpful comments. Elmes, in particular, marked gaps in logic or steps unexplained. In July 1995, another revised draft went out for review. A few of the stalwart reviewers above provided helpful comments on the corrected product. Ferko Csillag, Jeremy Crampton, Derek Thompson, and Michael Goodchild gave very helpful suggestions. I received encouragement and advice from unofficial reviewers Elizabeth Kohlenberg, Gail Langran Kucera, and Thomas Poiker. Richard Morrill provided help on location–allocation models. Geoffrey Dutton, Denis White, Daniel Karnes, Ilya Zaslavsky, and Francis Harvey all contributed comments on the definition in Chapter 1. Another draft emerged in February 1996 for another review cycle. Barbara Buttenfield, Michael Goodchild, Brian Klinkenberg, and Daniel Sui (for a third time!) provided the final review with useful alterations.

At John Wiley, Christopher Rogers, Executive Editor, put his faith in this project on the basis of a prospectus and an outline that still bears some resemblance to the final product. Frank Lyman, Geography Editor, helped in the early stages of the project. Nanette Kauffman took over this role and survived the blizzard of anxious email as the project came to completion. In the production process, Jeanie Berke has provided a complete education in all the steps that convert a raw manuscript into final form. Her sense of humor has overcome many seeming disasters. Harry Nolan provided the simple, elegant design. This project stretched the limits of the possible in production deadlines, and Mark Cirillo deserves special thanks for his role in manufacturing.

Most of the tables and figures were produced by John Wiley from rudimentary sketches by the author. Edward Starr interpreted the raw material with an air of calm assuredness. All the weird electronic formats did not deflect him and Rolin Graphics from the schedule. Sources and credits for all graphics not produced for this book appear at the back of the book.

While the assistance from all of these sources helped create this book, none of them hold any responsibility for the limitations that it may still contain. This project will require further refinements as the explorations in GIS continue. Please send comments or suggestions to "`chrisman@u.washington.edu.`"

Contents

PART 2: TRANSFORMATIONS AND OPERATIONS 89

Chapter 4: Attribute-based Operations 91

Chapter 5: Overlay: Integration of Disparate Sources 105

Chapter 8: Comprehensive Operations 185

Chapter 9: Transformations 207

PART 3: THE BROADER CONTEXT 233

Chapter 10: Evaluation and Implementation 235

Chapter 11: Social and Institutional Context 255

PART 1

BASIC BUILDING BLOCKS OF GEOGRAPHIC INFORMATION

Social and Cultural Context
Institutional Context
Transformations
Operations
Representation
Measurement

An information system organizes data into a structure, so that simple observations can be converted into more useful information. A geographic information system serves the distinctive needs when geographic descriptions play a central role in the observations. To understand this kind of information system, this book uses a series of nested *rings*. At the innermost ring, the process of geographic *measurement* requires choices that can be organized as *measurement frameworks*. Differences in these measurement frameworks best explain the technical choices of *representation* for geographic information; measurement and representation, in turn, strongly influence the *operations* that can be performed with the information. Finally, *transformations* can convert from one measurement framework to another. Thus, each ring builds upon decisions made at the simpler levels.

This book successively links measurement to representation and then to analytical operations and transformations. Part 1 provides a base for the rest of the book, covering the measurement and representation rings. Chapter 1 defines geographic information, reviews the conventional approach to measurement, and introduces reference systems for measurements of time, space, and attributes. Chapter 2 develops a revised approach to measurement based on measurement frameworks. Chapter 3 describes the translation of measurement schemes into practical systems of representation. In addition, this chapter describes the procedures for converting existing documents, such as maps, into digital form. Part 2 continues with the operations and transformations rings.

CHAPTER 1

MEASUREMENT BASICS

CHAPTER OVERVIEW

- Connect current developments to the history of the technology.
- Define a geographic information system and geographic information.
- Review conventional approach to measurement.
- Introduce reference systems for time, space, and attributes.

DEVELOPMENT OF GEOGRAPHIC INFORMATION SYSTEMS: THE CONVERGENCE OF MANY TECHNOLOGIES

Human societies have collected and processed geographic information for millennia. Some of the first written records from Mesopotamia and Egypt contain property boundary information as a part of legal transactions. Other records describe routes and characteristics of distant places. Maps and graphical representations seem to be at least as ancient as the written word (for reading on the history of cartography, see Harley and Woodward 1989). Demand for geographic information has historically exceeded the supply. In the earliest civilizations, the supply of geographic information was limited, probably because this kind of information was more difficult to collect, represent, and transmit than the other kinds of information scratched on ancient clay tablets. After all, a sale of five goats can be represented with the simplest icons, but a description of an agricultural field requires multiple steps of abstract thinking. In addition to the development of numbers, the real world had to be simplified into points, straight lines, and triangles to be able to estimate the quantity of land. Early advances of mathematics had their roots in these practical problems.

Technology, in its broadest conceptual and cultural sense, has been a driving force in the ability to construct and disseminate geographic information. My grandfather, Oscar Chrisman, learned civil engineering at the University of Missouri in the first decade of this century. While the instruments he used for surveying were carefully constructed using modern optics, the basic techniques (laying out a baseline and turning angles with a leveled device) would have been understandable to an Egyptian "rope-stretcher" or a Roman *agrimensor* (literally, "field measurer"). Many of the developments for geographic information in the twentieth century involve the combination of some new ingredient from outside the field with the basic concepts from the past. In the following cases, the equipment has become much more sophisticated, but the underlying trigonometry remains the same. The aerial photograph requires the development of aircraft and photography, and "total station" surveying equipment uses lasers to measure distances and angles, then records the results directly in digital form for computer drafting software in the office. The Global Positioning System broadcasts timing information from a constellation of satellites. These signals can be converted into distances, thus providing a huge set of triangles to measure position. Other satellites can capture digital images of huge areas at increasingly detailed resolutions in spatial and spectral terms.

The twentieth century has produced a vast change in the hardware used for geographic measurement. In fact, the technology used for geographic information may have changed more in the twentieth century than it did in the previous 4000 years. The most critical new hardware element is the computer, which has conquered the workplace in most of the developed world. Alongside the impressive development in computer hardware, the development of software has sparked a fundamental rethinking of the concepts underlying geographic information. The separate specialities known as digital cartography, spatial analysis, remote sensing, multipurpose land information systems, and a host of others have converged to create an interdisciplinary focus on geographic information systems. Each of these contributory technologies has expanded rapidly, but the combination has expanded even faster. A geographic information system can integrate information from different sources, thereby exploiting a variety of technical advances as they occur.

One of the most durable technologies for representing geographic information has been the map. A map consists of a set of symbols recorded in spatial relationship to each other, so that the position of the symbols acts as an integral part of the message. It is all too easy to forget that the map is part of a more complex system. The marks on the map lie latent until the map reader decodes them and interprets the message or meaning of the symbols. The map reader who grasps the spatial relationships and performs spatial operations (such as locating the nearest exit off the highway) is an integral part of the map system. Maps are so well established as a technology that they control our perception of geographic information. One of the greatest barriers to progress in geographic information systems has been the tendency to cling to the printed map as a model for digital developments.

In the modern geographic information system, the map is replaced by a database accessed through a software system. In some cases, the software simply reproduces graphic products that look like traditional maps and depend on visual interpretation. At this level, the technology only provides a replacement for the map, often at a much higher cost. The next level of technology requires more sophisticated software that can reorganize the raw data to discover new relationships. It is at this level that the real advantages of a geographic information system become apparent.

DEFINING A GEOGRAPHIC INFORMATION SYSTEM

There are dozens of definitions for the term *geographic information system* (GIS), each developed from a different perspective or disciplinary origin. Some focus on the map connection, some stress the database or the software tool kit, and others emphasize applications such as decision support (Maguire 1991). One of the most general definitions was developed by consensus among 30 specialists:

> *Geographic Information System—A system of hardware, software, data, people, organizations and institutional arrangements for collecting, storing, analyzing and disseminating information about areas of the earth. (Dueker and Kjerne 1989, pp.7–8)*

While this definition may seem bland, it encompasses all the characteristics, as long as the terms are expanded to their intended meaning. For example, the word *system* implies a group of connected entities and activities. An automated information system organizes a collection of data, computer procedures, and human organizations to serve some particular purpose. For a GIS, the purpose could involve a complex decision such as the policy for timber harvest or a routine decision to grant a permit or simply activity to maintain an inventory. Notice that the definition carefully distinguishes between the data in the system and the information that results from the system. Data provide the raw material for information much as map symbols convey a map message. For both maps and information systems, the raw data are not enough; additional relationships must be constructed from the context.

The most common understanding of a GIS emphasizes that a GIS is a tool. However, no tool is totally neutral; a GIS can be designed to be effective and efficient for a certain range of purposes. Tools are developed within a social and historical context to serve changing needs, but tools are also intended to change their environment. The perspective of this book can be summarized by the following definition:

Geographic Information System (GIS)—The organized activity by which people

- *measure* aspects of geographic phenomena and processes;
- *represent* these measurements, usually in the form of a computer database, to emphasize spatial themes, entities, and relationships;
- *operate* upon these representations to produce more measurements and to discover new relationships by integrating disparate sources; and
- *transform* these representations to conform to other frameworks of entities and relationships.

These activities reflect the larger context (institutions and cultures) in which these people carry out their work. In turn, the GIS may influence these structures.

BASIC COMPONENTS OF GEOGRAPHIC INFORMATION

Geographic information is commonly broken into the components of *space*, *time*, and *attribute*. Space, though an obvious component of geographic information, deserves careful definition. Our world of direct experience is basically three-dimensional. In fields such as geophysics, the entire planet must be treated as a solid, but for most geographic purposes, attention can be limited to the thin shell of the earth's surface. For much mapping (and GIS), the dominant spatial expanse is two-dimensional. Even though it is often convenient to use a simple Cartesian two-dimensional space, the surface of the earth is not a limitless plane, and the two-dimensional relationships are wrapped around the **geoid**.

Time often plays a silent role in maps, though there is always some implicit or explicit temporal reference. The most common map is a form of snapshot, valid for a specific point in time. Of course, time is far from pointlike; time extends and separates. The inexorable progression of time and the inability to turn back the clock lead to the most prevalent view of time as an arrow—a line with one direction. Sometimes, however, it makes sense to treat time as cyclical, such as when astronomical and climatic phenomena repeat. In these cases, the sense of cycles comes from the connection between the spatial realm and the temporal, not because the time actually repeats. When the earth returns to a particular position in its orbit, the stars return to the relationships observed in a previous year. The regularity of these spatial relationships had a profound influence on the early development of calendars.

In Newtonian physics, time and space provide a system of reference for phenomena. In this approach, these dimensions act as containers of the material world; thus, space and time are external or absolute. Applying this view to geography, Hägerstrand (1970) developed the highly graphic view of time as an additional spatial dimension. He created a three-dimensional diagram in which the life histories of people interact in time and space (Figure 1-1). Most GIS software packages adopt this model by implementing space and time as immutable external axes.

The Newtonian view of time and space as external absolutes is not the only possibility. Modern physics recognizes that space–time relationships depend on the arrangement of mass and energy within that space—a model of relativity. Similarly, the relationships between objects in geographic circumstances can construct a relative space that differs from the abstract dimensions. For example, the connections of mass transit in large cities limit relationships to a world of subway stops, not strict distance. Part 2 will demonstrate that both the absolute and relative views of space can be accommodated in the operations that can be performed on geographic information.

The third component of geographic information can range from observable physical properties to aesthetic judgments, all described by the term **attribute**. The theory of measurement provides some structure to the realm of attributes, usually without considering space or time. The next section will discuss the approach to attribute measurement that currently dominates social science instruction. This approach has limi-

Geoid: Three-dimensional shape of the earth defined by the surface where gravity has the value associated with mean sea level.

Attribute: The range of possible values of a characteristic; an attribute value is a specific instance of the characteristic associated with a geographic feature.

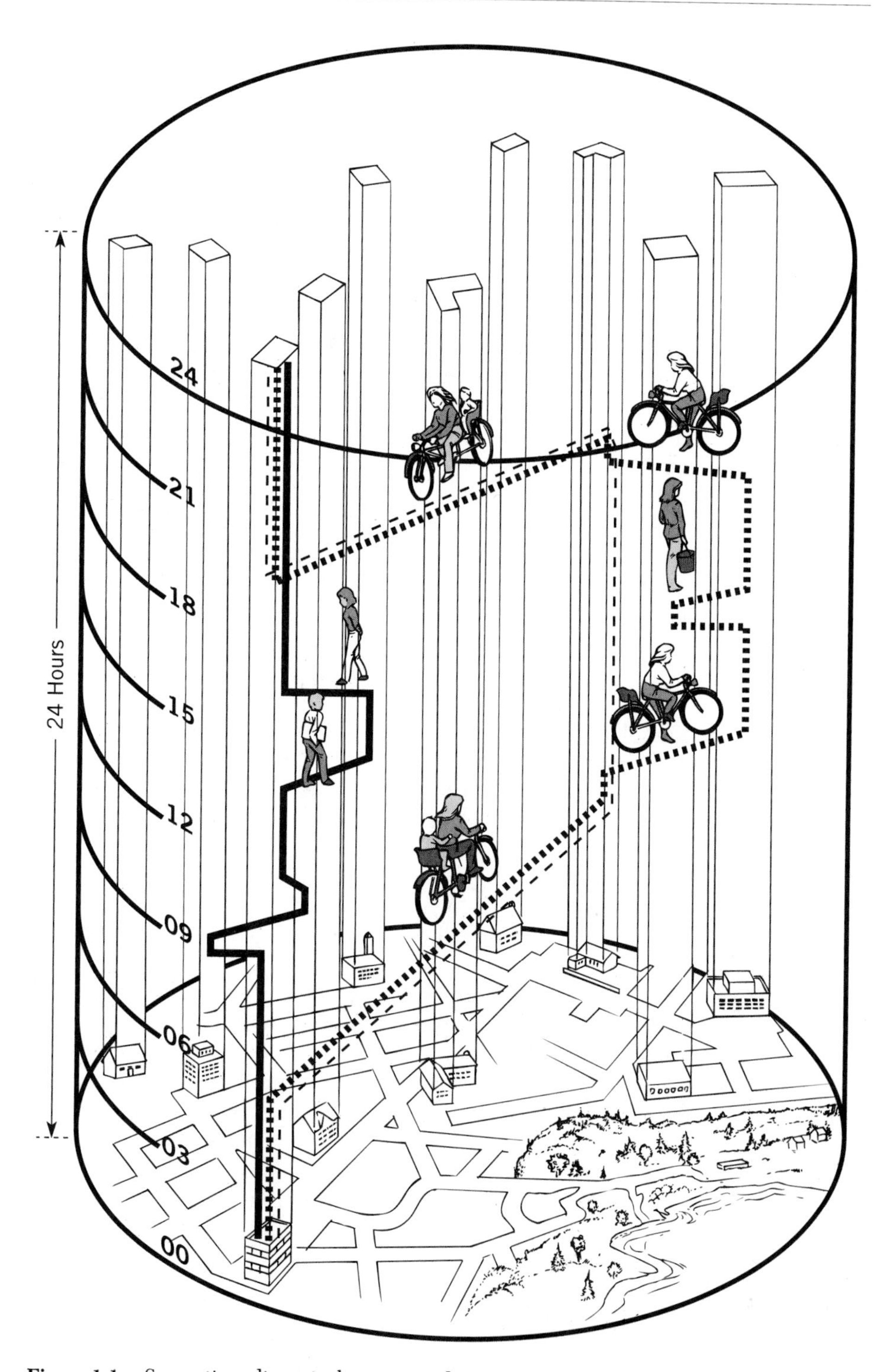

Figure 1-1: Space-time diagram showing residents in their daily lives. Redrawn from Parkes and Thrift (1980).

tations, but they can be resolved by formulating a more comprehensive scheme that combines the temporal, spatial, and attribute components. This more comprehensive approach is the subject of Chapter 2.

MEASUREMENT: THE CONVENTIONAL VIEW

Cartographers have usually focused on the communication process, and have considered the map reader the ultimate consumer. If we adopt an informal definition of **information** as "a difference that makes a difference," the human perception of the message is the final difference in cartography. By contrast, in a GIS, much of the interaction occurs within the system. The relationships that remain latent on the map must be accessible to the software instead of awaiting the human map reader.

The formal study of differences lies at the core of measurement theory. Measurement imposes an external human logic, a mental abstraction, onto the world. The body of possible frameworks for measurement has been the subject of mathematical study for many centuries. It is important to recognize that geographic information fits into the general framework for measurement, although it raises some particular issues and special problems.

Background: Development of Measurement Theory

Measurement, as viewed by classical physicists by the end of the nineteenth century, seeks to determine the numerical relationship between a standard object and the object being measured. These physicists developed procedures to measure length, mass, electrical charge, and more. Consider the attribute *length*. Every entity in space can be measured by comparing its length to some other length. If we place a "standard" measuring rod alongside the other object, mark where the end of the rod falls, and place the rod again beginning at the mark, this physical procedure mimics addition. The number we obtain represents the ratio of the length of the object to the length of the rod.

Nineteenth-century physicists constructed a rather complex model of the world with remarkably few of these fundamental properties. They were considered *extensive* properties because they *extended* in some way as length does in space. (Other properties, like density, were built up as ratios of the extensive properties and were thus *derived*.) Extensive measurement has been an element of commerce and centralized government in many societies. One of the ways that the first Qin emperor consolidated China around 210 B.C. was by standardizing weights and measures. In Europe, this standardization process occurred much later. The medieval European world had a different measuring rod and measuring container in each market square. Not until the

Information: Data (observations, measurements, etc.) placed in context of a system of meaning (a set of relationships and assumptions about those relationships). Information, built into larger context, constructs knowledge.

eighteenth century did Enlightenment thinkers conceive of universal mathematical values. Universal standards for measurement were a consequence. During the French Revolution, the Enlightenment concept of standardization led to the metric system, now called **SI**. In the final analysis, even a worldwide standard is simply a convenience, not a mathematical truth.

Extensive properties are rather restrictive. For example, having a universal standard measuring rod located in Sèvres, France, is not very practical for all the geographic attributes that must be measured. By the early twentieth century, physicists began to move beyond the classical concept that the meter was an intrinsic property of one particular rod. After all, they could develop predictive models of thermal expansion in the rod and other effects. They saw that the universality of the standard did not really depend on the rod at all. As science became more analytical, the *operation* of measurement became just as important as the physical standard; thus the object and the measurement result became separated. A twentieth-century philosophy of measurement called **representationalism** viewed numbers not as properties inherent in an object, but as the result of relationships between measurement operations and the object (Michell 1993). While this view redefined the philosophy, extensive measurement was still considered the basis for science.

Defining measurement exclusively in terms of extensive measures left almost no room for the social sciences to develop a measurement theory for less physical properties. Debate during the 1940s between physicists and social scientists led to a stalemate. The physicists could not consider the perceived loudness of sounds as a measurement, because it did not involve extensive operations like addition. In response, Stevens (1946), a psychologist at Harvard University, published a paper in *Science* proposing a framework based on what he called **levels of measurement**. This scheme for measurement adopts the representationalist philosophy, defining measurement as the "assignment of numbers to objects according to a rule" (Stevens 1946, p. 677). Stevens' schema has become a basis for social science methods and a framework for cartography and GIS. Because Stevens' classification is often misapplied and misinterpreted, the levels of measurement must be reviewed carefully. Some revisions and extensions must be considered to accommodate geographic information.

SI: Système International d'Unités; the system of weights and measures established by international agreement in 1875. The International Bureau of Weights and Measures in Sèvres, France, oversees the measurement standards. SI defines seven base units from which many others can be derived: meter–length, kilogram–mass, second–time, kelvin–temperature, ampere–electric current, mole–chemical quantity, candela–intensity of light.

Representationalism: A philosophy of measurement that defines measurement as the connection of numbers with entities that are not numbers. The numbers are not seen as inherent but as a representation of a defined aspect of the entity. (This philosophy is opposed to the classical theory that measurement is the numerical expression of one quantity relative to another; quantities were seen as inherent.)

Level of measurement: A grouping of measurement scales based on the invariance of certain properties. Measurement scales at a common level of measurement can be transformed into another scale at the same level without reducing the information content.

Levels of Measurement

Stevens' levels of measurement were organized by the principle of **invariance** under transformations, a concept imported from mathematics and physics. Invariance implies that a **scale** retains its information content through a certain group of possible transformations. One example of invariance involves temperature. A measurement in °F can be transformed into °C without loss of information; thus, these measurements are at the same level. By contrast, a scale consisting of {cold, warm, hot} cannot retain all the information recorded on either numerical degree scale.

The following sections explain Stevens' four levels of measurement and illustrate them with a common example. Imagine a marathon in which contestants become associated with certain attributes or measurements.

Nominal At the most basic level, Stevens described a nominal "scale," in which objects are classified into groups. Any assignment of symbols can be used, so long as the distinct nature of each group is maintained. A nominal measure is based on set theory, a quite simple level of mathematics. The use of the word "scale" for a nominal measurement may evoke the traditional number line, but there is no such ordering implied.

In the marathon example, each contestant gets a number to wear. What does this number mean? Is it a measurement? If the number is simply pulled randomly out of a box, it has to be considered an arbitrary symbol (like a word or an icon). Other nominal attributes could be determined, such as the set of contestants wearing red shirts (Figure 1-2). Another kind of nominal grouping might partition all contestants into either a women's event or a men's event. Any numerical label for these two categories (0 and 1 or 1 and 2 or 359 and 213) would be totally arbitrary.

Ordinal The ordinal level introduces the concept of an ordering. An ordinal scale arises when objects can be sorted in some manner using a pairwise comparison; such

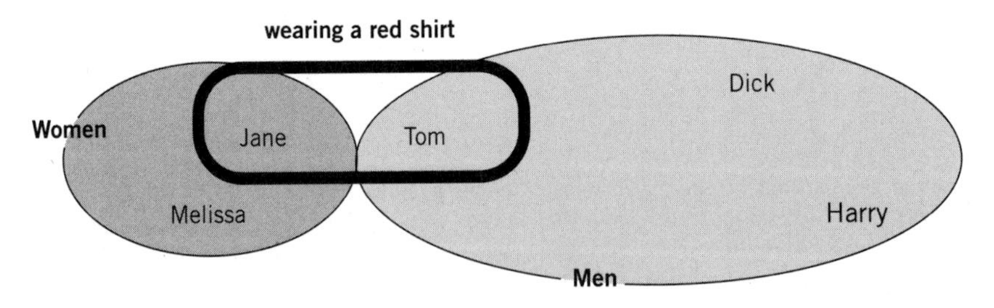

Figure 1-2: Nominal measures are not on scales at all. They create categories that can be treated as sets.

Invariance: Properties that remain unchanged despite transformations of the numbers used to represent a measurement.

Scale: A system used to encode the results of a measurement; typically a number line, but generalized to include a list of categories.

Ordinal

Order of arrival of contestants	*Women's race*	*Men's race*
First	Jane	Tom
Second	Melissa	Dick
Third	Leila	Harry

Figure 1-3: Strictly ordinal scales can arise from a total ordering, but ordinal scales may also arise from partial orderings.

a scale can exist in many forms. The most demanding form orders all objects completely without any ties (Figure 1-3), as in recording the order of finishing the race (first, second, third ...). It makes sense to use the word "scale" for such an ordering, because each successive element continues in the same direction. Similarly, the number on the shirt of each contestant in the race might also represent an ordinal measure, if the numbers are handed out of the box sequentially. Thus a seemingly nominal identifier might hide an ordering based on time or on some other property.

Not all orderings are as well behaved as this ideal model without ties. The more ties, the less scalelike the ordering becomes. For example, some ordinal categories use a semantic scale of words. Soils are ordered from "poorly drained" through "somewhat poorly drained" on to "well-drained" and "excessively drained." Opinion polls use orderings such as "agree" and "strongly agree." The ordinal level covers a wide range of possibilities; some behave in a nearly numerical way, whereas others are barely evolved from a nominal level.

As long as the ties behave symmetrically, the ordering still behaves as a scale. However, much of social science deals with messier circumstances. Respondents to a survey may create loops instead of linear orders: they might answer that they prefer A to B, B to C, and C to A, leaving researchers with a distinct problem creating a linear scale. Less severe problems occur when there is a tolerance involved in making comparisons. If A seems to be tied with B, and B tied with C, can we infer that C is also tied with A? Perhaps; the answer depends on how far C is from A on some underlying scale. C *may* be "just detectably different" from A, so B must fall between them.

Whatever the nature of the orderings, an ordinal measurement does not give much guide to numeric representation. It may be conventional to give out the numbers 1, 2, 3 ... to finishers in a race, but the numbers could have been any increasing sequence (0.5, 0.66, 0.75, 0.8 ...) or (1, 3, 597, 66,671 ...), because the difference between the pairs is not given. In the example of soil drainage, we do not know if the step from "poorly drained" to "somewhat poorly drained" is identical in distance to the step between "well drained" and "excessively drained." For different analytical purposes, the importance of each step in the scale might vary. Some orderings may relate to an underlying numerical scale, others may not; for example, it is hard to assume that "good" means the same thing for all respondents to an opinion poll. Ordinal values are essentially categories without the arithmetic properties usually ascribed to numbers. It is important to remember these limitations when some GIS user wants to "standardize" rankings on a scale from 1 to 9. The assignment of numbers does not automatically construct valid arithmetical relationships.

Interval In Stevens' scheme, numbers develop algebraic meaning by moving into the *quantitative* realm and becoming an interval scale. An interval scale places an object on a number line with an arbitrary zero point and an arbitrary interval (choice of a distance to be called one). Thus, interval data can be shifted around on the number line without changing the meaning of the measurement. For example, years can be recorded on the Gregorian calendar (A.D. or the "Current Era"), the Islamic calendar (1 A.H. = A.D. 622), or the geologists' Before Present (0 B.P. = A.D. 1950). In all these systems, the numerical value of a particular year has no particular significance. The year 2000 is not twice the year 1000 in any significant sense.

In the case of the marathon, we could assign arrival times to runners by simply noting the clock time on arrival (Figure 1-4). As long as some simple assumptions are valid, particularly that all runners departed at the same time, then these numbers capture all the ordinal results. In addition, the differences between arrival times can be interpreted. Some of the arrivals are closer to the next arrival than others, establishing a truly numerical measure of distance between values. A difference can be treated as twice as long as another.

Ratio Arrival times for a race provide a raw result awaiting further processing. Contestants would obtain a more useful measure by subtracting the time of their start. In fact, a difference between two interval measures becomes a measure on Stevens' next level: ratio.

In measurement theory, the ratio level gets the most attention. Ratio measures have a true origin (zero value) *and* an arbitrary interval (a meaning for the distance to be called one). These properties support the arithmetic operations of addition, subtraction, multiplication, and division. On a ratio scale, if a value is twice that of another, then it represents a doubling in the quantity. The easiest ratio measures to visualize are classical extensive quantities. In the race example, the elapsed time in running the race is a ratio measure obtained by subtracting two interval measures for the start and finish (Figure 1-5). The ratio measure of elapsed running time contains all the information of the ordinal scale for ranking winners, plus it adds the numerical properties that measure how fast each contestant ran. It is clear that these ratio measures convey more information and permit more analytical treatment.

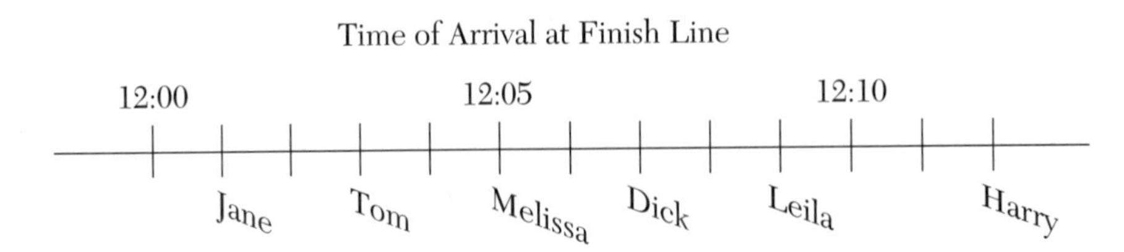

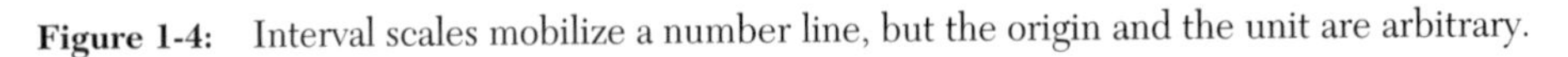
Figure 1-4: Interval scales mobilize a number line, but the origin and the unit are arbitrary.

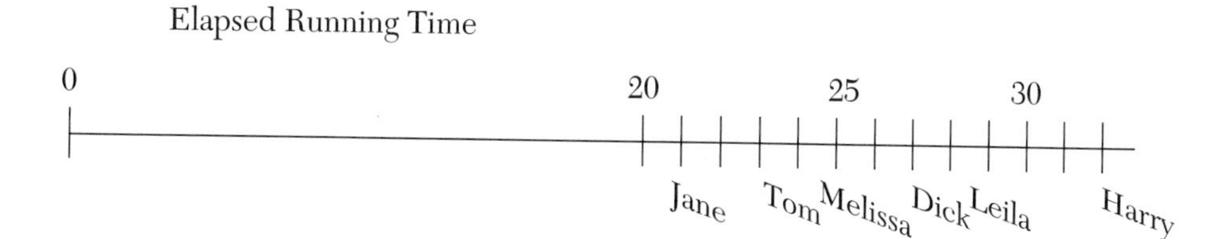

Figure 1-5: Ratio scales, the classical ideal for physical measurement, have a true origin and an arbitrary unit of measure.

Extensive and Derived Scales Stevens tried to combine extensive and derived (intensive) measurement into one level. Anything that can simply be rescaled by a constant he termed a ratio measure. This does recognize that most numerical measurements have an arbitrary *unit of measure*, so that a measurement in feet can be converted to meters with no real change in fundamental information. But commonality of ratio measures misses some of the important distinctions between geographic measurements. We can measure the total economic activity in a city in dollars. This value is effectively continuous (we could record money down to insignificant fractions of dollars) and arbitrarily scaled (we could use yen in place of dollars), with a true zero value (no money = zero). It could be combined with the value of other cities to obtain a measure for the total economy (by addition). Cartographic design principles would suggest that we use **proportional symbols** for such a raw (extensive) measurement (Figure 1-6 shows the total expenditure by Department of Defense for each California county in 1976). We could also obtain a per capita measure by dividing this economic activity by the number of persons in the county. Such a transformation removes the influence of size and concentrates on relative wealth. This value is just as "ratio" as the total value, but cartographic rules suggest a **choropleth map** presentation for these per capita figures (Figure 1-7). Per capita values of two counties cannot really be added together, since the denominators (populations) might be totally different. Notice that some counties with low total figures can have high per capita figures.

Stevens' system has been used to decide which statistical methods apply to a given measurement. Many introductory statistics books (particularly for the social sciences) connect the levels of measurement to a group of appropriate tools. Similarly,

Proportional symbols: A thematic mapping technique that displays a quantitative attribute by varying the *size* of a symbol. Typically, proportional symbols use simple shapes such as circles and are scaled so that the area of the symbol is proportional to the attribute value. Proportional symbols are located at a point, even when they represent information collected for an area.

Choropleth map: A thematic mapping technique that displays a quantitative attribute using ordinal classes applied as uniform symbolism over a whole areal feature. Sometimes extended to include any thematic map based on symbolism applied to areal objects.

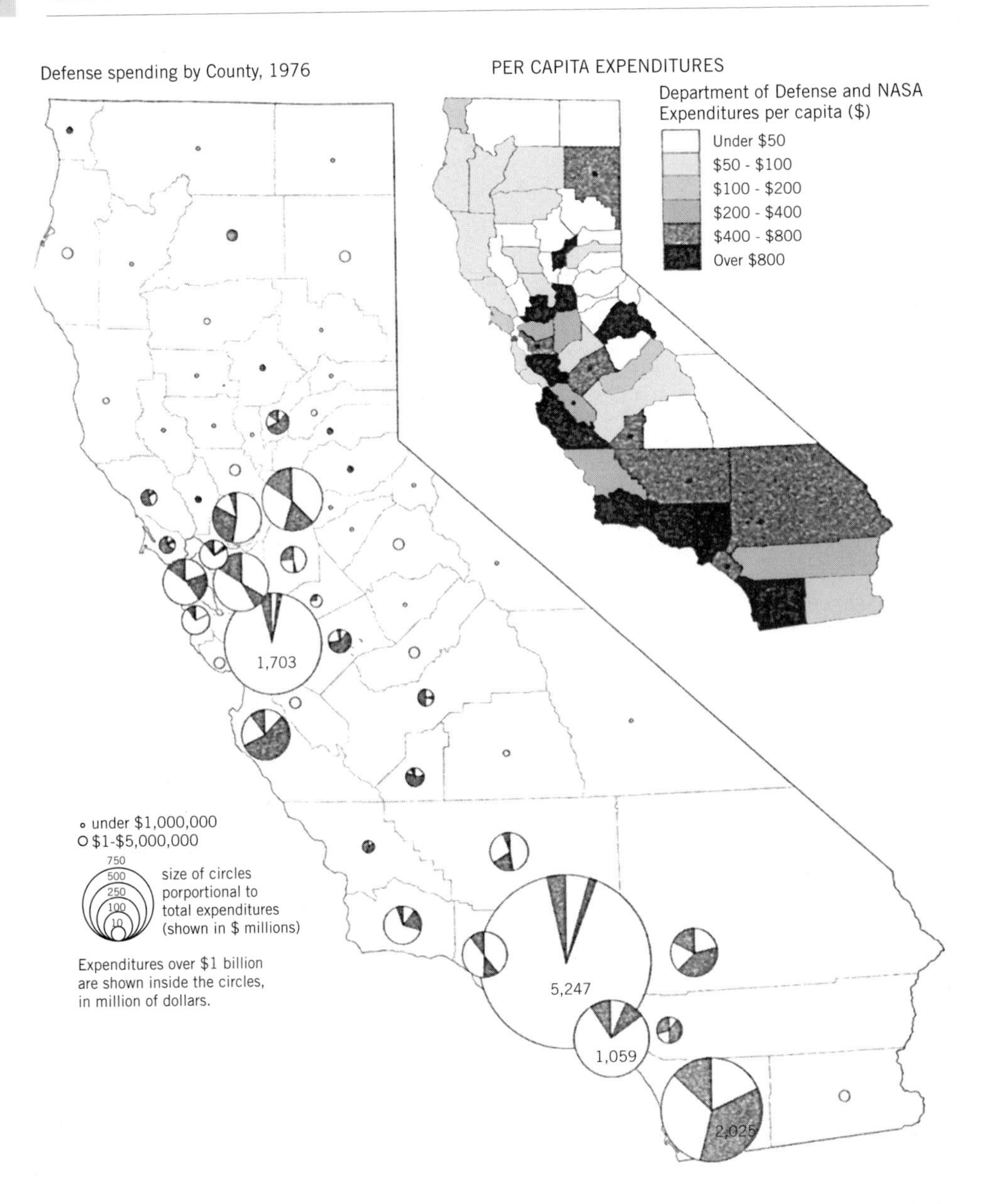

Figure 1-6: Proportional symbols use the graphic variable of *size* for a simple geometric symbol such as a circle. They are considered appropriate for raw measures, such as total population or total economic output. This map portrays defense spending for each county by scaling the area of the circle to be proportional to the dollar figure. Source: Donley and others (1979, p. 46).

Figure 1-7: Choropleth maps use the spatial object as the symbol. The graphic variable *size* cannot be used without a cartogram. Here, as in many cases, the graphic variable *value* (a gradation from light to dark) shows the range of the attribute. Since the area of the object is a part of the symbol already, this method is most appropriate for density measures, or a derived ratio such as dollars per capita. Source: Donley and others (1979, p. 46).

cartographic texts connect the cartographic tool kit of **graphic elements** to specific levels of measurement. For their intended purposes, these connections make some sense, but each of these decisions occurs inside a series of assumptions about the nature of the measurements. These assumptions must be examined carefully.

Perhaps the best way to explain the connection between measurement and operations is by a counterexample. Along highways, it is common to announce the towns and villages that the road passes through. Most highway signs announce the name with some extra information, such as population or elevation. One town in California has a sign that takes the spirit of local pride to an extreme (Figure 1-8). Adding these three numbers is clearly a joke; yet, professionals who work with geographic information often commit equally meaningless combinations with no humorous intent. The number 4663 measures nothing about New Cuyama because it combines people (on some census date), feet (above sea level), and date of establishment (relative to the Current Era). Having three numbers does not ensure that addition will produce any sensible result. Understanding the appropriate manipulations requires more than just the number scales.

What is Missing from Stevens Practical applications of geographic information require more distinctions than are available in Stevens' scales. Stevens' four levels are

Figure 1-8: Sign posted at the entry to New Cuyama, California.

Graphic elements: The characteristics of a symbol system that can be manipulated to encode information. For cartography, these include size, shape, hue, saturation, brightness, orientation, and pattern. (See Robinson and others 1995).

usually presented in the geographic literature as a complete set, but they are not enough. Even inside Stevens' invariance scheme, ratio is not the highest level of measurement. The ratio scale has one fixed point (zero) and the choice of the value of one is essentially arbitrary. A higher level of measurement would be achieved if the value of one were fixed as well. When the whole scale is predetermined or *absolute* (Ellis 1966), no transformations that preserve the meaning of the measurement can be made. One example of such an absolute scale is *probability*, where the meaning of zero and one are fixed and other restrictions prohibit rescaling. Even though it is common to report probabilities as percentages, the relationships of probability (such as Bayes' Law of conditional probability) operate correctly only when scaled from zero to one.

While Stevens' levels deal with an unbounded number line, there are many measures that are bounded within a range and repeat in some cyclical manner. Angles seem to be ratio, in the sense that there is a zero and an arbitrary unit (degrees, grads, or radians); however, angles return to their origin. The direction 359° is as far from 0° as 1° is. Any general measurement scheme needs to recognize the existence of such cyclical measures.

Another class of geographic measurements treats counts aggregated over some region in space (such as population). Counts are discrete, since there is no half person to count, but a count is certainly much higher in numerical power than the other discrete levels (nominal and ordinal). Since the zero is a fixed value, counts may seem to be ratios, but the units of a count are not arbitrary, so they cannot be rescaled as freely.

As a further criticism of Stevens' system, nominal categories are not always as simple as portrayed. Nominal measures apply the strict rules implicit in classical set theory. Pure equivalence relationships require all members to be equally central to the category in perfect symmetry. Many classifications, however, adopt more flexible rules; they involve some kind of *graded* memberships as formalized in **fuzzy set theory** or involve some kind of resemblance to a **prototype** member of the class. In both these situations, an object will have some degree of membership, rather than a perfect equivalence class. The practical application of categories may require bending the rigid rules.

Thus, Stevens' four levels of measurement are not the end of the story. The concept of a closed list of levels arranged on a progression from simple to more complex does not cover the diversity of geographic measurement. Still, Stevens' terminology provides a starting point for a more complete understanding of measurement.

Fuzzy set theory: An extension to set theory that permits an object to have a degree of membership (usually represented as a number between zero and one). Fuzzy membership values do not have to follow the rules of probability.

Prototype: An approach to categorization that defines a category by identifying a particular object as the typical example. Other objects assigned to this category may not share all characteristics with the prototype object. The degree of resemblance can be treated as a distance in a taxonomic space.

REFERENCE SYSTEMS

The concept of measurement levels is normally applied to attributes. To handle geographic information, it is important to consider the measurement concepts required for time and space. In contrast to attribute measurements, measurement of time and space naturally invokes the concept of a *reference system*. This general concept, in turn, can be applied to attributes to provide a perspective beyond Stevens' four levels.

Temporal Reference System

Time, with its strong sense of a linear order, is simpler than space. The linear axis of time—measured in units of seconds, hours, and years—orders our lives in many ways. Global activity is synchronized by a common reference time (Greenwich Mean Time) and a common calendar. Time seems to be an extensive phenomenon, except that it cannot reverse itself. A complete **temporal reference system** merely requires an origin (time to call zero) and a unit of measurement. By adopting a common reference system, time measurements can be compared and mathematical operations like subtraction become valid.

Some aspects of time have repeating or cyclical elements. The calendar year is frequently tweaked with leap days and even leap seconds to make it remain synchronized with the gentle slowing of the earth's revolution around the sun. In environmental studies of all kinds, the seasons play an important role. Stevens' scheme does not allow for objects that can be ordered either spring–summer–fall–winter–spring or fall–winter–spring–summer–fall to make a growing year or a water year in some particular climate.

In addition to continuous time measured on the interval scale, time can be treated on an ordinal scale. Some periods are simplifications of the continuous historical time scale, and other phases or stages simply indicate temporal order without reference to any historical origin. Administrative procedures often specify such an ordinal time; they lay out the sequence of events, perhaps with some guidelines for duration, without respect to any particular starting point.

Spatial Reference Systems

Space, with its independent dimensions, is more complex than time. Stevens does not treat multidimensional measurements, because he considered the invariances on a number line as the main distinctions. Quite to the contrary, spatial measurement

Temporal reference system: An agreed measurement scheme for time; involves a time to start counting (an origin) and a unit. International conventions establish a calendar and synchronized clocks based on Greenwich Mean Time.

requires a set of geometric assumptions to create a **spatial reference system**. These geometric assumptions involve far more than can be expressed in the concept of levels of measurement.

Each spatial reference system operates through multidimensional measurement, and certain basic geometric relationships remain invariant to various techniques of representing the measurements. For example, Cartesian coordinates on two orthogonal axes can be freely converted into a radial system (Figure 1-9). Even though the units of measurement are not comparable, these two representations are equivalent. The two orthogonal distances create a triangle, and the radial coordinates specify that same triangle using the hypotenuse and an angle. The theorems of geometry demonstrate that the two triangles are congruent, a finding that would not be apparent from their measurement scales.

The spatial reference system of latitude–longitude (Figure 1-10) uses the axis of the earth's rotation to orient one axis (north–south). The plane of the equator (orthogonal to the axis) provides the zero reference for angles north and south (latitude). Longitude measures the angle east–west around the equator, from an arbitrary zero set by international convention along a meridian through Greenwich, England. Thus latitude–longitude pairs are not simply numbers; rather, they involve relationships to a model of the earth. Other spatial reference systems usually create a simpler reference model (such as a plane or a cone) positioned with respect to the spherical model. Cartesian **coordinates** may seem to measure simple distances, but these distances operate inside a spatial reference system that orients the axes and provides meaning to the numbers.

A common system of spatial reference is a critical element of a GIS, since it brings different map "layers" into correspondence (Figure 1-11). Of course, using the same

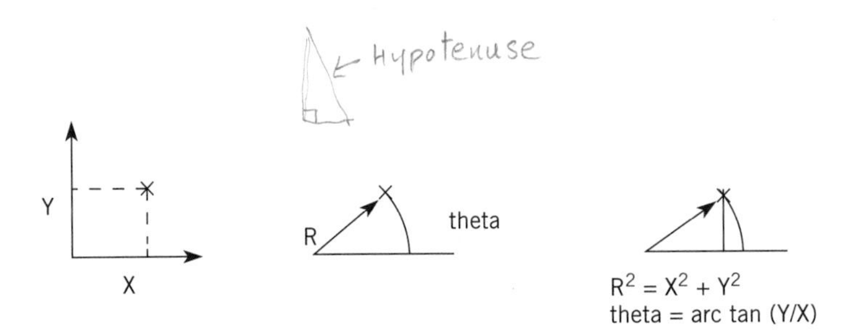

Figure 1-9: Conversion between measurements in two dimensions. Radial coordinates (angle and distance from an origin) are equivalent to Cartesian coordinates (distances parallel to two orthogonal axes). The angle measurement is cyclical, not extensive.

Spatial reference system: A mechanism to situate measurements on a geometric body, such as the earth; establishes a point of origin, orientation of reference axes, and geometric meaning of measurements, as well as units of measure.

Coordinates: A structured set of measurements related to a specific spatial reference system; usually applied to pairs of distance measurements (X,Y) on independent axes of a planar reference system or to angular measurements such as latitude–longitude pairs on a spherical reference model.

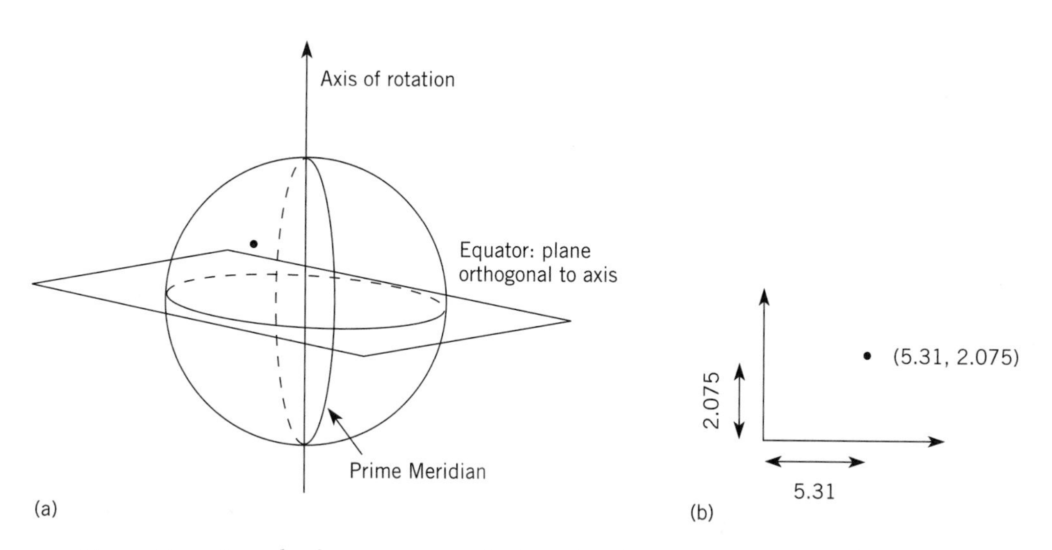

Figure 1-10: Spatial reference systems: Latitude–longitude pairs measure angles from the plane of the equator (orthogonal to the rotational axis) and the Prime Meridian (the meridian of Greenwich, England). The World Geodetic Reference System specifies a particular ellipsoid to use with this model of angular measurement. Other reference systems on earth may use a plane or a cone, connected to the geodetic reference system. Measurements on these flat projection surfaces are usually distances from a pair of axes (Cartesian coordinates).

spatial reference system does not guarantee that the rest of the measurements become compatible. Further description of these reference systems appears in Chapter 3.

Attribute Reference Systems

Just as reference systems apply to time and space, the same concept also applies to attributes. Each particular attribute scale requires its own reference system. Stevens' four levels provide a guide to the information content for such a reference system. For a ratio scale, the unit of measure must be given. For interval, the units and the zero point are required. The categorical levels require more information, because each category has its own definition. Thus, the levels of measurement provide a useful starting point in defining the reference system for attributes.

Some attributes require other kinds of reference information. Absolute measurements can simply state that they measure probability, or proportions, because these imply a unit of measure. Counts require attention to the kind of object tabulated. There is no bounded list of all the possible attribute reference systems, but the discussion in this chapter delimits a basic scheme for the information content of these reference systems (Table 1-1). Geographic information involves three components (time, space, and attribute), each connected to its own system of reference, a system of meaning.

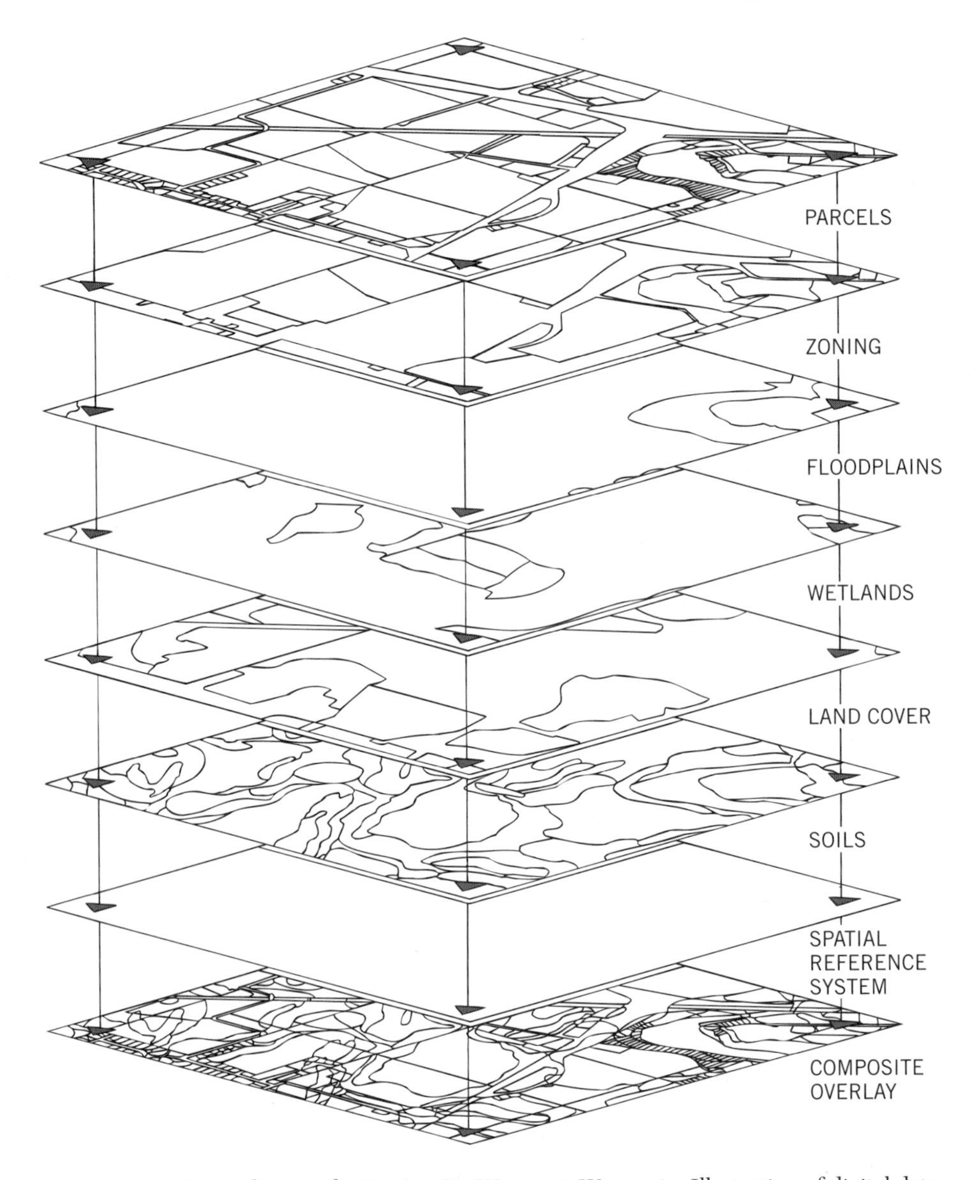

Figure 1-11: Layer diagram for Section 21, Westport, Wisconsin. Illustration of digital data registered to a common spatial reference system, through a connection to the National Geodetic Reference Network. Composite information products can be produced because the spatial references match.

TABLE 1-1: Information content for attribute reference systems.

Level of measurement	*Information required*
Nominal	Definitions of categories
Graded membership	Definition of categories plus degree of membership or distance from prototype
Ordinal	Definitions of categories plus ordering
Interval	Unit of measure plus zero point
Extensive ratio	Unit of measure (additive rule applies)
Cyclic ratio	Unit of measure plus length of cycle
Derived ratio	Unit of measure (ratio of units; weighting rule)
Counts	Definition of objects counted
Absolute	Type (probability, proportion, etc.)

To use measurements effectively, additional distinctions must be made. These distinctions do not come from the numbers, but from a larger framework surrounding the measurements. A spatial reference system provides a mechanism to construct a more integrated structure, but coordinates by themselves do not ensure compatibility of diverse information. Time and attribute also have reference systems, but these three systems just provide the basic axes. A more comprehensive framework must include the interactions of these three components. The next chapter develops such a framework for geographic information.

CHAPTER 2

MEASUREMENT FRAMEWORKS

CHAPTER OVERVIEW

- Present a comprehensive taxonomy of rules that govern measurement of geographic phenomena.
- Provide examples of major measurement frameworks.
- Explain basic principles of two-dimensional topology.

As the discussion of levels of measurement and reference systems in the previous chapter demonstrates, measurement depends on rules. Stevens tried to construct a system based upon "assigning numbers to objects according to rules," and many social science disciplines found this framework helpful. In most situations, the identity of objects was not a question. Psychologists study people, economists study businesses, and so on. When dealing with geographic information, however, the nature of the object is not simple at all. In fact, much of the diversity in geographic information arises from the objects themselves and their relationships. A set of rules for measurement will be called a **measurement framework** in this book. Differences between frameworks explain many of the seemingly incompatible viewpoints within the GIS community.

In the database literature, the term **data model** applies to the logical structure of

Measurement framework: A scheme that establishes rules for control of other components of a phenomenon that permit the measurement of one component. Geographic information has three components: time, space, and attribute.

Data model: In the database literature, general description of sets of entities and the relationships between these sets of entities (Ullman 1982); collection of object types, collection of operators on those object types, and a collection of integrity constraints (Codd 1981). In a GIS, composed of a measurement framework and a scheme for representation.

entities and relationships. In GIS, however, the term data model has been attached to a more primitive role in the implementation of systems of representation, which are discussed in the next chapter. What are termed data models usually developed directly from map display techniques, which are often the final results of a series of operations, each with its own assumptions. Because a measurement framework precedes any choices about the storage or symbolization of the data, it will avoid potential confusion when separated from the issues of representation.

There are a number of ways to develop a systematic description of a **database**. Many collections of data are assembled with little attention to fundamental assumptions; thus, the distinct steps of measurement and representation might be very hard to discern. The database schema may be inherited from some other application or developed to suit technical issues of representation. To adopt the more formal approach of a mathematical theory, a model contains symbols, relationships, and **axioms**. Codd (1981) recognized the same three components in structuring a data model and used the terms object types, operators, and integrity constraints. In a mathematical context, axioms are asserted without proof. Similarly, a geographic database makes some assumptions about the nature of the landscape. Establishing a reference system for time, space, and attributes is not enough. The most basic step in creating a geographic database identifies the objects that can be measured and the nature of relationships between the objects. These decisions form the basis for a measurement framework.

EXAMPLE OF A SIMPLE MEASUREMENT FRAMEWORK: THE GEOGRAPHICAL MATRIX

One framework for geographic data has been in circulation for at least the past thirty years; it simply asserts that objects have attributes. This *geographical matrix* (Berry 1964) has rows for places and columns for attributes of those places (Table 2-1). Structurally, there is no real difference between the geographical matrix and the *flat file* of **cases** and variables in social science statistical packages. Both approaches could be implemented in the matrix metaphor of a spreadsheet. The geographical matrix approach takes the identity of the objects for granted, and then expects the attributes to be measured for each object. In the case of cities, the definition of each city is open to challenge. The political boundaries used in official statistics often fail to capture the social and economic life of a city. More important, the rows of the matrix just define

Database: Structured collection of data with software to provide access in different ways; has a data model, a data structure, and an implementation (representation).

Axiom: A proposition accepted as true without proof; an assumption that is formally recognized.

Case: In statistics, an individual unit of observation. Selecting the *unit of analysis* (discrete unit of control) establishes the measurement framework for statistically based studies.

TABLE 2-1: A geographical matrix for the cities of the United States

City Name	*Population 1990*	*% Office Vacancy*	*Debt/ Person*	*Rainy Days*
New York	7,072,000	18.1	$4778	111
Los Angeles	3,485,000	14.3	2296	35
Chicago	2,784,000	22.1	2160	114
Houston	1,631,000	19.1	2,430	90
Philadelphia	1,586,000	18.9	2,418	153
San Diego	1,111,000	22.7	1,482	42
Seattle	516,000	15.0	2,074	155

Source: All figures from Statistical Abstract of the United States, 1994 (Washington DC: Government Printing Office); % Vacant from Table 1228; percentage vacant for existing office buildings, June 1993; Municipal debt in dollars per capita, 1992 from Table 489; Average days per year with precipation of .01 inch or more from Table 385.

different instances of the same type, with no provision for their spatial relationships. One may add additional variables to the matrix for location, but such spatial measurements require different treatment compared to other attribute columns. Although quantitative geographers, such as Haggett (1965) and Berry in his later work, recognized that geographic objects are not such "independent spaceships floating in the void (Berry 1973, p. 18)," the simpler matrix vision still dominates the view of quantitative geography.

A geographical matrix for cities can be explained in terms of the three elements: entities, relationships, and axioms. Together, these define a measurement framework—a set of rules to structure the process. In the simple matrix form, cities exist and have attributes such as a tax rate and a population. This list of facts hides the relationships and assumptions behind each statement (Figure 2-1). Some attributes, like the tax rate, belong to the corporate entity of the city and apply uniformly everywhere within. Other attributes are not so directly attached. It is much more constructive to recast the measurement of population as a relationship between the city and another class of entities: people. A person can have the relationship "lives in" with respect to a city (Tomlinson lives in Ottawa, Goodchild lives in Santa Barbara, Dan-

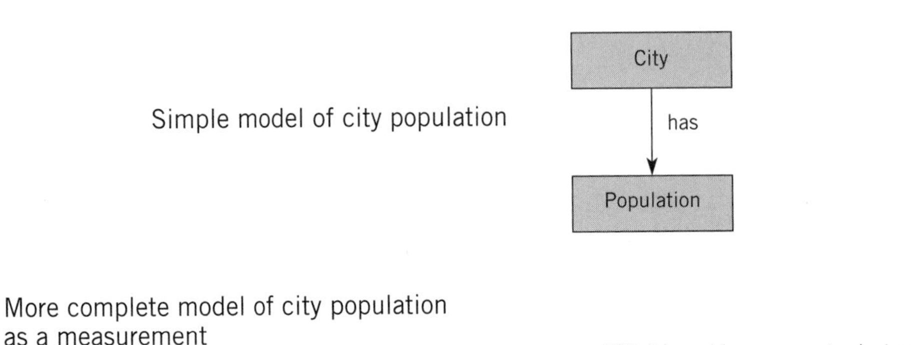

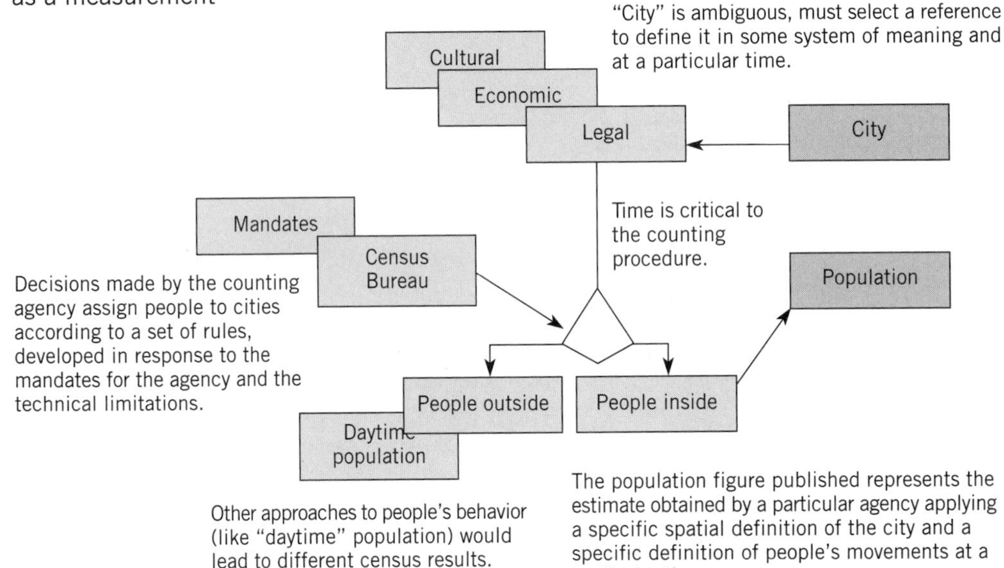

Figure 2-1: Two versions of city data model. In the simple formulation, a city "has" a population, as a kind of inherent property. In a more careful examination, a city is a spatial unit, and people move in and out of the city. A measurement procedure like a census, applied at a specific time, produces a particular result as a sum of the number of people found in the city. This procedure requires a number of assumptions and definitions.

germond lives in Redlands, etc.). The population measurement attached to a city comes from a rule to aggregate these relationships into a number.

Such a simplified framework serves important functions, but each step in the process deserves attention. Various definitions of population can best be described as assumptions about the relationship between people and cities. For example, we might say that a person can only live in one city at a time. But which city do we count: the one they work in, the one they are in at 11:15 on April 1, or the one where they registered to vote most recently? The choice between such axioms for geographic information is not really a mathematical issue. The institutional, social, and cultural context for the information system motivates the basic assumptions. The case and variable format of the simple geographical matrix fails to reveal the geographic questions of time, space, and attribute because it makes everything simply an attribute. Thus, a full understanding of GIS requires a richer set of measurement frameworks.

CONTROL AND MEASUREMENT

Geographic information involves three components: time, space, and attribute, as introduced in the previous chapter. Sinton (1978) described three possible roles for these components. In his scheme, in order to measure one component, one of the others has to be *fixed* while the third serves as **control**. To give an example, an old tide gauge has a strip recorder, one pen, and a long roll of paper that can be pulled across a drum (Figure 2-2). The location of the gauge is fixed (has no variability), and the speed of the paper is firmly regulated or controlled. Control of speed means that distance along the paper directly mirrors the passage of time. A mechanical linkage connects the pen to a float, so that the pen on the paper reproduces the movements of the float in the tube (probably at a reduced scale). Then, the recorder can measure the float height (the intended attribute) as it varies over time. In the digital era, the control would consist of periodic samplings (discrete events in time) that measure the height of the water. This form of environmental sampling may seem totally obvious; are there any alternatives? Yes. Instead of using time as the control, a sensor could be designed that recorded the exact time that the water passes a certain fixed height. This mechanism could be designed to ring an alarm or to shut a floodgate, as well as to record the time. In this case, height becomes the control and time is measured.

The role of control is an inescapable part of obtaining a measurement. In the tide gauge example, some of the variability in either time or water height has to be sacrificed to measure the other. Control can come in different forms, but it always imposes an element of discrete divisions on one component to serve as a basis for a measurement operation on another component. The discrete element of control puts a limit on the **resolution** of the information. In turn, the resolution sets some bounds on the **accuracy** of the measurement system. These connections begin with the mea-

Control: A mechanism of restraint on the variation of a system to permit measurement of one component of a phenomenon while other components vary only within the limits of the control.

Resolution: Least detectable difference in a measurement; in a geographic context, resolution applies to all components (time, space, and attribute) according to the measurement framework. Not identical to accuracy.

Accuracy: Closeness of a measurement to a value thought to be true; repeatability can be estimated by repeated measurement, measured by variance for continuous measures; accuracy of classification for categories can be summarized by a misclassification matrix when compared to a survey of greater accuracy.

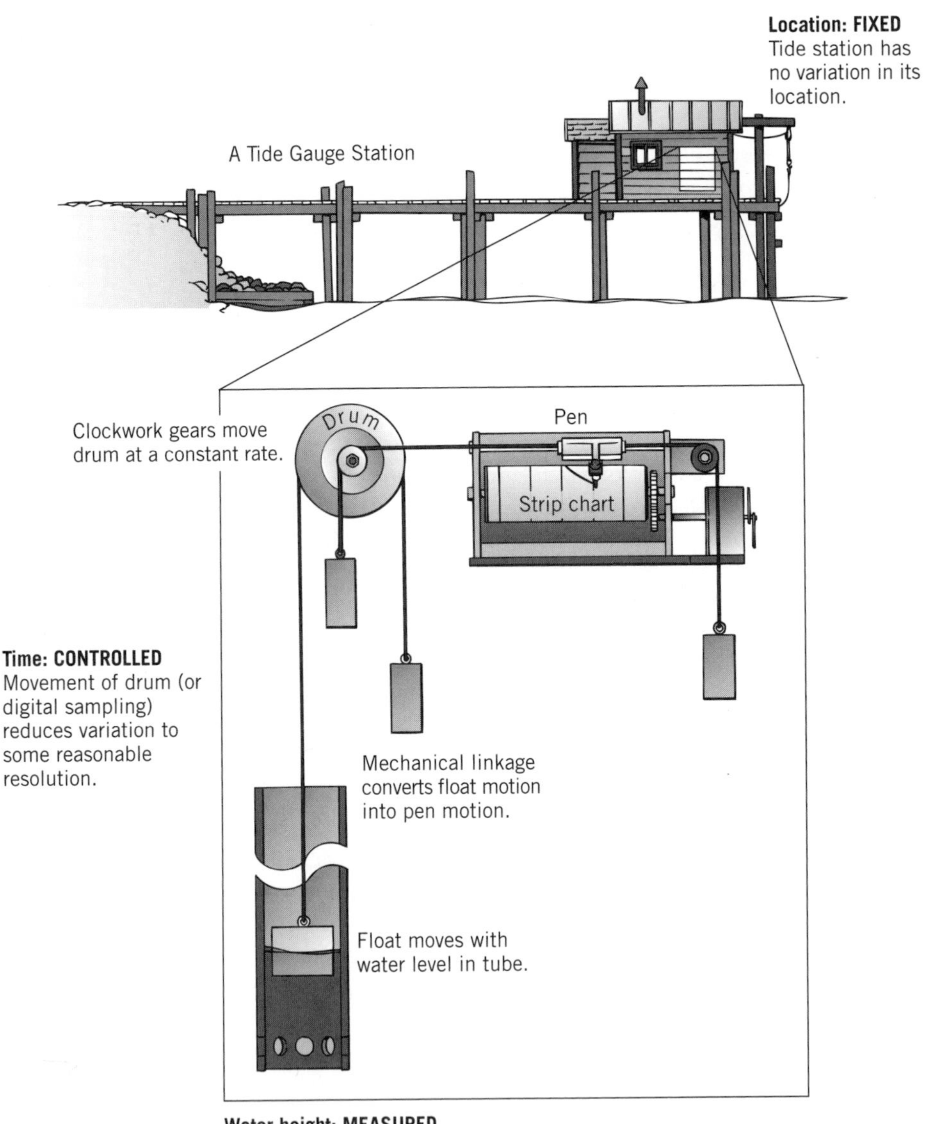

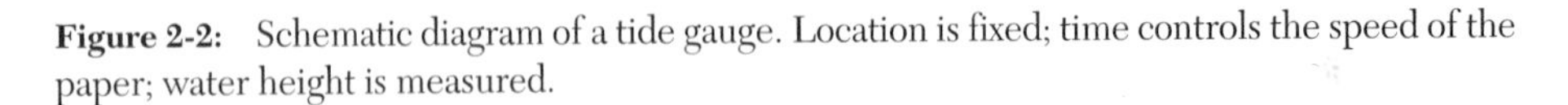
Figure 2-2: Schematic diagram of a tide gauge. Location is fixed; time controls the speed of the paper; water height is measured.

surement framework, but they become more prominent in understanding the representation process in the next chapter.

Unlike the tide gauge examples, most maps are conceived as snapshots in which the time element is fixed. This illusion of an instantaneous view from above predates aerial photography. Fixing time leaves the two components normally described as the basis for thematic cartography: space and attribute. In some abstract sense, we may conceive of a totally continuous **field** with the attribute value changing continuously as we move continuously through the space. Cartographers simplify the mathematical field to a fairly concrete concept of a **surface**. Operationally, a surface must be approximated by a measurement procedure. This chapter covers a range of different methods, starting with the two opposite approaches: control by space and control by attribute.

A spatial reference system introduced in Chapter 1 provides a structure for spatial measurement, but the spatial reference must not be confused with the choice of control. Part of this confusion comes from association of the term *geodetic control* with a spatial reference system. Geodetic measurements help establish the connection between local spatial measurements and a worldwide spatial reference system, but relating the coordinates to a global system does not determine which components will be used as control to permit measurement of the other.

In some cases, space is used as a control. The most classic instance of spatial control involves a systematic spatial sample, such as the TM sensor on Landsat that records a value for each **pixel** spaced at 30 meters. These pixels might be carefully aligned to the Universal Transverse Mercator (UTM) projection, a spatial reference system, but it is the choice to sample every 30 meters that provides the control, not the geometric reference. In Table 2-1, the political city limits serve as control for the geographical matrix. In both cases, the attribute is aggregated over a spatial object—pixel or city. Forms of spatial control will be discussed later in this chapter.

In other cases, the attribute serves as the control. If we have a map of the forests in France, the control is the binary simplification into forest and nonforest. Certainly there are variations between forests, but these distinctions are simplified to provide the control. Given this division, the lines on the map make the best attempt, under the circumstances, to measure the location of that distinction. These boundaries can be measured using UTM as a spatial reference system, without altering the role of the forest category in acting as control.

This general discussion contrasts the opposite extremes. Now, each measurement framework will be developed in detail, starting with the use of attribute as control.

ATTRIBUTE AS CONTROL: ISOLATED OBJECTS TO CONNECTED COVERAGES

The role of control may be most apparent when categories are used to chop the attribute into discrete pieces. Then, in principle, the spatial position of the boundary

Field: An abstract construct of a mathematical relationship viewed as a spatial structure. In physics: a region of space subject to a force (as a magnetic field).

Surface: A spatial distribution that associates a single value with each position in a plane (technically a field of a single-valued function), usually associated with continuous attributes.

Pixel: Smallest resolvable unit in an image; an area (usually rectangular) forming a part of a systematic, uniform division of a study area. Contraction of *pic*ture *el*ement.

TABLE 2-2: Summary of attribute control frameworks

Isolated Objects	
Spatial Object	Single category distinguishes from void
Isoline	Regular slices of continuous variable
Connected Objects	
Network	Spatial objects connect to each other, form topology (one category or more)
Categorical Coverage	Network formed by exhaustive classification (multiple categories, forming an exhaustive set)

between categories can be freely adjusted (measured). Two basic approaches can arise from the decision to use the attribute as the control: *isolated objects* and *connected coverages*. Each of these two approaches has two variants (Table 2-2). If a single category is taken in isolation, then the geometric objects are described in isolation, a framework termed *spatial object*. A collection of spatial objects can be developed by a systematic division of a continuous attribute into slices (or **contour** intervals). To conform to cartographic terminology, this will be called an *isoline framework*. Alternatively, objects may have relationships to each other. A *network framework* recognizes that linear objects have relationships of connectivity. If the attribute partitions space exhaustively into categories, then the geometric result is a connected network of nodes, chains, and polygons that carry out that exhaustive partitioning. This special form of a network will be termed a *categorical coverage*. Each of these frameworks is explained in the following sections.

Since all of these frameworks rely on attribute control, the attribute changes in discrete jumps where the boundaries of objects fall. While there is some unity to this *object view*, the similarity distracts attention from the very real differences in the relationships implied within each framework.

Isolated Object Frameworks

Control by attribute is most directly implemented by a single category surrounded by the void—everything else. There are two variants: spatial objects derived from categories and isolines derived by slicing a continuous surface.

Contour/isoline: Contour line: a line connecting points of equal elevation on a topographic surface. An isoline generalizes this concept for any continuous distribution.

Spatial Object Framework In the simplest case, there is a single category, like an airport, to be located. Each occurrence of an airport is effectively surrounded by a void (everything that is not an airport). Because the category reduces to a simple binary yes/no, edges are sharp. Also, because the objects are isolated, objects need no relationship to other objects of the same kind. This spatial object measurement framework lies at the core of many mapping programs and virtually all computer-aided design (**CAD**) packages. This framework allows the spatial description for **features** such as oil wells, navigation buoys, highways, and wetlands using simple geometric **primitives** (point, line, or area) for each object (Figure 2-3). Each object is described as a geometric whole, since it will forcibly occur in isolation. The message of the object framework is: "Here is an airport"; "Here is another airport," and so on. In the pure form of this framework, the only relationship is between the object and a position; there are no relationships between objects. Linear objects depart from this to some extent, thereby creating the need for the network framework.

Isoline Framework The spatial object approach can be extended relatively easily to handle a classified version of a continuous attribute. For example, the continuous variation of elevation can be simplified into the query for all points having an elevation of exactly 1000 meters. These points connect to form a contour on a surface. Each closed loop of a contour becomes an isolated polygon, sharing the cartographic form of the object framework, though it is generated by an entirely different process. A collection of contours is then created by a regular spacing of discrete intervals (Figure 2-4).

Isolines make a clear case for the role of control. The original attribute is continuous, but it is reduced in measurement level to a discrete set of specific values. The variation on the surface between contour intervals is simply lost. It is quite common to treat an isolated contour as a simple category: above or below the threshold. Such simplification risks losing sight of the continuity of the original surface. Given a collection of isolines, the human visual processing system can construct an impression of intermediary shapes, but these relationships come from sophisticated image processing in the human brain, not from the isolated contour lines themselves.

Connected Coverage Frameworks and Topological Relationships

Both the previous frameworks assume that objects remain distinct. Points, by definition, cannot merge with other points. Areas in an isolated object framework are also forcibly distinct from other areas of the same type; otherwise, the area would simply

CAD: Computer-aided design, software packages designed to automate drafting of mechanical drawings.

Feature: Cartographic feature: an instance of a defined class of objects that cannot be divided into objects of the same type.

Primitives: Basic components that are sufficient to build a larger system; the primitives of two-dimensional geometry are points, lines, and areas.

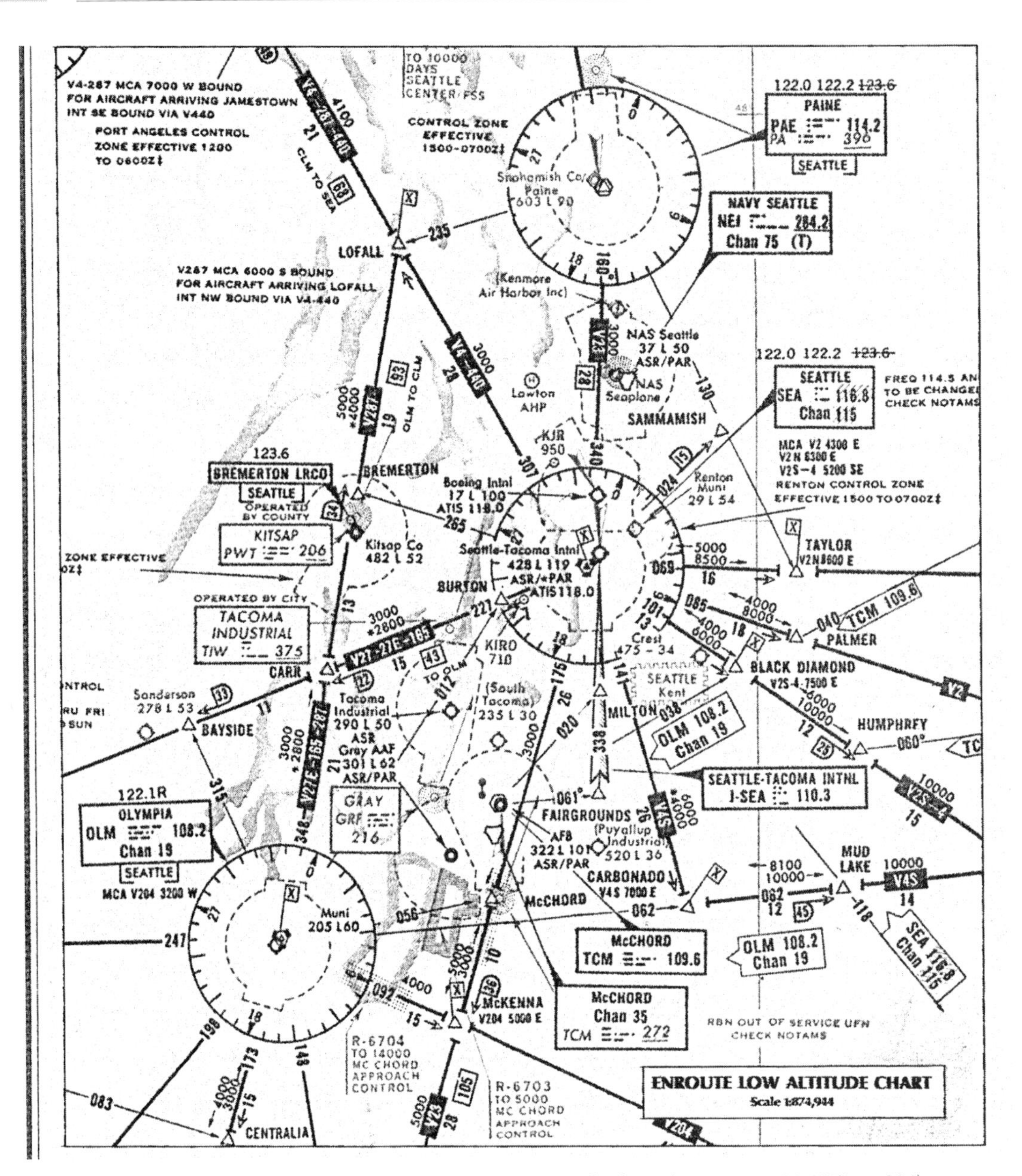

Figure 2-3: Examples of object view: Enroute low altitude chart. (Source: USGS 1970, p. 304)

get larger. As mentioned earlier, line objects are more complicated. Most single categories of linear objects, like "road" or "river," will connect to each other, forming a network. In analyzing these connectivity relationships, it is important to recognize that

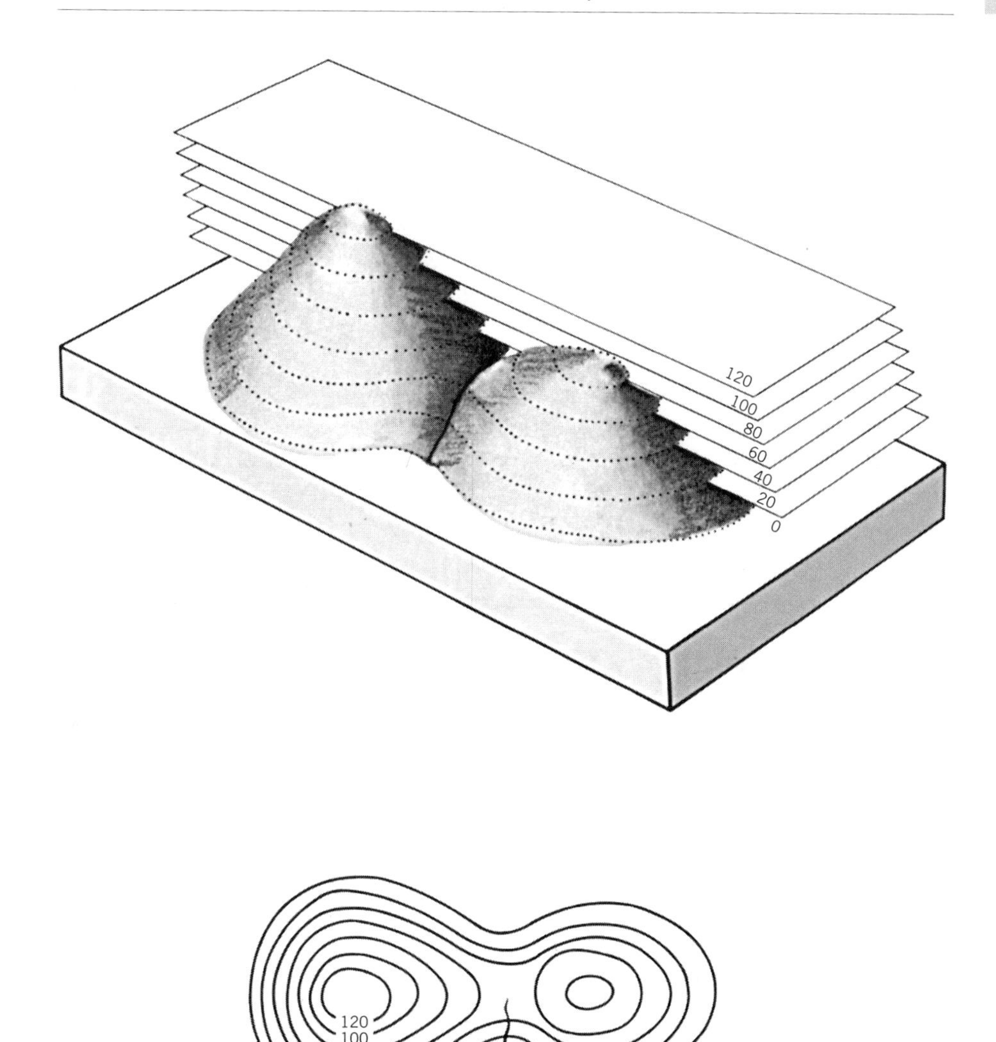

Figure 2-4: Isolines are created by controling the variation in the attribute to certain values then measuring the location of these slices. Redrawn from Robinson and others (1995) p. 509.

streets (lines) connect at intersections (points). The streets also divide the area into a set of blocks or polygons. Thus, drawing lines creates points and areas as well. An isolated object framework does not recognize these relationships and thus misses much of the spatial structure of a network.

The geometric roles of points and lines and polygons can be explained most simply in terms of **topology**. Topology, as used in cartography, concerns those characteristics of geometric objects that do not depend on measurement in a coordinate system. For a mathematical topologist, the network formed by streets or rivers on a plane is hardly very exciting, but an understanding of the relationships between dimensional objects has been quite important in the development of geographic information technology. Using the terms adopted in the Spatial Data Transfer Standard (Figure 2-5), **nodes** bound **chains** and chains bound **polygons**. (Note: Various software packages use *arc* or *edge* or *segment* for chain.) The relationship of boundary also implies its

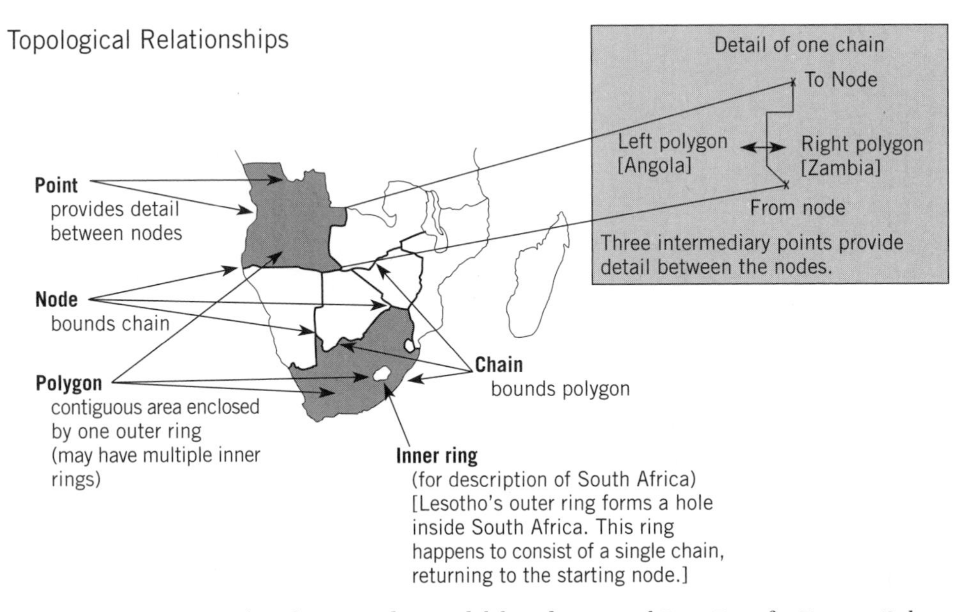

Figure 2-5: Terminology for vector data model, based on Spatial Data Transfer Format. Relationships to other primitives create a topological structure.

Topology: A branch of mathematics concerned with those properties of geometry that are independent of a distance metric and are unchanged by any continuous deformation. In cartography, topology refers to *combinatorial topology*. The other branch, *point-set* or *algebraic topology*, "emerged in the twentieth century as a subject that unifies almost the whole of mathematics" (Boyer and Merzbach 1991, p. 622) and is used as the basis for relational databases.

Node: A zero-dimensional object that is a topological junction (or end point) and a geometric location.

Chain: A directed set of nonintersecting line segments with nodes at each end and reference to left and right polygons.

Polygon: An area (bounded continuous two-dimensional object) consisting of an interior area, one outer ring, and zero or more nonintersecting nonnested inner rings.

inverse (mathematically termed *co-boundary*, but also rendered in the passive voice *bounded by*): polygons are bounded by chains and chains are bounded by nodes. From these direct relationships, more complex neighborhoods can be built; for instance, the adjacent polygon is the other object bounded by each bounding chain. The *qualitative* relationships of connectedness and contiguity are thus more fundamental than the *quantitative* properties of length and area.

Some topological relationships need careful explanation. A closed line in the isolated object framework is frequently called a "polygon," but this line acts as a boundary of an area, not the area itself. In an isolated approach, there is no way to know if there is a hole inside the area. The terms for cartographic topology (Figure 2-5) are designed to handle these tricky situations. A polygon is an area that must be bounded by one **ring** (its outer ring). In the simple case, the outer ring suffices, and the ring looks like an isolated object. If there are holes inside the area, they are represented by additional *inner rings*. In some cases such as the land–water coverage of a glaciated region, there can be thousands of inner rings in a single polygon. The area inside the inner ring is not a part of the enclosing polygon, so there is no need to construct infinite recursions of nested structures. Figure 2-6 shows a series of maps of Isle Royale National Park in Lake Superior. Lake Superior is one polygon with a number of inner rings for islands. Isle Royale is one polygon with many inner rings, one of them for Siskiwit Lake. Siskiwit Lake, in turn, has an inner ring for Ryan Island, at the most detailed scale. Each of these polygons has attributes (such as land or water), but there is no topological need to record how deeply nested the inner rings are, because the polygon is defined by being continuously connected, not based on the attribute of land or water. All the land may belong to the state of Michigan and to the national park, but these groupings are political, not geometric.

The relationships of topology apply to any geometric information, whether these interactions are recognized or not. Checking topological integrity provides a powerful means to verify data quality, a topic developed in the next chapter. Despite the commonality of topology, two distinct measurement frameworks emphasize different aspects of these relationships. A network framework focuses on the connectivity of linear elements by themselves. A coverage framework uses a linear network of boundaries to define a set of contiguous areal units. Both of these frameworks, though they are unified by a similar topological model, make different assumptions about measurement.

Network Framework A network framework is based on identifying a particular category, much like an isolated object framework, but the network recognizes the interaction of the identified objects in a coherent structure. A category such as river or road is still surrounded by the void, but the meaning of a river or a road requires connection to other elements in the network. These relationships require a more comprehensive approach than an isolated object.

Ring: A sequence of nonintersecting chains that close.

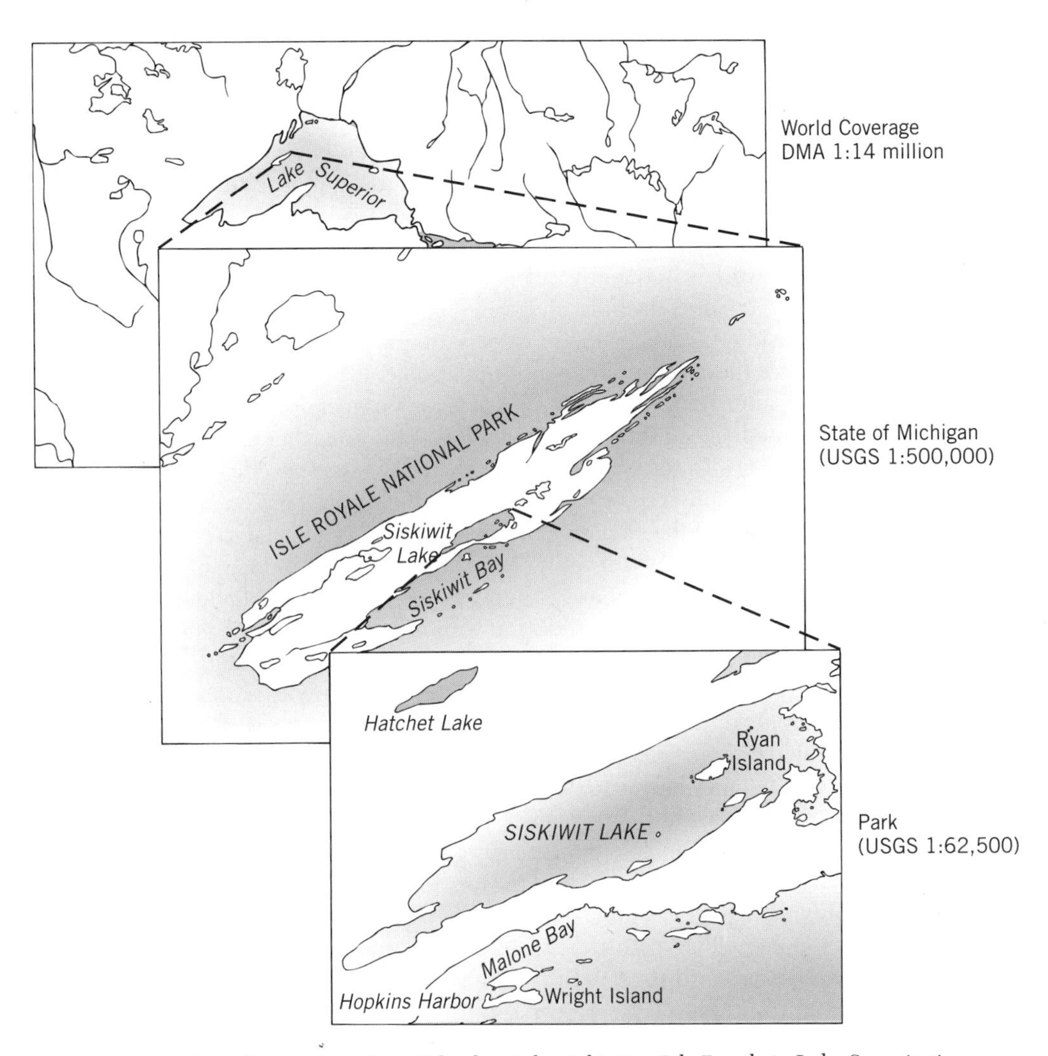

Figure 2-6: Nested inner rings: Ryan Island in Lake Siskiwit in Isle Royale in Lake Superior in North America.

The relationships in a network create additional distinctions. Some networks can be treelike (dendritic) if they have only one path to each node (Figure 2-7a). Conventionally, streams are considered to fit this model, but there are all sorts of exceptions in practical application (Figure 2-7b). Lakes and reservoirs form areas that do not behave as simple points in a tree network. There are also stream networks that diverge into deltas. At a detailed scale, braided stream channels, with islands inside the stream bed, become most confusing to the assumption of treelike behavior.

The model of a **planar graph** applies to most road networks, but the complexities of overpasses, tunnels, one-way streets, and various three-dimensional structures violate assumptions of planarity (Figure 2-8). A highway network requires a non–planar graph model that relaxes the strict interpretation of some nodes. Nevertheless, topological relationships are still required to ensure connectivity throughout the structure, even if it is not on a single plane.

Categorical Coverage Framework The other connected coverage framework uses another form of control by attribute. Whereas the spatial object approach depended on a single category surrounded by the void, a more general approach to nominal measurement recognizes that the whole space must be classified into mutually exclusive categories. These exhaustive categories ensure that the attribute has a value at all points. The discrete nature of the categories provides the control, differentiating a categorical coverage from the methods of surface measurement that provide a continuous value at all points. Geographers have offered a variety of terms for this circumstance. Some insist on keeping alive Bunge's use of *area-class map* (Bunge 1962; Mark and Csillag 1989), but this term is somewhat ambiguous; it could refer to a

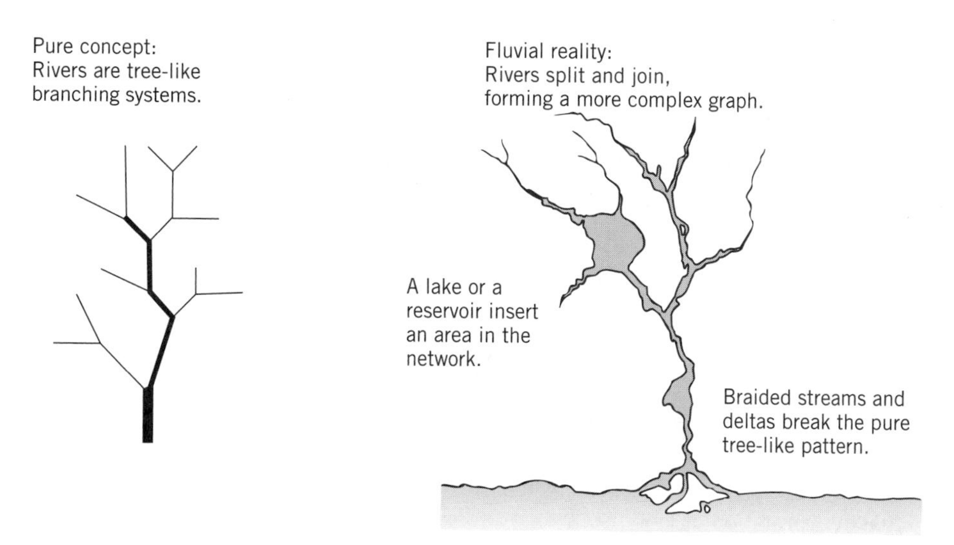

Figure 2-7: Comparison of pure forms and geographic application I — Dendridic networks: (a) Pure dendridic model exhibits tree structure; all branching in one direction; (b) river systems may include cycles, and branch in both directions.

Planar graph: An arrangement of nodes and lines such that the lines intersect only at nodes, thus remaining embedded in a plane (or a surface topologically transformable onto a plane).

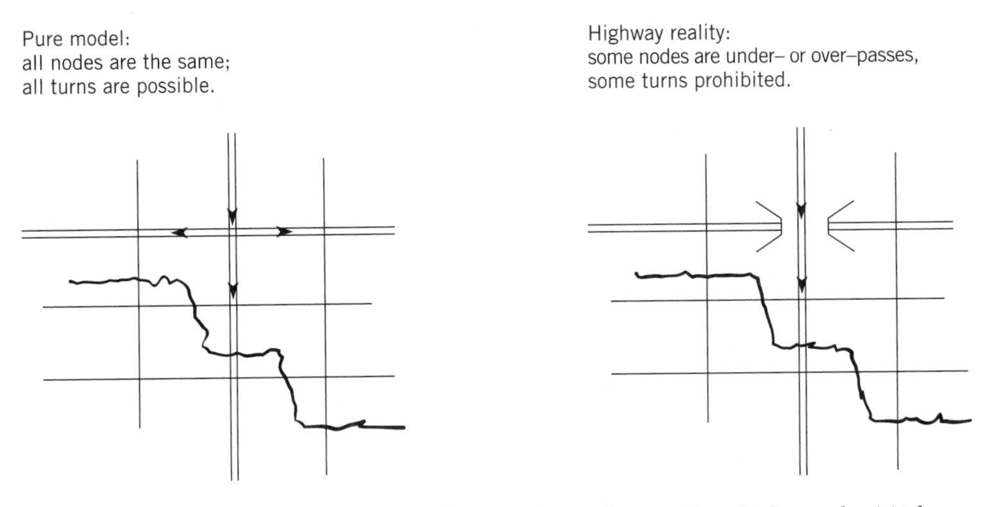

Figure 2-8: Comparison of pure forms and geographic application II — Cyclic graphs: (a) Planar graph model permits cycles, but all intersections are the same; (b) most highway systems are built on the earth's surface, but some connections bypass each other in the third dimension.

choropleth map as well. I find it more apt to call these **categorical coverages** (Chrisman 1982a) to focus on the measurement, not the graphic appearance. Because the different categories must abut, a boundary is required to separate them. Thus, each boundary measures the location of the transition between a specific pair of categories. The topological model applies particularly to this framework.

For example, a network of boundaries distinguishes a set of land use categories around Cwmbran in Wales (Figure 2-9). The basic logic of this land use map started with the list of classes, serving as control. This particular list was derived from the Anderson codes (Anderson and others 1976), with specific interpretations for the British landscape. Each point on the map was placed in the category that fit best, within certain limitations of resolution. Each boundary line acts as the break between the adjacent categories.

The research literature on geographic information is full of challenges to the measurement assumptions made in constructing such categorical coverages. Some of these come from the mismatch between Stevens' view of nominal categories and the actual interpretation of land use from prototypes (mentioned in Chapter 1). Many of the boundaries in Cwmbran are quite sharp, where a land use category stops at a fence or a property parcel boundary. Some boundaries at this scale are rather wide, because they follow roads or hedges. Some categories in the system are less well defined, such as *heath*, which represents various vegetation types including gorse, bracken, and tundra-like sedges. The transition from improved sheep pasture to heath may not be

Categorical coverage: An exhaustive partitioning of a two-dimensional region into arbitrarily shaped zones that are defined by membership in a particular category of a classification system.

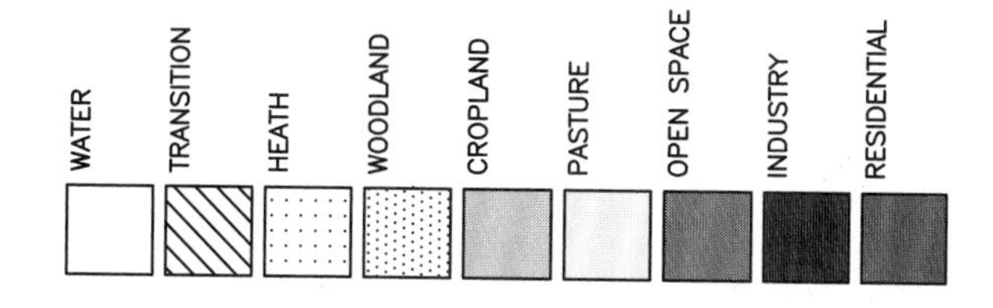

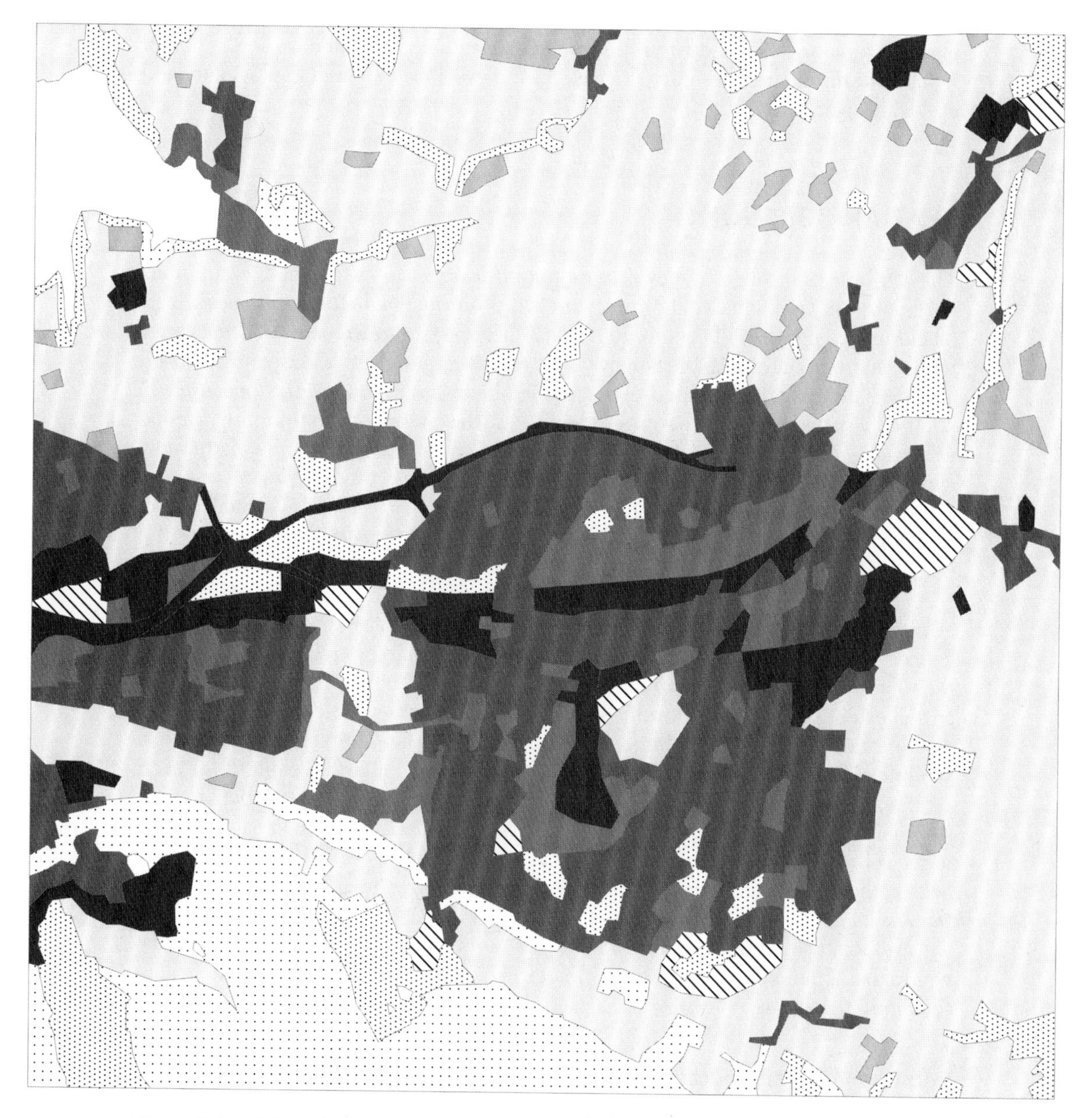

Figure 2-9: Categorical coverages partition the whole plane into an exhaustive set of categories. Example shows land use for Cwmbran, Gwent, UK, during May 1981. Region covered is 8 km by 8 km on the UK National Grid. Land use classification by Chrisman (1982a) using field methods compiled on OS 1:25,000 topographic map ST 29/39.

sharp. In addition, the process of making this particular map was far from exact. I sketched some of the lines in pencil onto a copy of the 1:25,000 topographic map. These imperfections are quite typical in implementing this measurement framework. The representation issues discussed in the next chapter will analyze the generalization inherent in geographic representations. A categorical coverage intentionally simplifies by choosing a single category to represent each location. This framework is common in many forms of land inventory such as soil surveys, forest stand mapping, and land use/ land cover mapping.

SPATIAL CONTROL

The attribute-controlled frameworks do not offer direct tools to handle continuous attributes. The traditional contour map limits the information to the position of selected elevations, not the full continuum. To handle a surface with continuous measurements, another kind of measurement framework is required. In a sense, positions on a surface are less critical than they are in the object-based frameworks because any point on a surface can be measured. Ideally, a surface takes on continuously varying values—an infinite number of values between any given pair of points, no matter how close. Of course, it is impractical to measure each of these infinitely many points; a subset of points must be used to characterize the variation in the rest. A spatial control framework, then, controls spatial position to obtain measurements of the attribute (the height of the surface, or any other attribute distribution).

Several distinct possibilities exist to obtain a measurement of a continuous attribute. An ideal field can be observed at a point, a location with no extent. We could conceive of moving an ideally tiny thermometer or barometer around and recording the value obtained at each sampling point. In practice, though, instruments have some spatial extent. At the other extreme from the pure point, a sample could be an average or a sum over an area as large as the spacing between samples. For example, a satellite sensor records the number of photons in a particular wavelength for each position (cell or pixel) it observes. If the cell is small enough, then the particular measurement rule applied is not very important. There will not be much variability within the cell, and any rule would tend to come up with the same value. As cells become coarse, the distinctions between the rules become more critical to interpretation of the results.

Just as attribute control subdivided into a number of distinct frameworks, spatial control has a number of variants. These alternatives can be organized first according to whether they use points or areas, then based on the rules applied (Table 2-3).

Point-based Frameworks

A point-based measurement does not require much consideration about measurement rules. The main issue concerns how to choose the points. There is one primary

TABLE 2-3: Summary of spatial control frameworks

Point-based Control	
Center point	Systematic sampling in regular grid
Systematic unaligned	Random point chosen within cell
Area-based Control	
Extreme value	Maximum (or minimum) of values in cell
Total	Sum of quantities (e.g., reflected light) in cell
Predominant type	Most common category in cell
Presence / absence	Binary result for single category
Percent cover	Amount of cell covered by single category
Precedence of types	Highest ranking category present in cell

option—center point—and some other possible techniques rarely encountered in practice.

Center Point Framework The purest form of a space-controlled method records the measurement at a regular lattice of points. If drawn as a grid of parallel lines, the point samples can be taken at the intersections (Figure 2-10), or these points can be seen as a *center point* of the cell. This method could also be labeled "systematic aligned" (Berry and Baker 1968), because the points are all lined up. The center point system creates a regular arrangement of points and applies a simple rule at each point (simply recording the value).

Digital Elevation Matrices (**DEM**) are based on this framework, though they may not be measured directly as point samples. At the US Geological Survey (USGS), for instance, DEMs can be constructed from contours (isoline frameworks), image matching (a kind of area-based control), and profiles. The profile information is the closest to a regular point sample. In all these cases, some form of interpolation may be involved to produce the point values (see Chapter 9 for a coverage of interpolation and other transformations between frameworks).

The center point measurement framework may also be applied to measure categorical attributes. In principle, the category at the point is recorded, though it may be hard to determine certain categories at an infinitely small point. For instance, if the point falls on a leaf of a tree, does that imply a forest? The tree could be in a garden of

DEM: A framework for recording spot elevations in a regular rectangular grid (matrix); an acronym originally created from Digital Elevation Model at the US Geological Survey. To avoid ambiguity, DEM will be used exclusively for a grid framework, so it can be read matrix.

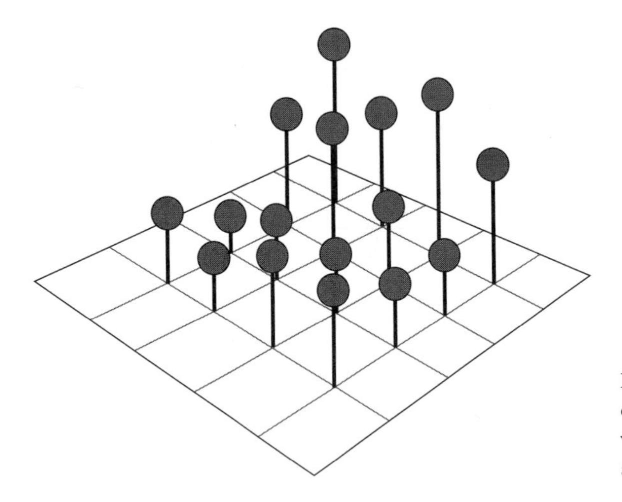

Figure 2-10: Point-based space controlled measurement framework: Attribute value recorded at a regular lattice of points.

a residence, in a parking lot of a huge factory, or in many kinds of nonforest land use. The Swiss Arealstatistik project conducted its land use inventory based on a 100 meter spacing of sampling points. Despite the point sample, there is usually an area component to land use determination. The Swiss Federal Bureau of Statistics (1992) developed a rule book to examine the surrounding 100 meters and determine the land use attribute assigned. Figure 2-11 shows one of the rules to distinguish two forms of agricultural land. Large fields that can support mechanical agriculture belong in category 81. If obstructions to agricultural machines occur within 50 meters of the point, the sample point is downgraded to the lower quality category (82). In practice, it becomes difficult to distinguish point-based methods from the area-based methods.

For both continuous and categorical attributes, the spacing of the grid may interact with the spacing of the phenomenon. For instance, in the American Midwest, the

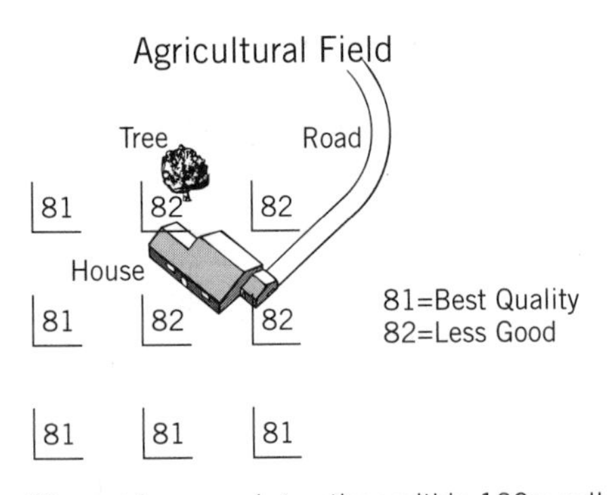

Figure 2-11: Example of a rule to classify land use in the vicinity of a point sample. Drawing simplified from Category 81 Günstiges Wies- und Ackerland & 82 Übriges Wies- und Ackerland (Swiss Federal Bureau of Statisics 1992).

landownership and road networks are laid out on a rectangular grid following the Public Land Survey System. The rural roads are very likely to occur at one-mile intervals. If a regular grid is laid over this terrain with the traditional cardinal orientation angle and with a spacing between points that divides evenly into a mile, then the chance of hitting a road is totally dependent on the location of the initial point (Figure 2-12). If

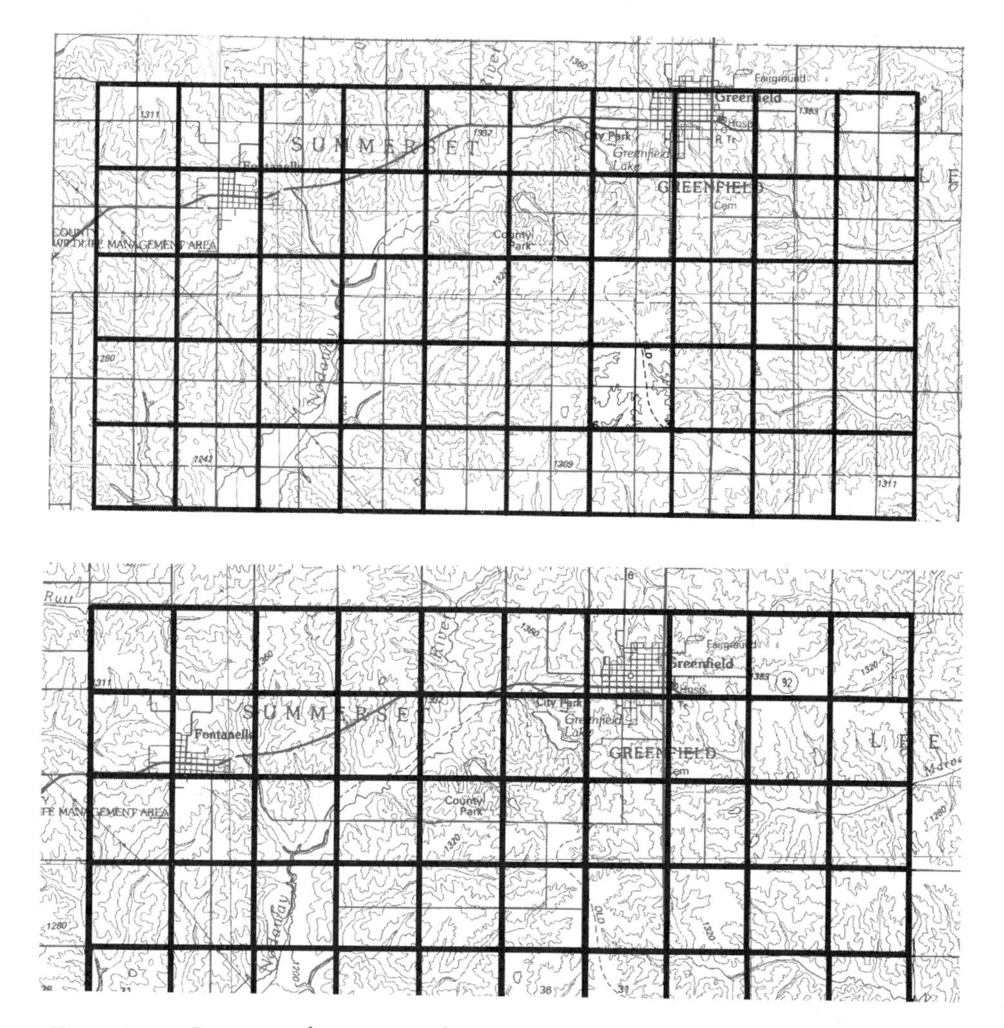

Figure 2-12: Interaction between regular point sampling and regularities in the landscape. Example based on USGS 1:100,000 topographic map 41094-C4-CF-100 for Adair County, Iowa. Grid spacing of 1600 m approximates the spacing of section lines of the Public Land Survey. (a) Initial point located in center of section rarely falls on a road, (b) initial point on road intersection often falls on road.

the initial point just misses the road, then the whole grid will avoid the road. Not all landscapes are as rigorously rectangular as the roads of Adair County, Iowa, but there are periodic elements to many land systems.

Systematic Unaligned Theoretically, a sample of points that is not so rigidly aligned would provide a more robust description of variation over the region. Random location of the point within the cell would respond more flexibly to the periodic landscape of Adair County. However, a "systematic unaligned" sample is rarely, if ever, used in cartographic applications. For continuous attributes, the irregular spacing would complicate many vital neighborhood operations (such as slope calculation, presented in Chapter 7). For categories, this form of spatially stratified sampling has been applied for some kinds of natural resource surveys.

Area-based Measurement Frameworks

A regular system of spatial control can also be seen as an exhaustive partitioning into regular geometric figures (a tessellation). In most cases, the basic cells are rectangular, or nearly so. (Satellite sensors see trapezoids due to the motion of the platform relative to the earth.) Systems of equilateral triangles or hexagons have been proposed, but they are not used in very many applications. For ease of discussion, the spatial unit will be called a *cell*. Within this spatial control, the measurement of the attribute depends on the attribute values present within the cell. The area-based frameworks apply different rules to decide which value to record using those available.

Extreme Value One simple rule takes the highest (or lowest) value available inside the cell. For example, on aeronautical charts, there is a lowest flight altitude given for rather crude cells (Figure 2-13). This value is based on the safe clearance over the highest obstruction in the cell. For flight safety this is a reasonable rule, but for other purposes it overestimates the elevation. Extreme value is an example of a rule that can be applied to ordinal and higher levels of measurement because they establish an ordering.

Total Compared to the extreme value rule that simply selects one value, other rules can use all the values available to contribute to the final result. Various arithmetic functions can be applied to measurements at the ordinal or higher levels, though there must be some assumptions about spatial resolution to identify the candidate values.

Measurement hardware often imposes the rule for assigning a value to a cell. A satellite sensor records the total number of photons detected over a certain period while passing over a certain cell (Figure 2-14). This is an area-based value, not a spot reading. The total photons reflected or emitted depend on the mixture of materials on the surface, intervening atmosphere, and other factors. Certain highly reflective objects can contribute the bulk of the light, even though they cover a small portion of the cell. These sensors actually use time as the control, because they sum the energy

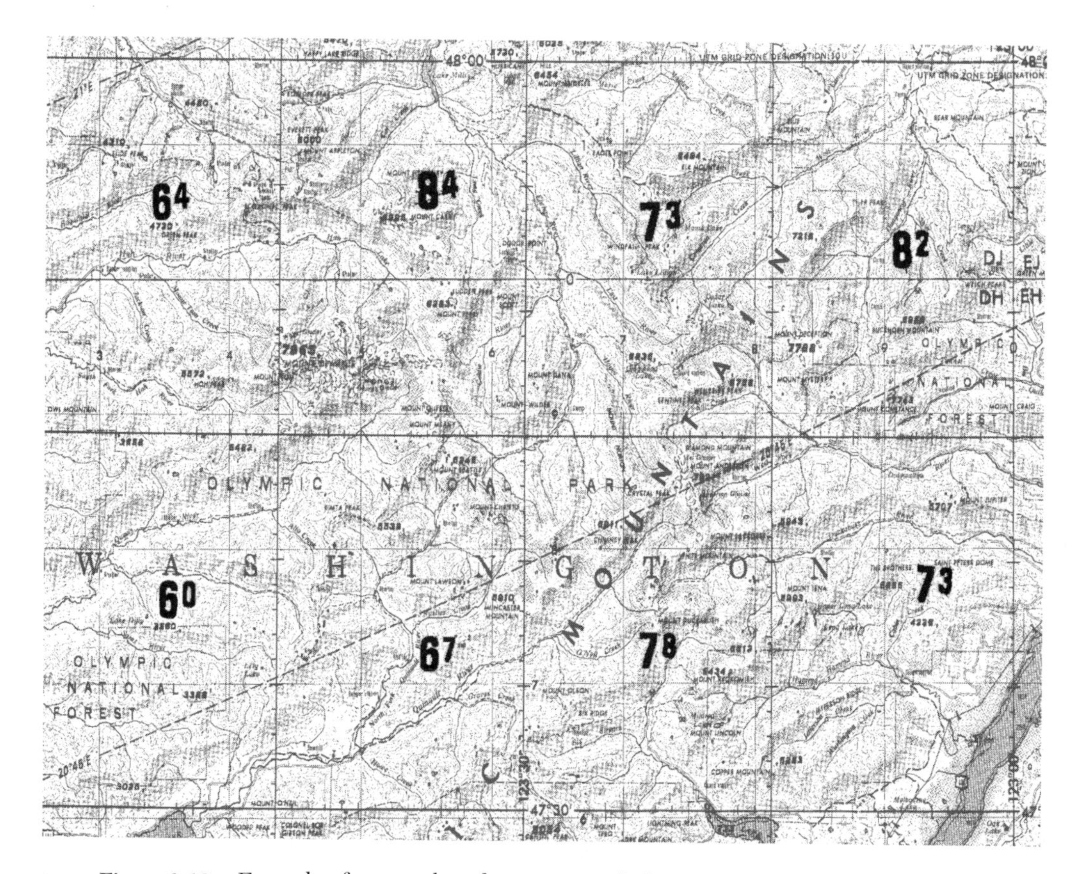

Figure 2-13: Example of an area-based space controlled measurement framework for elevation data. Aeronautical chart shows minimum safe flying altitude based on maximum obstruction within a cell (in this case, 15 minute quadrangle). Taken from DMA Joint Operations Graphic, published at 1:250,000. Figures given in 1000 foot units with superscript for hundreds: 14^8 means 14,800 feet.

as the satellite moves. Coupled with the geometry of the flight path, the time sample converts to a ground position, though it is not easy to recover the cell on the ground to high precision. For practical purposes, satellite images have to be treated as area-controlled sources.

Predominant Type For categorical attributes, the rules of arithmetic do not apply to the attribute values. Instead, there are a number of methods to characterize the distribution in a cell. All of them depend on measuring the amount of the various categories within the cell. This information is assembled, the cell value is decided, and most of the detailed information is discarded.

The *predominant type* rule assigns the cell to the largest category within the cell. This amounts to the *mode* of the distribution within the cell (the most frequent cate-

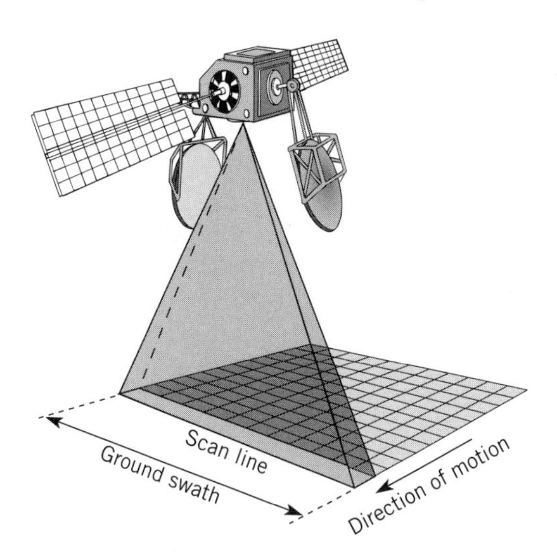

Figure 2-14: Satellite sensor records light reflected from an area.

gory). When the cells are fine and only one category appears in most cells, the choice of rule will make little difference. While predominant type seems to ensure the best result for the particular cell, this rule will underrepresent categories that tend to be thin relative to the cell width. Predominant type is the closest approximation to a categorical coverage, given the choice of spatial, not attribute, control.

Presence/absence and Percent Cover For categorical attributes, an isolated object approach is an alternative to the comprehensive coverage. Similarly, with spatial control, there are frameworks for isolated as well as comprehensive viewpoints. If a particular category has been selected, the cell can record whether or not that category is present. Such a presence/absence rule ignores all other categories, a step that may be necessary to encode linear networks such as streams or highways without losing them in surrounding categories. Separate grids for each category can be stored in binary layers as a "bit map." Of course, there will be no way to tell which category was more important within a cell that contained multiple categories.

One way to salvage more information would be to record the percent of cover, that is, an estimate of the proportion of the cell covered by the category. This measurement seems to retain more information about the cell, but without the spatial distribution. The percent cover measurement framework applies particularly to coarse cells, such as the one degree resolution world soil and wetland data (Matthews 1993). The spectral mixture approach to remote sensing (Adams and others 1986) also adopts this approach to measure graded membership. Rather than classifying a cell into one category or another, this procedure estimates the percentage mixtures of various spectral templates in the observed remote sensing measurements.

Precedence of Types If categories of a coverage can be ranked by importance, then another coding method could select the most important category within the cell—precedence of types. For example, one natural precedence occurs in treating the highway network, which has an ordering based on importance and traffic. A cell is usually coded for the highest class of road present. Once the cell has been given its single value, it is not possible to reorder the priorities.

Although spatial control simplifies the measurement process, there are still many decisions in choosing what value to record. Much of the regularly sampled geographic data currently available comes from automated equipment such as satellite sensors or scanned aerial photographs. The measurement rules for these situations are built into the hardware logic or the postprocessing of these data streams. These rules will influence the use of the data source.

RELATIONSHIP CONTROL

So far, the presentation of measurement frameworks reinforces the opposition between the object and grid methods. Sinton's original purpose was to show the different motivations behind these two common approaches to control. That observation is helpful, but the choice of control by either space or attribute does not exhaust all the possibilities, even discounting the difficulties of handling time.

When controlling space or attribute, the variation in that component is deliberately limited to discrete units. These units permit measurement in the other component. Another way to make a measurement is to define the unit through relationships between objects. This creates a new class of measurement frameworks, which includes some alternatives that do not fit into the orthodoxy of spatial or attribute control. Information organized in these frameworks also will not fit very neatly into software developed around the classical measurement frameworks.

Measurement by Pair

Measurements can apply to pairs of objects (Goodchild 1987). For example, trade statistics measure the flow of products and finance between countries (Table 2-4), but not the physical path taken by these flows. Similarly, traffic demand connects homes to workplace or other destinations. These interactions have formed a major topic in economic geography for many years (Ullman 1954), but they are hard to represent with simple object-attribute models and to portray with standard cartographic techniques (Figure 2-15). Interaction tables can potentially have as many cells as the number of sources times destinations. The measurement does not belong to either the source or the destination, but to the combination. Thus the control comes from the relationship between two objects.

These measurements could be attached to lines between the two endpoints, but

TABLE 2-4: Example of measurement for pairs of objects

Exports From/To	*North America*	*Europe*	*Japan*	*Australia & N.Z.*
North America	22,550	10,954	4,592	914
Europe	11,056	101,485	2,673	961
Japan	8,954	6,016	0	680
Australia & N.Z.	495	644	1,015	349

Average monthly trade between countries belonging to the Organization for Economic Cooperation and Development, by region, 1992. In millions of US $.

Source: OECD Statistics Directorate: February 1995 *Monthly Statistics of Foreign Trade*. Each region contains multiple countries, except Japan, thus the diagonal (shaded) shows export trade between countries within the region.

these lines do not necessarily follow the geometric path on the ground. Since the interaction pairs do not have a direct spatial meaning, current GIS software provides few facilities to manage them.

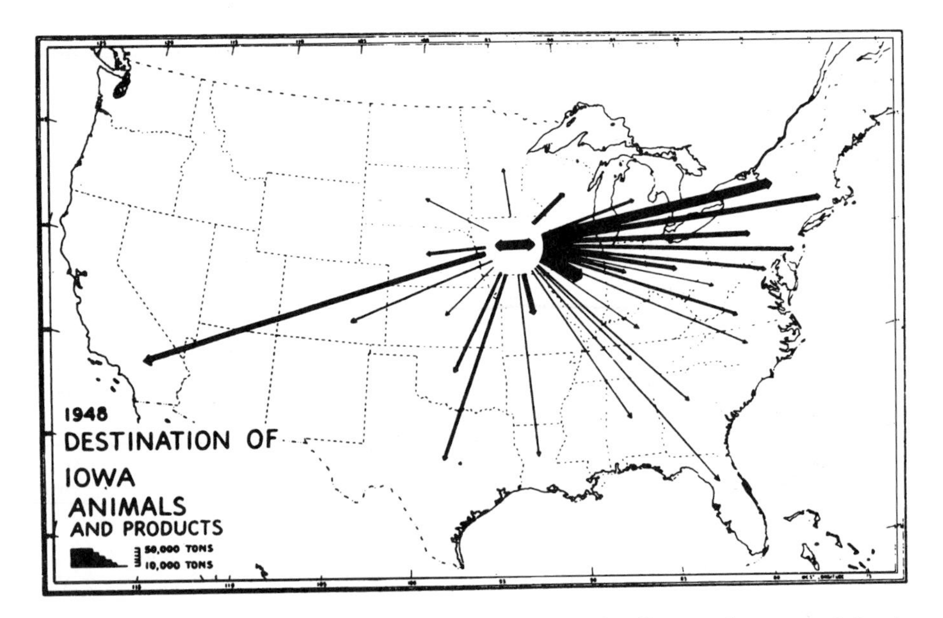

Figure 2-15: Example of cartographic treatment of commodity flow. Trade in animals leaving Iowa by rail in 1948. Reproduced from Map 5.3 in Ullman (1980).

Triangulated Irregular Networks (TIN)

Another example of a relationship control framework is the system of terrain representation called triangulated irregular network (**TIN**). A TIN provides a way to represent a surface (Peucker and Chrisman 1975) without controlling either elevation or space as one must with the isoline or center point frameworks, respectively. Each triangle in a TIN connects three neighboring points so that the plane of the triangle fits the surface sufficiently (Figure 2-16). A TIN can be formed from a scattered collection of points, but it works best when they include surface-specific points like peaks, passes, and samplings along ridgelines and stream courses. These points are measured using an object framework that provides a spot height measurement. The triangle establishes a relationship between three points, asserting that the surface in that region can be approximated by an inclined plane with a particular slope (gradient and aspect, see Chapter 7). The whole region is covered by triangles, so every point can be associated with an elevation value.

A TIN approximates the surface with the planes of triangles in three dimensions. These planes are not controlled by an attribute value, as in the isoline framework. Nor are they spatially controlled as in a grid of points. A TIN is much more than the elevation values at points. The triangles serve as a form of control that permit the measurements at the vertices of the triangles to be used to estimate the measurements for all

Figure 2-16: A Triangular Irregular Network (TIN) draped over a terrain. Photograph of Pilot Range, New Hampshire.

TIN: Triangulated Irregular Network: a system of terrain representation that builds triangular facets to connect point heights. The points and triangles are chosen to represent a surface within some limits.

neighbors unambiguously. The relationship between the three corner points of each triangle provides the constraint on the variation of the surface. Because a TIN uses such a distinctive method of encoding elevation, it is not surprising that TIN data is hard to integrate with other more traditional systems of control.

TIN has been presented as a data structure, a system of representation, but it must be recognized as a distinct form of measurement as well. Chapter 7 discusses TIN representation and surfaces in greater detail.

COMPOSITE FRAMEWORKS

In the physical sciences, measurement is rarely direct. Temperature is measured by the expansion of mercury up a glass tube (a distance), or it is measured by the difference in thermal expansion of two metals wrapped in a spiral. These measures of distance are converted into temperature using various physical laws. While geographic relationships may not be as deterministic, indirect measurement is quite common. Indirect measurement operates by observing one quantity that can be linked, by a set of assumptions, to another quantity. Under these conditions there may be multiple stages of control, which produce a composite of measurement frameworks. The accumulation of control introduces discrete elements at each step, thus limiting the resolution and accuracy of the final result.

Associating Attributes—Indirect Measurement

Based on a categorical coverage, many different thematic interpretations can be constructed using a lookup table form of indirect measurement. For example, soil surveys contain large tables with all kinds of attributes for the soil mapping units. The SOIL-5 database from the US Natural Resource Conservation Service lists dozens of such attributes, each with a different unit of measure—nominal, ordinal, and ratio. A selection of variables gives a sense of this diversity (Table 2-5). Of course, the soil mapping units were not designed to portray permeability, erosion hazard, corn yield, slope, and suitability for playgrounds with equal accuracy. Soil boundaries are drawn to distinguish a given set of categories, subject to a set of cartographic and pedological limitations. These are not necessarily the boundaries that one would obtain in a direct measurement of each attribute.

The simplest form of indirect measurement takes the basic categories of a categorical coverage and ranks them with respect to some particular purpose. This process upgrades the nominal categories to an ordinal scale by importing the ordering related to the purpose. In the soil survey report, each soil mapping unit is ranked "Slight," "Moderate," or "Severe" in terms of its limitation for a host of activities. Of course, the ranking orders the whole soil class, not the specific instance. Each polygon of the class gets an identical rank.

TABLE 2-5: Selected attributes for soil mapping units

Attribute	Measure
Estimated Soil Properties (each attribute by soil horizon)	
Texture	Classified by USDA, AASHTO, "Unified"
Clay content	% range
Moist bulk density	grams/ cm^3
Permeability	inches/ hour
Available water capacity	inches of water/ inch of soil
Soil reaction	Ph
Salinity	mm halides / cm
Shrink-swell potential	Low/ Moderate/ High
Erosion factor (K)	proportion of standard soil loss
Wind erosion group	numerical class
Suitability for Sanitary Facilities (by slope class)	
Septic tank absorption fields	Slight/ Moderate/ Severe
Suitability for Building Site Development (by slope class)	
Dwellings without basements	Slight/ Moderate/ Severe
Local roads and streets	Slight/ Moderate/ Severe
Recreational Development	
Picnic areas	Slight/ Moderate/ Severe
Playgrounds	Slight/ Moderate/ Severe
Capability and Yields per acre of Crops [Figures shown for Irrigated and Nonirrigated by Slope Class]	
Corn	Bushels/ acre
Soybeans	Bushels/ acre
Alfalfa Hay	Tons/ acre
Windbreaks	
<species - up to 20 listed	height>

Source: Headings on Soil Interpretations Records for Christian County, Missouri, dated 1983, given to landowners on request at the USDA Soil Conservation Service (SCS) District Office. These are extracts from a larger database and may be regionally specific. Original table lists 110 attributes.

A soil survey in the Corn Belt lists an estimate of corn yield for each soil class. This ratio figure may be obtained from a few sample fields or from expert opinion. When this value is applied through indirect measurement to the soil polygons, many repetitions of the same value occur all over the map (Figure 2-17). The abrupt nature of the categories remains, although yield would likely be more continuous. The shaded map maintains the illusion of a continuous attribute by showing ranges of possible values, yet there are only 19 distinct values of yield given in the lookup table for the 113 soil classes in this particular county. Thus, indirect measurement cannot hide the abrupt change between classes implicit in the measurement framework. The map is a composite that can only be understood by explaining the two processes: first, control by attribute (soil class) to measure boundary locations and then conversion to another attribute (corn yield) on a seemingly higher scale. In addition, indirect measurement makes the assumption of homogeneity despite the careful recognition of **soil inclusions** in the text of the soil report. Fisher (1991) has suggested solutions to the representation of inclusions using Monte Carlo simulations.

Assigning a ratio measure to a category might seem rather innocuous, particularly if it is merely a guide to expected yield for farmers. However, information has a way of moving into many unanticipated applications. In Ohio, Indiana, Missouri, Iowa, Nebraska, and neighboring states, various forms of expected crop yield assigned to soil mapping units are used in the assessment of agricultural land for taxation. This kind of use of indirect measurement for important public decisions raises many questions about the tradeoff between special-purpose measurement and general-purpose surveys (a topic revisited in Chapter 10).

Choropleth Framework

One of the most common geographic measurement frameworks, the **choropleth framework**, cannot be explained as a direct measurement. Consider the classic choropleth map of counties (or provinces, census tracts, or some other jurisdiction) with the title "Population Density by County, 1960" (Figure 2-18). It is well understood that county boundaries do not necessarily conform to regions of uniform population density. County boundaries seem to be taken as a fact of life—a base map rather than any kind of measurement. But even base maps must be measured. Counties fit perfectly as categorical coverages because an exhaustive set of categories (counties) serve as control to delineate (measure) a network of boundaries. In the next step, the census aggregates its raw tabulations for these spatial divisions. The county now serves as a *spatial* control for the attribute measurement process. Thus, at the most immediate level, the choropleth framework has an irregular form of spatial control that permits a continuous measurement of an attribute. This spatial control arises from the base map with the reverse combination of attribute control and spatial measurement.

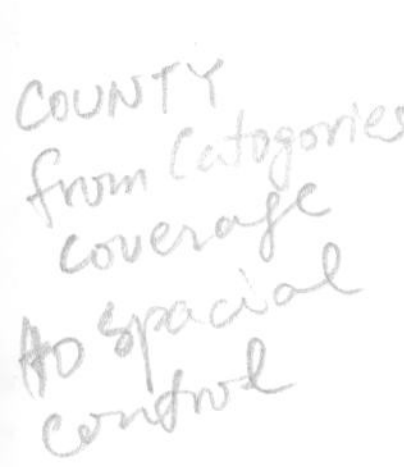

Soil inclusions: A category of soil expected to occur inside the units mapped as another category.

Choropleth framework: Measurement framework whose spatial units (derived from a categorical coverage of named objects) serve as control for attribute measurement (e.g., census tabulation).

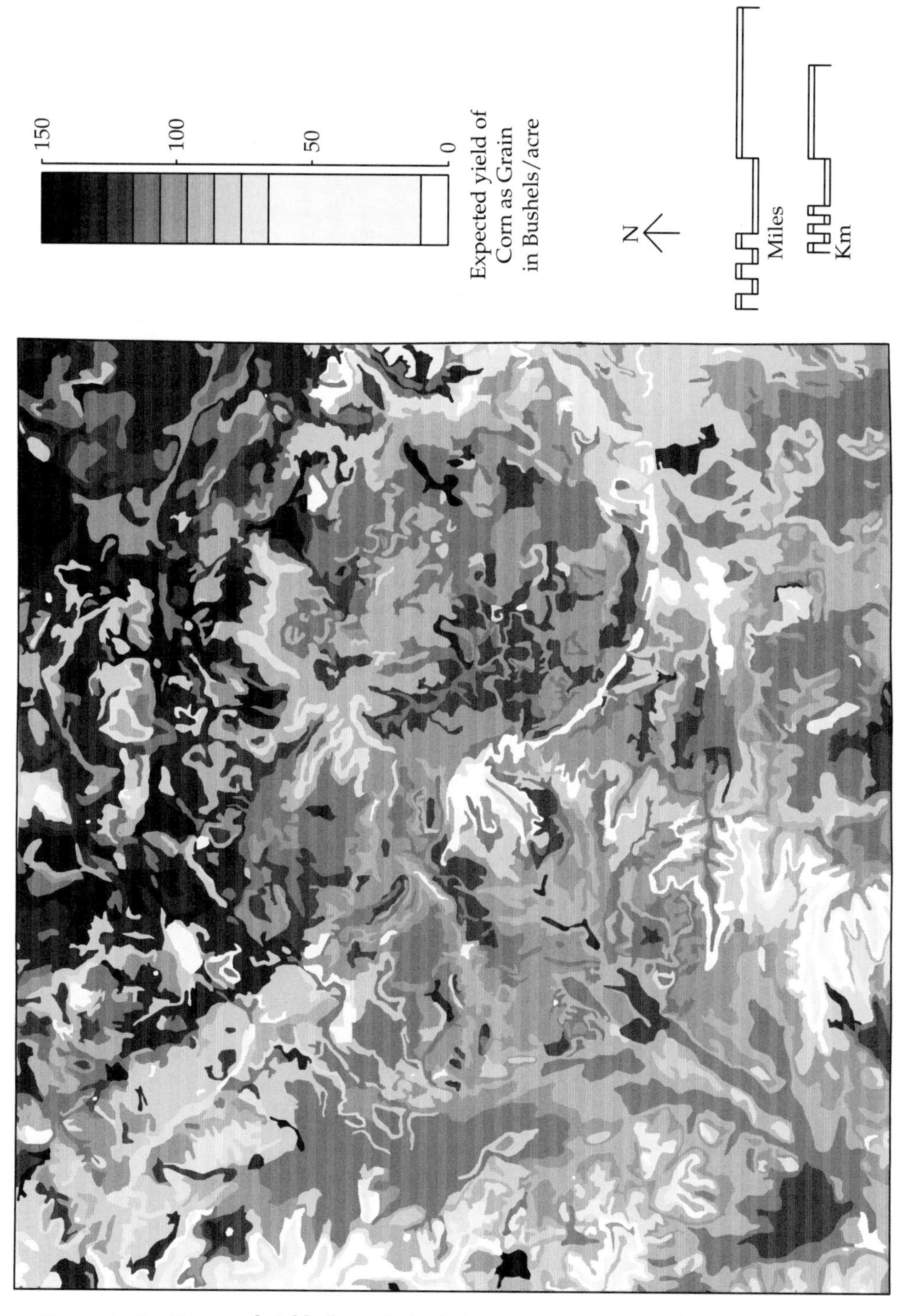

Figure 2-17: Estimated yield of corn in bushels per acre. Area covered: Oregon, Wisconsin approximate dimensions 10 km by 10 km. Digital data developed by Dane County Land Records Project (Niemann and others 1987) in cooperation with USDA Soils Conservation Service Wisconsin Office. Attribute table from Dane County Soil Survey, 1976.

The boundaries on Figure 2-18 do not relate to the attribute shown, but to the categorical coverage of political jurisdiction.

Examples of tabulation by collection zones include the obvious cases such as census returns that are aggregated by objects from blocks and census tracts up to municipalities and states (as shown in Figure 2-18 at the county level). The census phenomena such as population, housing, and income do not necessarily change character crisply on the edges of the tabulating units. A phenomenon tabulated by one set of choropleth zones may be quite different from the same variable tabulated for another set of zones. The real units of measurement, such as households in the case of the census, are so many that it makes little sense to treat each one as a distinct spatial unit. Furthermore, the protection of privacy may require deliberately reduced resolution in both space and attribute.

Tabulation by collection units extends beyond the classic case of the census. Various kinds of areal collection zones are used in different applications; such zones range from school attendance areas to television viewer areas to watersheds. Not all spatial aggregations are areas, however. In dealing with highway statistics, the records of accidents and other activities are often tabulated by a variety of levels from intersection to highway route number. Most commonly, the highway is segmented into tabulation units that are homogeneous according to some property such as traffic flow or pavement type. The process of designing different segments for different purposes has

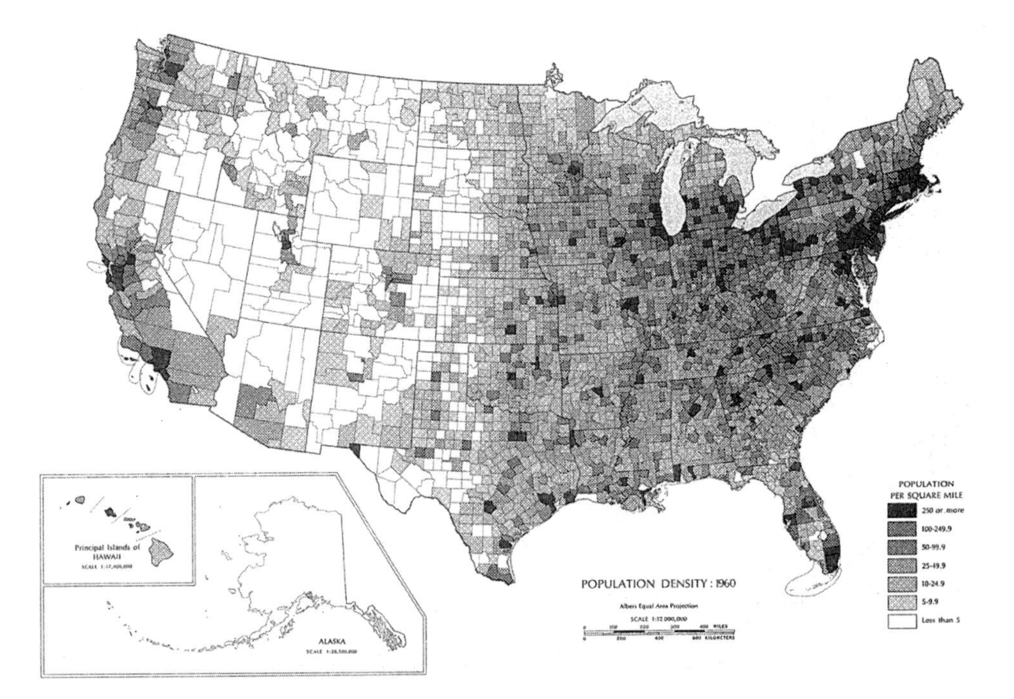

Figure 2-18: Choropleth map: Population density of the US by county 1960. Source: USGS (1970, p.241).

come to be called **dynamic segmentation**, though it has no particular temporal nature (Nyerges 1990; Dueker and Vrana 1992).

Despite sharing the indirect measurement process, the choropleth framework differs from the indirect measurement of soils discussed earlier. In the soil case, the ranking or attribute was attached to the whole class that had provided the control for the original coverage. In some sense, the corn yield by soil class is an attribute of the attribute. In the choropleth case, the attributes have been aggregated from a more detailed distribution, thus imposing a form of spatial control. The choropleth boundaries have little to do with the variability in the attribute; by contrast, boundaries retain most of their meaning after indirect measurement. Although they might look similar, particularly when displayed using choropleth maps, these measurement frameworks do differ.

TEMPORAL FRAMEWORKS

Time lies at the heart of many geographic problems. For example, the ultimate purpose of an analysis might be to explain the processes behind some changes or to predict future configurations of the landscape. Yet, time remains the weakest part in the measurement frameworks used for geographic information. The snapshot model, in which time is fixed, lies at the base of most geographic information models, just as it did for printed maps.

The classic approach to change detection places a series of snapshot maps into a kind of jerky motion picture. Time graduates to a kind of control, as long as the measurement structure of the map has not changed. The calculation of change requires the overlay techniques presented in Chapter 5. Of course, errors in the maps may become inextricably confused with changes. It becomes technically challenging to retain adequate control over the spatial and attribute components, so that the repeated snapshots really repeat the same measurement procedure.

A form of control through temporal relationships may sidestep this problem. Most nongeographic databases are not structured as snapshots; rather, they center on transactions. The working world of GIS already links the daily flow of permits and inspections to revisions in the database, but tragically all the temporal information gets lost as our current software throws out the past to remain current. As geographic measurement moves away from the limitations of maps, databases will be able to reflect change in the landscape more flexibly. Some conceptual advances have occurred (Langran 1991), but they have not yet become incorporated into the regular GIS tool kit.

Environmental modeling has always had a strong temporal component. This kind of model provides a system of relationships that link particular measurements into a larger structure, eventually seeking to predict future configurations. These relationships come from an understanding of environmental processes, not just from the map patterns. Still, the elements of the model do not escape the basic trade-off described in this chapter, that is, some components must serve as control to allow measurement of another.

Dynamic segmentation: A method for referencing attribute information along a network that does not divide each segment of the network wherever any attribute changes.

SUMMARY

The measurement frameworks for geographic information introduced in this chapter are not intended to exhaust all possibilities, though they do cover the bulk of current applications. Each framework involves choices in recognizing objects and relationships inside a system of axioms (Table 2-6). These frameworks then serve as the guiding concepts for the systems of representation used to implement the information system—the topic of the next chapter.

TABLE 2-6: Summary of measurement frameworks

Control by Attribute	
Isolated Objects	
Spatial object	Single category distinguishes from void
Isoline	Regular slices of continuous variable
Connected Objects	
Network	Spatial objects connect to each other, form topology
Categorical coverage	Network formed by exhaustive classification
Control by Space	
Point-based Control	
Center point	Systematic sampling in regular grid
Systematic unaligned	Random point chosen within cell
Area-based Control	
Extreme value	Maximum (or minimum) of values in cell
Total	Sum of quantities (e.g., reflected light) in cell
Predominant type	Most common category in cell
Presence / absence	Binary result for single category
Percent cover	Amount of cell covered by single category
Precedence of types	Highest ranking category present in cell
Control by Relationships	
Measurement by pair	Control by pairs of objects
Triangular Irregular Network (TIN)	Control by uniform slope
Composite Control	
Choropleth	Control by categories (name of zones) then by space

CHAPTER 3

REPRESENTATION

CHAPTER OVERVIEW

- Introduce the primitives used to represent spatial and attribute measurements.
- Describe basic representation models: vector and raster.
- Follow steps that convert existing documents into digital databases.
- Introduce components of data quality evaluation applied to digitizing.

Representation involves a symbol acting in the place of some entity. A system of representation can choose to suppress some details in the world to emphasize others. The clearest precedent for geographic representation involves the graphic symbols used on traditional maps. A map populates a small space with the representation of a larger space, using map symbols to stand in place of things in the world. Thus, a small star serves to encode that a particular city is the capital of its state. While the direct representation of space facilitates many kinds of visual processing, the transformation to a smaller space creates many dilemmas. Traditional map symbolism often dictates a measurement framework to suit the technical limits of graphic reproduction. Rather than acting as a flexible presentation of geographic information, symbol becomes confused with measurement. This legacy still influences the understanding of geographic information.

Digital representations are far less direct than maps in structuring spatial relationships. Digital symbols are extremely simple until built into more complex structures. Despite their simplicity, digital representations improve on the amazing capabilities of the printed map. This chapter reviews the primitive elements used to represent geographic data and then describes the basic data models and **data structures** that orga-

Data structure: Arrangement of data entities that permits the construction of relationships through software operations; implements a data model.

nize these primitives into a useful representation. The final portion of the chapter traces the various steps used to convert data sources into an information system.

PRIMITIVES FOR REPRESENTATION

Computers provide a few methods to represent numbers built into the hardware. The basic unit of storage is a bit, a single binary digit. These bits are grouped into larger units to represent numbers. Most computers provide three "word" formats for numbers: *integer, floating point,* and *double precision* (Figure 3-1). At one time, the number of bits in a word varied, but these days most computers store integer and floating point values in 32 bits and double precision in 64 bits. Thirty-two bit integers can represent any whole number from –2147483647 to 2147483648. With integers, results of any division that involves a remainder will be truncated to the next lowest integer.

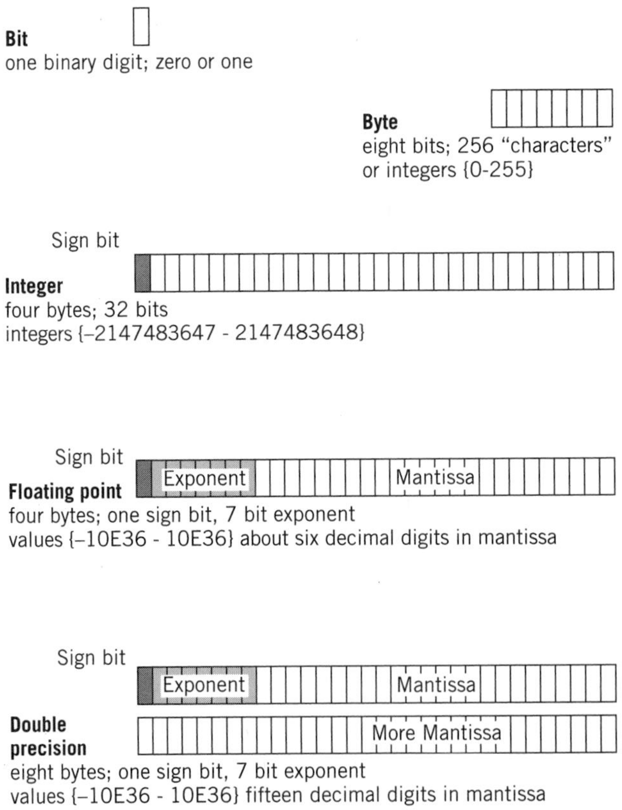

Figure 3-1: Computer storage at the most basic level (bits, bytes, words—integer, floating point, and double precision).

Floating point uses a logarithmic notation with an exponent to establish magnitude and a mantissa for the value. This format provides about six decimal digits of resolution, scaled over a much larger range (typically -10^{36} to 10^{36}). Calculations in floating point are more forgiving than the truncation associated with integers, but ultimately a floating point number is still represented with finite resolution. Double precision provides about 15 digits, essentially an extended form of floating point, but storage is doubled and speed of calculations is usually decreased. Choosing between these formats has consequences in handling spatial information.

Besides numbers, computer representation can manage information coded as characters. Most storage media work in eight-bit *bytes*. International standards provide a coding for certain basic characters, though the full range of accents and diacritical marks remains far from standardized. Since characters are nominal symbols, the hardware can do little other than copy them. Fixed-length text strings can be stored in words. A variable-length *string* structure, embedding a count followed by the text, is required to handle text flexibly.

Primitives for Attributes

Attribute values in a digital system are directly encoded using the units of computer storage. The case and variable framework of the geographical matrix encourages the data structure of a two-dimensional array. One storage location is allocated for each attribute value. Numbers are often encoded in floating point, because six digits are usually adequate for the significant portion of an attribute measurement. Counts are usually encoded as integers to make use of the increased range, although a 32-bit integer is no longer adequate to represent the sum of the world population of humans. When storage bulk becomes a concern, as with remotely sensed imagery, continuous measurements may be rescaled into a range from 0 to 255 and encoded as an integer in a single byte. Most users forget these details and simply treat an attribute as a "number"; they assume that the computer can handle all the intuitively obvious arithmetic operations. Computer hardware manipulates the storage units without recognizing different levels of measurement; thus, the validity of calculations depends on external meaning.

Nominal attributes must be coded as some kind of number as well. The coding rules become a part of the data structure schema that must be preserved to retain the meaning of the information. Categories may be encoded in a single byte with no loss of information content, as long as there are only 256 possibilities.

Primitives for Space: Coordinates

Representation of space requires a spatial reference system as defined in Chapter 1. On a map, location is encoded by relative position on the piece of paper. In a computer database, location is based on analytical geometry. For practical purposes, the spatial component of geographic information is represented in the form of coordinates, that is, ordered measurements inside a spatial reference system. These measurements may be angles on an **ellipsoid** (latitude, longitude) or orthogonal distances on a **projection** plane (Figure 3-2). As introduced in Chapter 1 (Figure 1-9), there are different ways to encode equivalent measurements, even on a plane. Each GIS software package will implement a certain range of alternative spatial reference systems as specific representations of the mathematical possibilities.

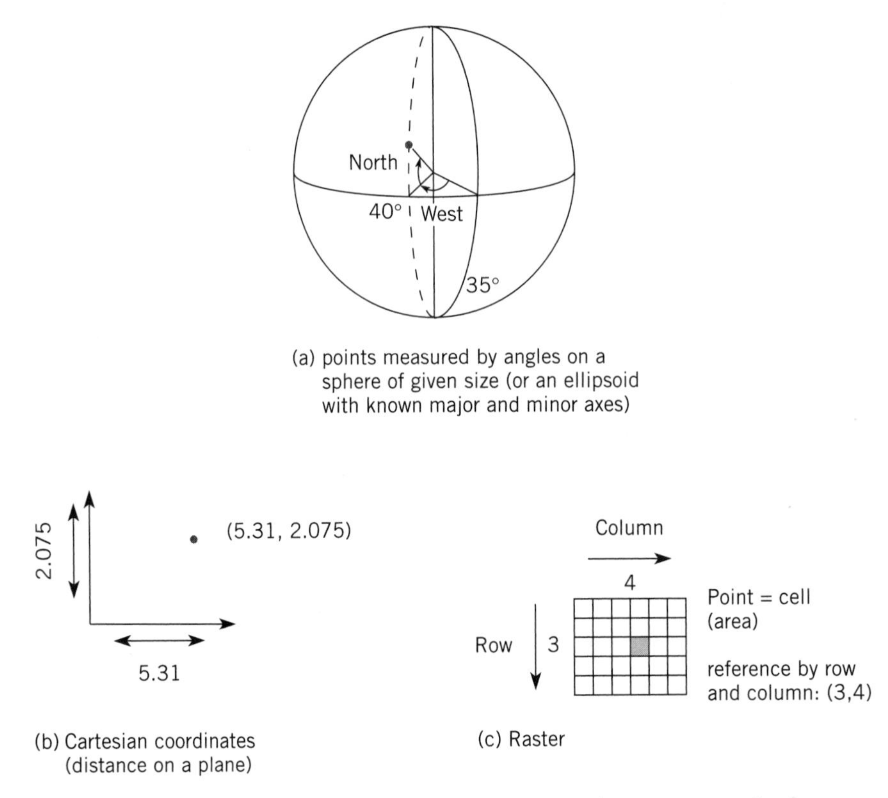

Figure 3-2: Coordinate reference systems: Measurements relative to a spatial reference system.(a) Spherical coordinates measure angles on a solid surface of a given radius. This system can also handle location on an ellipsoid (a distorted sphere); (b) Cartesian coordinates based on distances along two axes. Note that the convention of (X,Y) is somewhat arbitrary; some Cartesian coordinates such as UTM were given as (northing, easting); (c) Cellular coordinates counted as integers. Raster hardware tends to sweep from the top of screen to the bottom, so the convention is more likely to place Row 1 at the "top" or north of an image.

Ellipsoid: Three-dimensional object formed by rotating an ellipse around its minor axis; an oblate ellipsoid approximates the shape of the earth (geoid), computed by the best fit to geodetic observations. See Table 3-2.

Projection: Coordinate transformation that converts latitude–longitude measurements into planar coordinates. Projections can be based on a developable surface (such as a plane, cylinder, or cone) or on a mathematical function.

Coordinates are almost invariably stored as pairs of the basic computer words, with interesting consequences (Chrisman 1984a). All number systems supported in computer hardware have limited resolution, because they are effectively integers. When used for coordinates, integers cannot represent all intermediate locations along diagonal lines (Figure 3-3). Most diagonal lines from one whole number pair to another pass through locations that cannot be represented in the integer system without bending the straight-line segment (Egenhofer and Herring 1991). In order to limit these geometric problems, the representation systems for coordinates must have excess resolution. Integer storage also means that no coordinate representation can offer truly continuous variation. The size of the discrete jumps in resolution can be made so tiny that they seem effectively continuous, but only at some expense.

Resolution has a consequence in setting the maximum extent of the area represented. Single word floating point with its six digits can circle the equator (just over 40,000 km) so that kilometers can be resolved, but not much more. On a county scale, these words may resolve centimeters, but only by placing the origin close to the county. Otherwise, resolution is wasted in measuring each coordinate relative to some distant origin. With a local offset, the database of the adjacent county could not be appended without overflowing the range of a floating point value. Current workstation software has opted for double precision, paying the penalty of slower speed and increased storage to avoid the troubles of single–word storage. With double precision, there is enough resolution to distinguish the left and right side of every virus particle on the planet. For mapping purposes, this resolution is spurious and must be filtered back to some reasonable level. It is particularly odd to use such profligate storage for measurements obtained from inaccurate maps by fallible digitizing systems.

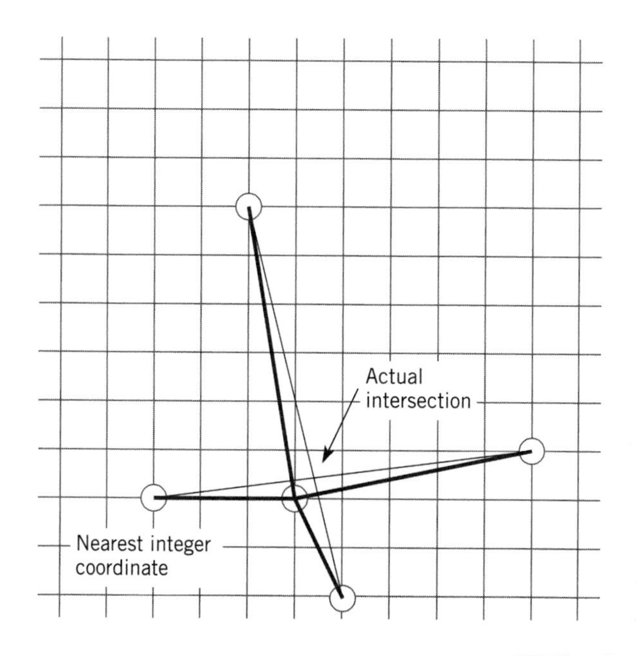

Figure 3-3: Representing intermediate locations in an integer coordinate system. Thicker grey lines represent the result of intersection of two straight lines once rounded off into the integer space. Note that the new segments have different slopes from their parent segments. Coordinates stored as floating point numbers do not avoid this problem.

REPRESENTATION MODELS AND DATA STRUCTURES

Representation of geographic information proceeds by organizing the primitives into more complex structures. These data structures often provide the key technical differences between competing software packages. The specific details of a data structure act as instances of a more generic data model of entities and their relationships. Much of the substance of a data model comes from a measurement framework, as discussed in the previous chapter. Many alternative data structures are possible, but there are relatively few generic models behind them. This section will concentrate on the two dominant models of representation in GIS, vector and raster.

Vector Model

Measurement frameworks based on attribute control are implemented most directly using the **vector** model. Based on analytical geometry, a vector model builds a complex representation from primitive objects for the dimensions: points, lines, and areas. These primitives have a nested dependency: areas are described by boundary lines, and the location for a line can be approximated by a string of line segments connecting a series of points. At the base, points are represented by coordinates. Cartographic data structures usually do not provide more complicated options between points. This simplicity contrasts with the richness of different curves in drafting or illustration software, where the paths between points may be **Bézier curves** or **splines**, not just straight–line segments. Of course, these software packages are oriented toward display, not analytical operations using the data. In engineering practice, many highway features are laid out with circular arcs or conic spirals, but most natural features do not have a preferred mathematical curve. For general cartographic representation, segments provide a simple, versatile approximation. Adding more complex curves would not really change the fundamentals, though it would make many relationships much messier to compute.

Representing Isolated Objects The spatial object framework translates into a simple vector representation with each line and polygon defined by a string of coordinates (Figure 3-4). Any string can represent a line, but a polygon should close. As long as the objects remain isolated, which is an axiom of the measurement framework, the representation serves its basic purpose. To represent an inner ring inside a

Vector: A spatial data model based on geometric primitives (point, line, and area), located by coordinate measurements in a spatial reference system; from mathematical term for a direction, or a directed line segment.

Bézier curve: A smooth curve that passes through specified points with a given direction (tangent) at those points.

Spline: A smooth curve that models the behavior of a thin spring (with a given modulus of elasticity) constrained to pass through specified points.

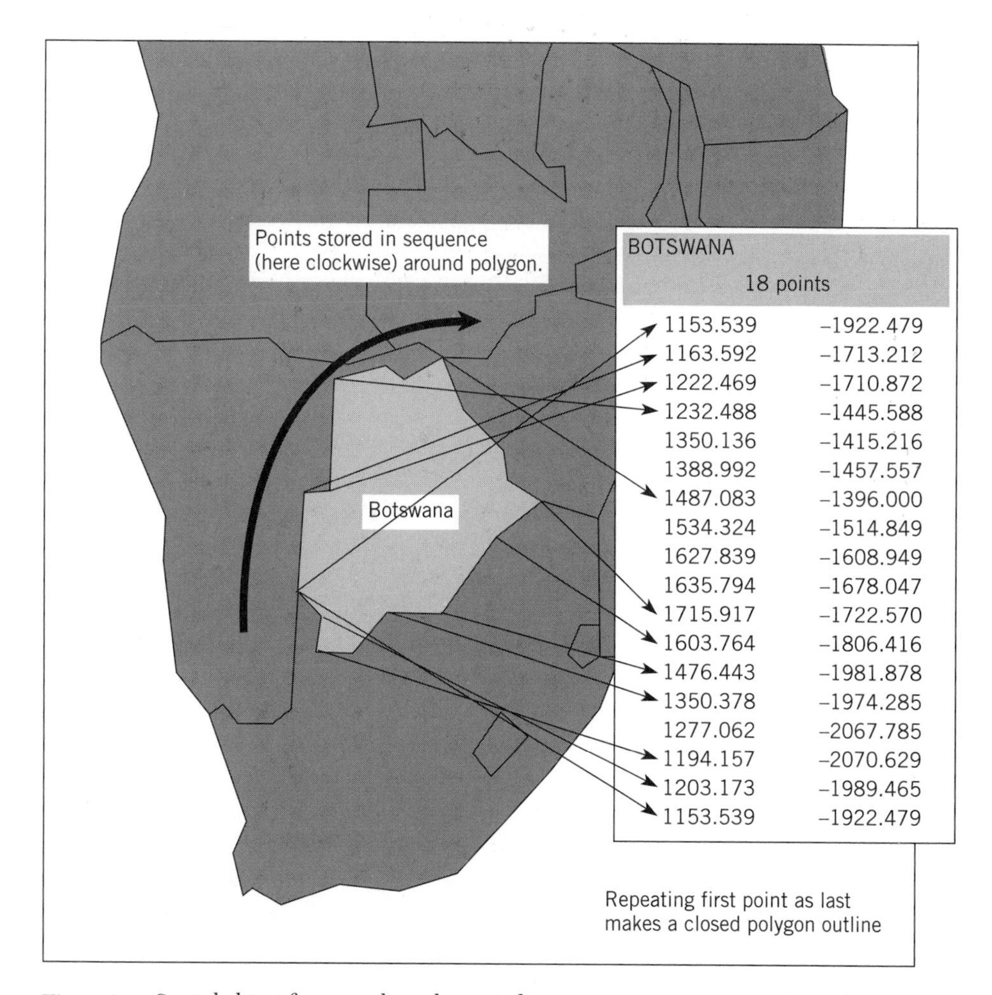

Figure 3-4: Spatial object framework implemented in a vector representation. Each object contains its own list of points (coordinates). No relationships are stored.

polygon, the isolated data structure often adopts a convention of a **retraced line** connecting the outer ring to the inner ring. This inelegant solution adds a burden of checking for duplicates, which diminishes the advantage of the simple isolated structure. The isoline framework, since it uses isolated contour lines, also translates directly to this simple vector representation. Any nesting structure to relate each closed loop to the next contour up or down creates relationships that diminish the isolation.

There is considerable potential for inconsistency if an isolated object representa-

Retraced line: A technique applied to simple vector data structures to embed inner rings within outer rings. A line that is repeated (drawn twice) should not appear graphically.

tion is applied to a connected network of polygons. Since each object is independent, each boundary is represented twice. A gap or overlap could easily occur without any easy method to detect it (Figure 3-5).

Representing Topological Objects The topological data model is more commonly used in software that implements a full range of operations on vector representations. The topological model incorporates network relationships along with the coordinate measurements (Figure 3-6); thus, it can handle the requirements of the connected coverage frameworks. This model centers around the boundary and provides explicit connection to nodes at each end as well as to polygons on left and right (as introduced in Chapter 2). These relationships can be implemented in a number of different specific data structures, particularly in relating polygons to their boundaries and inner rings to their outer rings (Gold 1988). One common structure creates a variable-length list of the chains around each ring, with notation for direction, as shown in Figure 3-6. However they are implemented, the data structures provide access to the same relationships.

The important characteristic of all the vector methods is that they permit essentially free placement of point locations and of boundary lines to represent categories. Thus, the vector model directly implements the intent of attribute-controlled measurement frameworks. Because of this linkage, the measurement framework and the representation may seem inseparable, but representation remains a distinct choice. The vector model can be applied to other measurement frameworks, such as the tri-

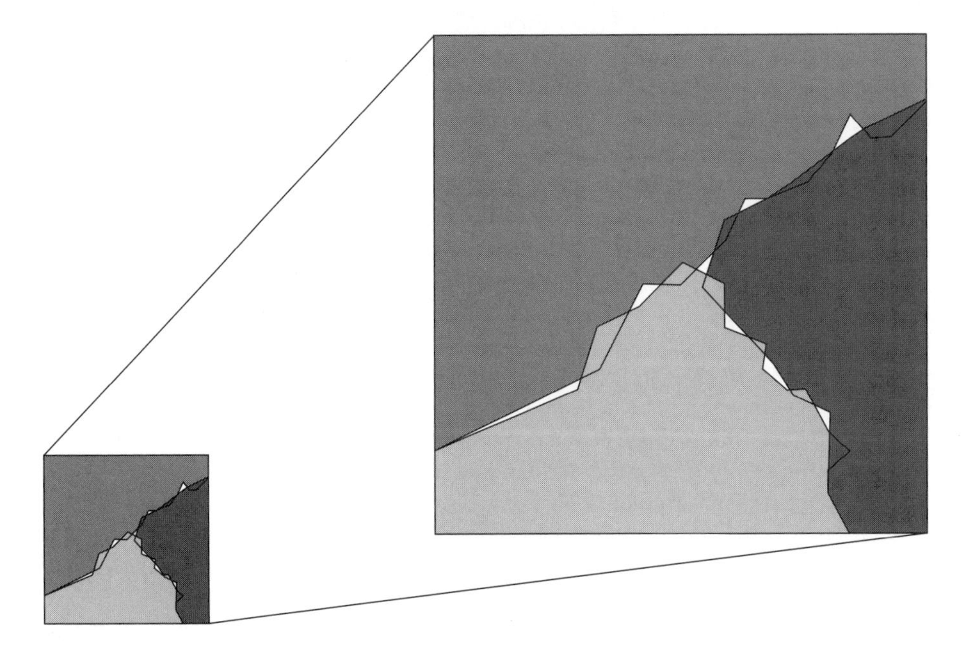

Figure 3-5: Adjacent polygons represented by isolated boundary lines may create slivers (gaps and overlaps).

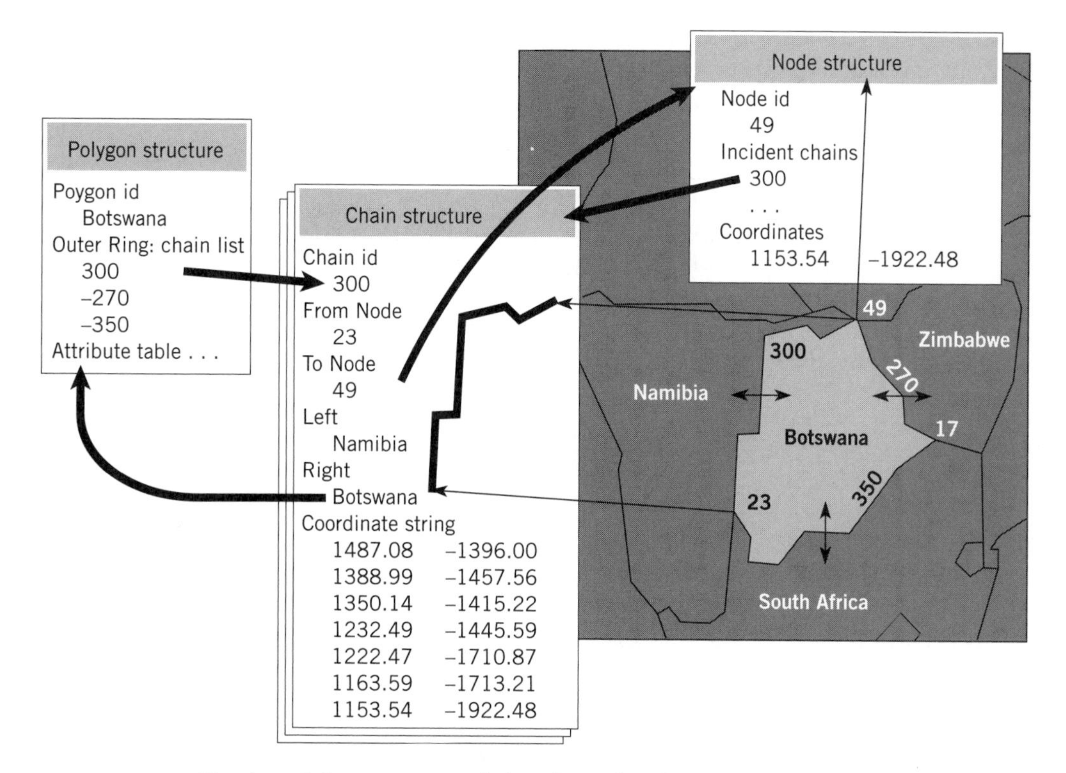

Figure 3-6: Topological data structure includes relationships between the components of a connected network. The polygon representing Botswana has three neighbors (Namibia, Zimbabwe, and South Africa) at this scale. (Actually, the node on the Zambesi River includes a 100 meter border with Zambia, not representable at this scale.) Each neighbor requires a border to separate the two countries. Each border begins and ends at a node. This diagram shows a polygon with a list of chains around its outer ring (using a sign for reverse direction).

angles of a TIN. Vector data structures are the method of choice for choropleth mapping of continuous attributes. As long as geographic data entry is difficult and expensive, there is a strong incentive to use one geometric description for many attributes, the approach of the geographical matrix. The evaluation of competing approaches will be revisited in Chapter 10.

Raster Model ⇒ Control Space

The other major family of representation models is called **raster**. Whereas the vector model is constructed from geometric primitives as a logical structure, the raster model has close links to the physical layout of computer graphics hardware. Raster derives from a word used in mechanical engineering for a tool that advances in a sweep back and forth. Television technology uses just such a mechanical sweep gun, so the term

Raster: A spatial data model based upon a regular tessellation of a surface into pixels or grid cells.

came to refer to the rows on the screen and eventually the cellular nature of a cathode ray tube (CRT) display. The hardware structure of remote sensing sensors, of line printers, and of CRT displays all contributed to the popularity of the raster structure. In addition to the hardware connection, raster structures could be implemented easily with the elementary data structures available even the earliest programming languages.

The raster model divides the region into rectangular building blocks (grid cells or *pixels*) that are filled with the measured attribute values. The raster approach is directly related to the frameworks that control space in order to measure attributes. The raster cells may be located within a spatial reference system, but they deliberately limit resolution to act as control. In many cases, the raster geometry is specified by the original sensing hardware (Figure 3-7). The raster representation is not restricted to

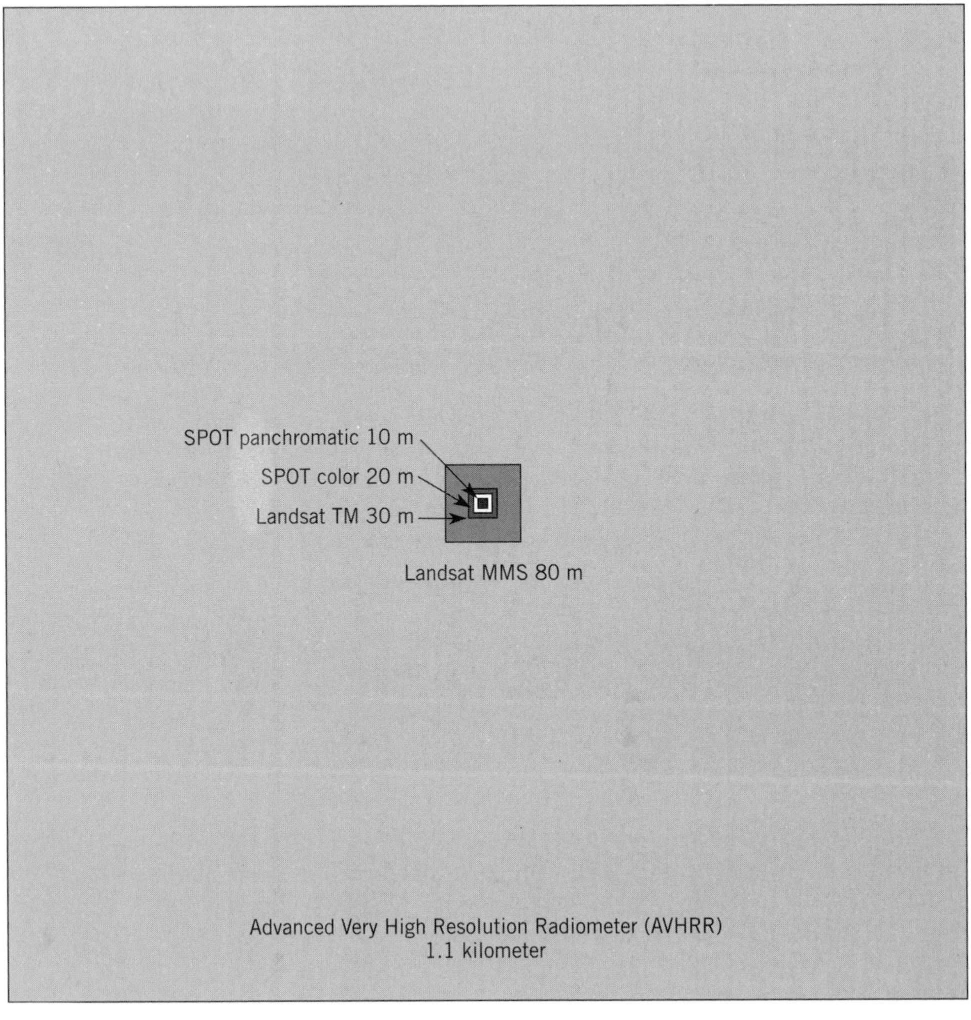

Figure 3-7: Pixel sizes from common satellite sensing systems.

its related measurement framework; for example, attribute-controlled measurements can be represented in a raster, though the process creates a composite using both attribute and spatial control successively.

An array in computer storage (random access memory or disk) provides the most direct implementation of a raster representation, but an array of fine resolution often contains redundant values. There are many methods of **compression** possible. Overall, compression can be divided into *loss-less* methods, which preserve all the data, and statistical methods, which might change some data to simplify the message. The later group effectively uses a form of cartographic generalization (usually implemented using the neighborhood operations described in Chapter 7), followed by one of the loss-less methods.

There are a number of loss-less compression methods; a few are diagrammed in Figure 3-8. If a raster contains chunks of cells with identical values, *run length encoding* can compress storage. Instead of storing each cell, each component stores a value and a count of cells with that value (Figure 3-8a). If there is only one cell, the storage doubles, but for three or more there is a reduction. In the special case of a binary (black/white) image, the compression need only store the cell where the value changes. These methods work along one row. More advanced methods store differences between rows instead of treating each row separately (Figure 3-8b). Compression algorithms form a routine part of the **TIFF** standard used by facsimile machines to reduce telecommunications costs, but methods that work for a digital transmission may not be best tuned for geographic access. Various forms of **quadtree** try to take advantage of two-dimensional character of spatial data for compression. A quadtree works by iterative division of a region into four square subunits. A *region quadtree* stores a categorical attribute and will not subdivide a square that is homogeneous (Figure 3-8c). The pattern of this compression also provides a measure of spatial variability (Csillag and Kummert 1990).

In addition to compression, a key issue with a raster system is the size of the cells. Smaller cells permit the raster to approximate the flexibility of the vector system as closely as required, but at a price in the storage consumed. A coarse cell system is sometimes distinguished from raster and may be called a *grid cell* system. A grid cell is sufficiently large that it can no longer be treated as a point, so one of the area-based rules discussed in Chapter 2 must be used to make a measurement. Though all raster pixels actually occupy some space, they should be so small that there is no internal detail at the resolution of representation. For tiny pixels, the measurement rule may not matter very much. Many of the original systems used for automation of site suitability analysis in the 1970s used quite crude cell sizes. For example, the Land Use and Natural Resource project in New York State used 1 km squares; the Maryland

Compression: A software procedure that encodes a data structure so that its storage occupies less space (under certain conditions); may preserve all the information (loss-less) or deliberately simplify.

TIFF: Tagged Image File Format: a family of image encoding formats that can vary the resolution and the number of bits used to represent each cell.

Quadtree: A spatial data structure that organizes a hierarchical structure of square cells through iterative division into four daughter cells (Samet 1990).

(a) **Compression by Run Length Encoding** (along rows)

3	3	3	3	3	3	3	3
3	3	3	3	3	3	7	7
3	3	3	3	7	7	7	7
3	3	3	3	7	7	7	3
3	3	3	3	3	3	3	3
7	7	3	3	3	3	3	3
7	7	7	7	3	3	3	3
3	7	7	7	3	3	3	3

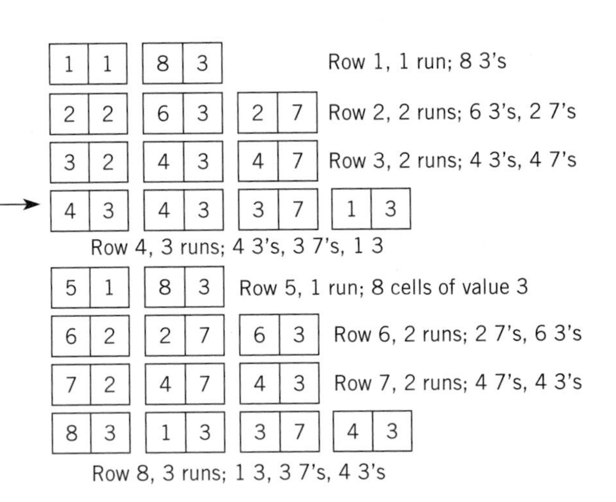

(b) **Compression by Row Differences**

1	1		1	8	3
2	1		7	2	7
3	1		5	2	7

Interpretation: Row 1, 1 change section; from cell 1, 8 cells change to 3 (whole row)

Interpretation: Row 2, 1 change section; from cell 7, 2 cells change to 7

Interpretation: Row 3, 1 change section; from cell 5, 2 cells change to 7

and so on...

(c) **Compression by Quadtrees**

3	3	3	3	3	3	3	3
3	3	3	3	3	3	7	7
3	3	3	3	7	7	7	7
3	3	3	3	7	7	7	3
3	3	3	3	3	3	3	3
7	7	3	3	3	3	3	3
7	7	7	7	3	3	3	3
3	7	7	7	3	3	3	3

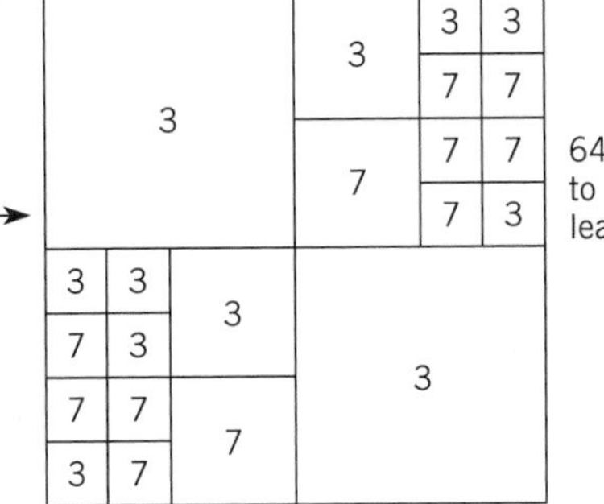

64 cells compress to 22 quadtree leaves.

Figure 3-8: Compression of raster representations. The redundancies of adjacent cells can be reduced along the rows, between the rows, or by quadtree.

Automated Geographic Inventory used 2000 ft cells. Crude cell sizes may have disappeared from state planning agencies, but they remain quite common in global environmental models. Cell sizes of 1° by 1° or even 5° by 5° are quite common, even though they are hardly square or uniform as they go poleward. For cells this large, the measurement rule becomes quite critical.

In the late 1970s, a serious and heated debate occurred over the virtues and disadvantages of raster versus vector models of representation. At that time, the debate often revolved around efficiency in implementation of data structures rather than the fundamental models. The choices made in the early days of GIS were often driven by technology, not application. In retrospect, the debate made the choices seem needlessly exclusive; each representation serves a measurement framework, and each framework has its appropriate uses. It is much more important to tailor the representation to the axioms of the measurement (and ultimately to the purposes of the enterprise) than to let the technology drive the decisions. As computing power has dropped in price, software has been able to deliver a much closer approximation of the conceptual models.

CONVERSION OF EXISTING DOCUMENTS (DIGITIZING)

The process of representation goes beyond the formalities of designing a computer data model. Each data source involves many practical decisions in converting the raw resources into the appropriate structure for further processing. As an example of all the possibilities, the rest of this chapter will consider one common source of geographic information: digitizing existing maps using manual as well as scanning technology.

In the earliest period of GIS, there were no resources of existing digital data. Any project had to begin with the process of building the database from scratch. In some cases, a digital source such as a remote sensing system could provide the data resource, but otherwise the only recourse was the conversion of existing information, typically in the form of paper maps. This process is called *digitizing*, because it converts physical representations into digital data. Even now, the process of database construction can be so protracted that a whole career can be spent in developing a database. The conversion industry currently consumes a large amount of the annual budget spent on geographic information worldwide. As the stock of materials becomes converted or fades into obsolescence, this level of prominence will decline. Yet, the process of digitizing provides an excellent organizational tool for confronting the issues of representation.

A geographic information system is hungry for data. Existing information resources, mostly in the form of maps and attribute tables, are often pressed into service without much reflection on their content. The technology of map construction, symbolization, and distribution impedes the representation of spatial phenomena, mostly by imposing certain measurement frameworks. Yet, no matter how limited they are, these information resources are often the best that can be found within the budgets of time and money. Digitizing has usually consumed the largest slice of project budgets—often over half the total funds. These costs can best be controlled by understanding the nature of the input material and the relationships that should be created in the database.

This description of digitizing will start with the two main technologies, vector and raster, and then cover the necessary geometric transformations. The next section deals with verification and data quality control, followed by some of the procedures used in database construction.

Vector Tracing

The most direct method to extract measurements from a map is to measure the motions of retracing the points and lines. The early vision of automated cartography (sometime between 1958 and 1962) began with an automated drafting table where the machine could record the movement of a pen (Coppock and Rhind 1991). The prototype of what we now call a digitizing table moved a huge magnet around under the table surface, homing in on the signal emitted by the pen. Modern **digitizers** use different techniques, usually based on detecting the position of a cursor relative to a grid of wires fixed into the surface of the table. The accuracy of the position depends on the fineness of the grid of wires and the signal–detection hardware. The best digitizing tables may be accurate to 0.075 mm (0.003 inch), though they usually record in integer units of 0.001 inch (Jackson and Woodsford 1991). Lower–cost tablets can still provide accuracies around 0.25 mm or 0.5 mm (0.02 inch) for smaller areas. These figures reflect the pure measurement capability of the device, not the combined accuracy of an operator trying to repeat measurements from map materials. Too often, the unsuspecting take the 0.001 inch resolution as a measure of accuracy.

In raw form, the digitizing system can measure the location of the cursor either when triggered by a button *(point mode)* or on a continuous basis *(stream mode)*. Stream mode can record a location after a fixed time or a given displacement. Each method has its disadvantages. Digitizing operators must position their cursor over the symbols on the map. The accuracy of the result depends on the width of the graphic symbol, the stress imposed on the operator by pushing the cursor across a large table at odd angles, and many other factors. Point mode clearly applies to point objects, and it works well for rectilinear features such as property boundaries. Stream mode may provide a more replicable result for complex linework, though it requires filtering to cut down oversampling. Stream mode may force an operator to move slowly since the cursor must remain on the line at all times.

In manual digitizing, the human operator interacts with the software to encode features from a map. Human pattern recognition is particularly important when the map includes many different symbols or when only a portion of the information is relevant. Manual digitizing is less accurate than good–quality scanners, and more forms of error are possible. It is fairly difficult to motivate digitizing operators to function under such exacting conditions. Recent developments of hardware and software have begun to address this difficulty with the "heads–up" system described in the next section.

Digitizer: A manually controlled machine that records a spatial measurement, usually on the surface of a tablet.

Raster Scanners

Since the earliest days of GIS, there has been a search for a way to avoid the drudgery and inaccuracies of manual digitizing. In the 1960s, **CGIS** designed its flow around a scanner that consisted of one lightbulb, one photocell, a drum to rotate the scribed map, and a motor to move the sensing equipment. The result was a raster scan of the original line map. Similar technologies have been developed explicitly for maps, some introducing high-precision lasers on rotating drums for color separation work, others attempting direct line following. However, the greatest breakthroughs came from the consumer marketplace. The demands copiers and facsimile machines have required cheap and accurate "push–broom" raster scanners. A modern fax machine has a charge–coupled device (CCD) array of photodetectors, often 5000 elements across. Multiple arrays can be arranged to cover the large widths of maps and other drawings. Each detector reports a gray-scaled value (or a color value). Coupled with a mechanism to move the paper (or the detectors), these detectors become a scanner. Resolution of the CCD detectors continues to get finer; resolution of .025 mm (1000 dots per inch) or better can now be bought at fractions of the price of a .2 mm scanner ten years ago. Accuracy of these scanners may still be closer to 0.5 mm (Jackson and Woodsford 1991). According to the basics of sampling theory, a scanner must sample at half the width of a line to ensure detection. A line as thin as 0.1 mm (0.004 inch) would require scanning resolution of .05 mm, well within the range of current devices. Cruder maps can be scanned with coarser sampling, such as the scanners sold for desktop publishing, though these scanners do not tend to be large enough for full sheet maps.

A scanner captures a picture of the graphic symbolism. Using the near neighborhood operations that will be described in Chapter 7, the skeleton of the linework can be isolated and converted to a vector representation. Junctions of lines, however, remain tricky for automated detection. The extraction of lines depends on the resolution of the scanner pixels; but even more, it depends on the nature of the original map.

The combination of a scanner with an interactive workstation permits an operator to read the scanned graphic and direct the line–detection algorithms. This combination is sometimes called **"heads–up" digitizing**. Alternatively, a scanned photograph can be used as a backdrop to construct other interpretations, thereby replicating the traditional environment of map compilation. The screen of the workstation becomes a flexible viewport to the graphic material, offering the ability to pan and zoom over a large virtual graphic. A simple mouse pointing device can select the elements of these

CGIS: Canada Geographical Information System, perhaps the earliest, certainly one of the most ambitious, prototypes of a modern GIS (Tomlinson 1967). CGIS created a digital coverage of land suitability maps for the areas of Canada with agricultural potential.

"Heads-up" digitizing: A digitizing station that provides a graphical user interface on the screen of a workstation (hence sometimes called *on-screen digitizing*). The operator uses a pointing device (mouse or trackball) to navigate on the scanned image of the original source, without having to look down at a digitizing tablet.

images just as accurately as on a large digitizing table. These techniques have already begun to make manual digitizing tables obsolete.

Transforming Digitizer Measurements into Coordinates

Measurements from a digitizer or a scanner are integers relative to the spatial reference system of the hardware. For most purposes, these numbers must be converted into a more meaningful spatial reference. The process starts with registration and involves issues relating to projections and geodetic surveying.

Registration on Device Whether derived from a digitizing table or a scanner, the units of spatial measurement are based on the hardware. It is quite difficult to reposition a map at exactly the same position on a table or to pass it into a scanner at exactly the same orientation. The raw measurements must be connected to some external frame of reference just for the conversion process, not to mention the later analytical stages. This process is called **registration**, a term derived from printing technology where the color plates have to be printed "in register."

The registration solution used in map printing can be applied to the digitizing process. A set of tick marks outside the area of the map serves as a reference, much as pin bars keep negatives in place for printing presses. If the marks are located arbitrarily, then they can be used just to provide orientation. If they are placed at a known distance apart, then they can be used to correct for systematic effects.

Most software for manual digitizing provides for registration, usually by digitizing a few points whose position is known. The points are called **control points** in photogrammetry and remote sensing (Lillesand and Kiefer 1994, p. 325), though they are only involved in establishing a spatial reference system, not a framework of "control" in the sense of a measurement framework in Chapter 2. The software fits these digitized positions to their intended locations. Such a transformation makes some assumptions about the map and the digitizer. In the simplest case, the map is rigid, and the points may simply be rotated, translated, and scaled (Figure 3-9) in the digitizer space. The *similarity* transformation for this case can be written as an equation for the output (X,Y), given input (x,y) (Equation 3-1).

Equation 3-1: Similarity transformation

$$X = A + Cx + Dy$$
$$Y = B - Dx + Cy$$

where

C = [scale factor] × cosine ([rotation angle])
D = [scale factor] × sine ([rotation angle])
A and B = offsets for the center of rotation in output coordinates

Registration: The process of connecting a spatial representation to a broader spatial reference system.

Control point: A feature whose location can be established in an external spatial reference system (ideally, the Geodetic Reference System) and on the source material to be digitized.

The assumption that the scale is the same on the two axes may not be justified in many situations, particularly when the scanner hardware moves differently in the two directions. Also, paper maps tend to expand and contract at quite different rates in the directions with or against the grain created by the rollers in the papermaking process. The *affine* transformation provides for different scales on the two axes (Equation 3-2).

Equation 3-2: Affine transformation

$$X = A + Cx + Dy$$
$$Y = B - Ex + Fy$$

where

C, D, E, F combine differential x and y scales and rotation
A and B = offsets for the center of rotation in output coordinates

A similarity transformation, with its four unknown parameters, can be fit to two points (four coordinate values). An affine, with its six parameters, requires three points (six coordinate values). If these minimal numbers are used, the results can be easily influenced by measurement error, thereby introducing unknown distortions in the product. More complex transformations can be developed to model the barrel distortion in a photographic lens or a higher order polynomial, if justified. The more parameters, the more points are required.

If extra points provide enough redundancy, a **least-squares** procedure can evaluate how closely the measurements fit the assumptions. Ideally, the transformation equation should model the geometry of the expected errors. Some experimentation may be required to select from alternative models. Usually, software implements one equation, and a user may not know which it is. Digitizing operators must pay attention to the measures of fit produced in calculating the transformation. Increasing the number of control points will only add a small amount of time to the job, while providing much improved confidence. If "known" coordinates are available, 10 or 20 control points are worth the effort.

Errors made in registration tend to be of two distinct types. Small variations are unavoidable in any measurement system. By taking additional measurements, this kind of random error can be controlled easily. A procedure such as entering control points is also prone to what photogrammetrists call *blunders*. These are large mistakes that come from entering entirely wrong points, putting the control in the wrong order, transposing digits, or other quite human events. If the transformation is fit using least squares, any blunders exert undue influence on the result and the transformation can be distorted. One common procedure to obtain the best transformation examines the differences between the fit and the intended values (residuals), eliminates large values (outliers), then refits the transformation. Some alternatives to least-squares from the realm of robust statistics can ignore outliers automatically. For example, the least-

Least-squares: An estimation procedure that minimizes the sum of squared deviations between observations and a numerical model for those observations; used in ordinary regression analysis and many other statistical procedures.

median-squares procedure will estimate the correct transformation even if the control points are contaminated with up to 50% blunders (Shyue 1989).

Transformations from Document to Projection Normal practice is to transform from the coordinates of the digitizer device directly into the projection selected as the spatial reference system for the project. This procedure may not be appropriate in some cases because it obscures the intermediary steps. Any source, be it a map or an aerial photograph, is already in some kind of projection. If this projection is not the same as the target projection, then the two systems will differ. Locally, the difference may simply consist of a rigid similarity transformation (Figure 3-9), but the differences can become substantial on larger areas. If the transformation from digitizer coordinates to the map is rigid and the transformation from the map to the projection is rigid then a direct transformation from digitizer to projection will also be rigid. In this case, there is no reason to pass through the intermediary steps. However, if there is some correction required (for example, to adjust for scanner geometry, lens distortion, differential shrinkage, or projection convergence of existing sources), then a single transformation may not be able to model the combination of geometric distortions.

Some countries limit the difficulty by ordaining a single projection as a common spatial reference system. All maps in Switzerland or the United Kingdom are referenced to their respective national grid systems. Larger countries and even large American states cannot cover their territory with a single projection zone without distortions beyond the tolerances required for many applications. There are hundreds of possible projection equations and an infinite number of parameters that control the final geometry (Snyder 1987). In practice for a given area, a small number of projections will be used.

The selection of map projections is often a matter of tradition, linked to the historical development of a given set of map users. Nautical charts, for example, are nearly universally drawn on the Mercator projection, with the equator as central meridian. This projection is conformal (it preserves angles) and it has a special property, namely, that corrected compass bearings are straight lines. Manual plotting of courses and positions influenced this choice. For topographic mapping, most of the world is cov-

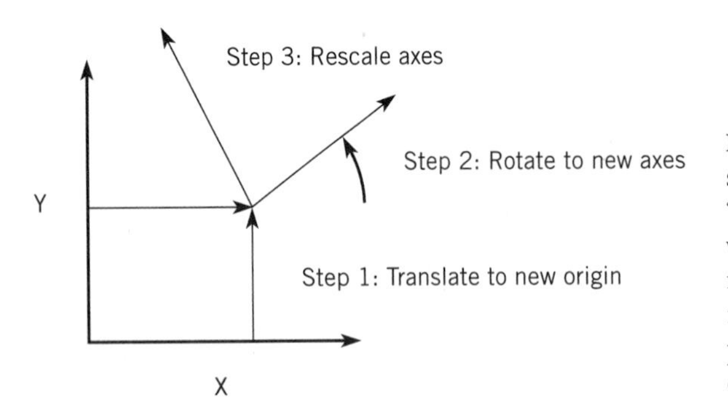

Figure 3-9: Components of simplest transformation: A rigid "similarity" transformation involves: translation (offset to a new origin), rotation (movement by some angle around a point), and a common scale change.

TABLE 3-1: Common projections used as spatial reference systems

Transverse Mercator systems	
worldwide	Universal transverse Mercator (6° strips)
United Kingdom	National Grid
Germany	Gauss-Kruger (3° strips)
Switzerland	National Grid (Gauss-Kruger)
State Plane System (USA)	North–south states, e.g., Illinois in two zones
Lambert Conformal Conics	
France	Grille Lambert, three zones
State Plane System (USA)	East–west states, e.g., Washington in two zones
Other Projections	
Malaysia (peninsular)	Malayan skew orthomorphic (oblique Mercator)
State Plane System (USA)	Alaskan Panhandle; oblique Mercator

The State Plane System for the US defines 125 zones, some as small as single counties.

ered at some scale in either a **transverse** Mercator such as Universal Transverse Mercator (**UTM**) or a conformal conic such as Lambert Conformal Conic (Table 3-1). The standard reference systems universally adopt a conformal projection, one that accepts a certain amount of scale error to preserve local angles. This choice was based on the needs of artillery and engineering users, who originally defined these systems. Equal-area projections might be more useful for many kinds of GIS analysis.

With computer representation of maps, projection equations are no real difficulty. A map in one projection can be converted to latitude–longitude (through the inverse of the projection function) then projected into some other form. Some calculation time may be required, but on modern computers it is no great burden. The US government has placed the General Cartographic Transformation Package (GCTP) in the public domain, and many commercial packages have incorporated this software.

Reference to Geodetic Surveys Digitizing must connect the local coordinates of the digitizing equipment to a permanent spatial reference system. Eventually, all projections are related to positions on the surface of the earth. For historical reasons, the

Transverse: A projection oriented at right angles to the equator. A transverse cylindric projection uses a meridian of longitude as its central meridian.

UTM: Universal Transverse Mercator; a spatial reference system using a set of transverse Mercator projections 6° wide that cover the earth (except for polar regions covered by two polar stereographic projections).

most common form of reference to geographic coordinates is latitude and longitude. These two angles combine with an assumption of a given radius for a sphere or a given ellipsoid to specify a location on a three-dimensional body. There are dozens of ellipsoids in use, though more precise measurement is leading toward some international standards for the shape of the earth (Table 3-2). One common source of incompatibility between spatial reference systems in the United States comes from the transition between the **North American Datums** of 1927 and 1983. This includes a change in reference ellipsoids plus the more local effects of readjusting the geodetic survey data. Ultimately, the spatial component of geographic information depends on the network of geodetic surveys to provide the framework for other measurements. The best points to use for registration are the geodetic reference markers used to construct the original map or some **well-defined points** whose coordinates are known from surveying measurements. Common practice in digitizing places too

TABLE 3-2: World geodetic standards: Reference ellipsoids

Name	*Equatorial (major) axis in meters*	*Flattening (1/f)*	*Region*
Airy 1830	6377563	299.325	Great Britain
Bessel 1841	6377397.2	299.153	Central Europe
Everest 1830	6377276.3	300.80	Indian subcontinent
Clarke's 1866	6378206.4	294.98	North America
Clarke's 1880	6378249.2	293.47	Africa; France
Krasovsky 1940	6378245	298.2	ex-Soviet Union
World Geodetic System 1972	6378135	298.26	NASA, US military
GRS 1980/ WGS 84[1]	6378137	298.257	GPS, new systems

These reference ellipsoids may serve as the best fit to the actual geoid in different parts of the world. A horizontal "datum" adopts a reference ellipsoid and locates geodetically surveyed points on that ellipsoid.

[1]At the resolution shown, Geodetic Reference System 80 and World Geodetic System 84 are the same.

Source: Snyder (1987) Table 1, p. 12

North American Datum: An adjustment of geodetic measurements that provides the horizontal reference for North America. The 1927 Datum held Mead's Ranch, Kansas, as a fixed point, while the 1983 Datum performed a simultaneous adjustment of all measurements. The 1927 Datum uses Clarke's 1866 ellipsoid, while 1983 uses the 1980 Geodetic Reference System. The plural of a geodetic datum is "datums," despite the word's Latin origins.

Well-defined point: A pointlike (isolated) feature that can be distinguished on the source and on the ground to sufficient accuracy; in US National Map Accuracy Standards of 1947, implemented as "plottable to .01 inch."

much faith in the corners of map sheets and other symbolism not directly related to the construction of the map.

In the past, geodetic surveying was very expensive, and the process of adjustment was complex. Observations formed a complex network of triangles to connect the areas of interest and to reduce errors. The Global Positioning System (**GPS**) has revolutionized many aspects of spatial measurement. Using this system of communication satellites, a receiver can observe four or more satellite transmitters, calculate the distance to each, and store the result as a coordinate projected into any system. With receivers built for surveying, not weekend yachting, and a second receiver on a known point, differential GPS surveying can provide geodetic-quality results without all the intermediaries required in the earthbound technology.

Mapping systems have been structured around the general assumption that geodetic control will remain expensive and rare. This may cease to be the case. The tools for integrating spatial data will become increasingly important if GPS equipment leads to many sources of information referenced to field-measured coordinates. This source of measurements should also diminish the need to digitize existing maps.

Before we become entranced by the possibility of geodetic-quality coordinates on everything, however, we need to remember that geographic information is often an abstraction. It is futile to record the location of a swamp symbol down to the centimeter, as it represents a region without clearly defined edges. The information content of a measurement does not require excruciating accuracy for all coordinates.

DATA QUALITY

Data Quality: Closing the Loop

Once a set of measurements has been organized inside a system of representation, it is prudent to consider how well the measurements do their job. The first word that may come to mind is *accuracy*. Maps, like measurements, are meant to be accurate. While accuracy evokes the spatial element of the map, there are a number of related issues that are better organized under the broader title of data quality. All measurements can be verified in some manner against the world or against each other. However, most geographic information involves purposeful simplification of the phenomenon to fit it into the measurement framework, so verification is not a simple process of checking each "fact" against an objective standard. Yet, the whole measurement process should be cross-checked to ensure that information is fit for its purpose.

In the ring system of this book (Figure P-1), the issues of data quality do not

GPS: Global Positioning System, a constellation of communications satellites that broadcast timing signals that can be converted into a distance measurement, permitting *trilateration* (surveying by knowing the sides of triangles).

fit as a ring in themselves, but as a reflection on the results of representing a given measurement. Issues of data quality are central to evaluating the results of digitizing and any other representation. Each chapter of Part 2 will consider how tools can be used to verify the information. These discussions will follow the framework for data quality adopted in the Data Quality Specification of the Spatial Data Transfer Standard (SDTS; Federal Information Processing Standard 173) (US National Institute of Standards and Technology 1992). SDTS defines data quality as the information a producer should present to a potential user to allow the user to make a determination of fitness for some particular use. The vehicle for this information is a Quality Report in five parts.

1. *Lineage* recounts the source materials and all the operations and transformtions to produce the product. This section should address the nature of the measurement (previous chapter) and the choices of representation (this chapter).
2. *Positional accuracy* deals with the spatial component of the measurements. Testing for well-defined points is well established, but testing the fidelity of other kinds of measurements is still a matter of debate within the research community.
3. *Attribute accuracy* covers the fidelity of the nonspatial elements of the measurements. Clearly, these depend on the measurement framework.
4. *Logical consistency* concerns the internal consistency of the representation. It verifies relationships that should be present.
5. *Completeness* also depends on the measurement framework. For isolated objects, the completeness concerns the errors of omission. For exhaustive frameworks, either raster or vector, completeness deals with the spatial rules, such as minimum width or minimum area, that may limit the information.

Verification and Quality Control

The measurement stage in digitizing is far from the end of the process. Verification and quality control occupy a major portion of the effort in building a GIS. The raw results from a digitizing procedure, whether a manual table or a scanner, create a set of geometric objects. The technical term for this mass of geometry is *cartographic spaghetti*, because it is "no more structured than spaghetti on a plate" (Chrisman 1974). A spaghetti file can be used to draw back the objects, though there are likely to be many little geometric flaws introduced by the digitizing system. These flaws can be removed by visual inspection and graphical manipulation, often the approach used with computer-aided design (CAD) systems.

The topological model (introduced in Chapter 2) was developed largely to reduce the effort in verifying the results of digitizing. By introducing a more complex set of axioms, the relationships between objects can be created with much less trouble. Specifically, the topological model prohibits lines crossing each other. All crossings must be declared as nodes, making the graph a simple planar one. This verification can be done automatically by the same software that performs polygon overlay or in an interactive setting.

Once a graph has been constructed, it can be checked for further requirements. For categorical coverages, such as parcels or soils, the purpose of the lines is to separate polygons. Thus, any chain ending at a dead-end node is a sign of some kind of error. Either a chain is missing or a chain has missed its intended node by an *undershoot* or an *overshoot* (Figure 3-10). Rather than having to inspect the whole map, the operator can direct attention to correcting these errors directly.

The next phase in any digitizing involves attributes. In some cases, attribute values are attached to features as the geometric measurements are taken. This may be efficient for point features and some line classes, but categorical coverages do not need to be coded on the left and right of each chain. If each polygon is identified by digitizing a single point inside the polygon, the creation of topological structure has a form of cross-check. All polygons should have one and only one "label point." If the topological processor finds two points inside one region, then either the points or the linework are in error (Figure 3-11). Typically, a chain is missing or an undershoot connects two polygons that should be separated. A polygon with no points may signal omission of a point, or it may signal that some linework inadvertently separated a polygon. These symptoms help direct the operator to fix the problems.

Once the input passes the test, it can be certified to be *topologically clean*, meaning that the software can depend on the structure. This kind of reliance is why Codd (1981) uses the term *integrity constraint* in describing the third (axiomatic) ingredient of a data model. Topological tests of logical consistency are particularly well developed in the current generation of workstation GIS packages. Like other tests of logical consistency, topological tests use the database against itself without external sources of information. A serious program of quality control must use the topological tests as a necessary beginning, but they are not sufficient as a total package.

Along with the topological tests, there should be some consideration of the other components of data quality. For example, some forms of internal evidence can be used to validate the range of values assigned to an attribute. Such a test does not really check the specific value for the particular object, but it can locate potential blunders in all components. More detailed tests of positional accuracy, completeness, and attribute accuracy require some source of external information. A source of information may be a database of higher accuracy that exists for a part of the region, a program of field inspections, or a number of other strategies. The procedures for testing geographic information are an area of substantial research interest (Goodchild and Gopal 1989; Chrisman and Lester 1991). Point samples are commonly used to test the accuracy of remote sensing classifications. The overlay process can also be used (see Chapter 5) to test one coverage against another. In both cases, the differences must be considered relative to the spatial resolution desired for the two products, the issues commonly termed *scale*. The testing program adopted by the French Institut Géographique Nationale for their BD-Topo project currently examines one sheet in five, allocating one person for one month to survey features, to examine attribute classification, and to produce a report for each sheet tested. Few agencies produce such a comprehensive documentation of data quality. Any testing program is expensive and must be designed carefully.

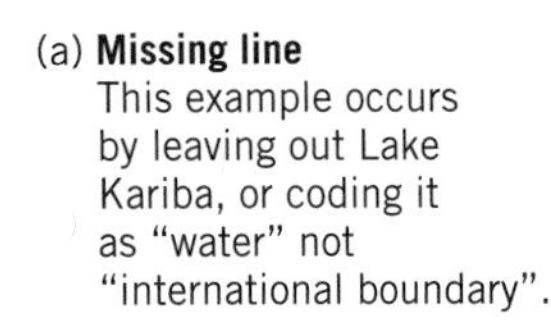
(a) **Missing line**
This example occurs by leaving out Lake Kariba, or coding it as "water" not "international boundary".

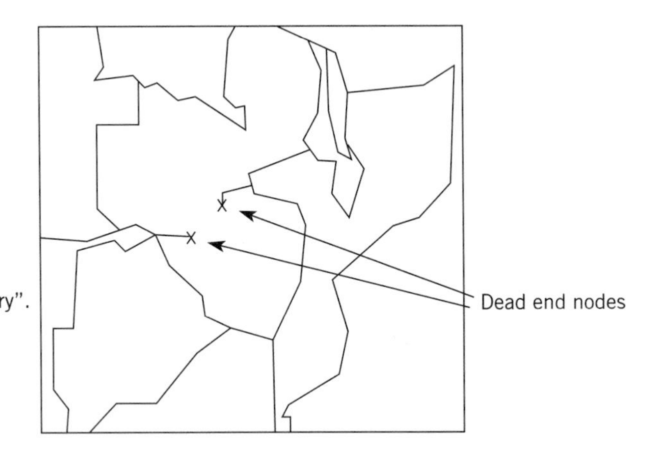

(b) **Undershoot**

(c) **Overshoot**

Figure 3-10: Examples of different digitizing errors flagged by a dead-end node: (a) missing line; (b) undershoot; (c) overshoot. Overshoots are much less trouble and may be deleted automatically. Undershoots require some correction, often on-screen.

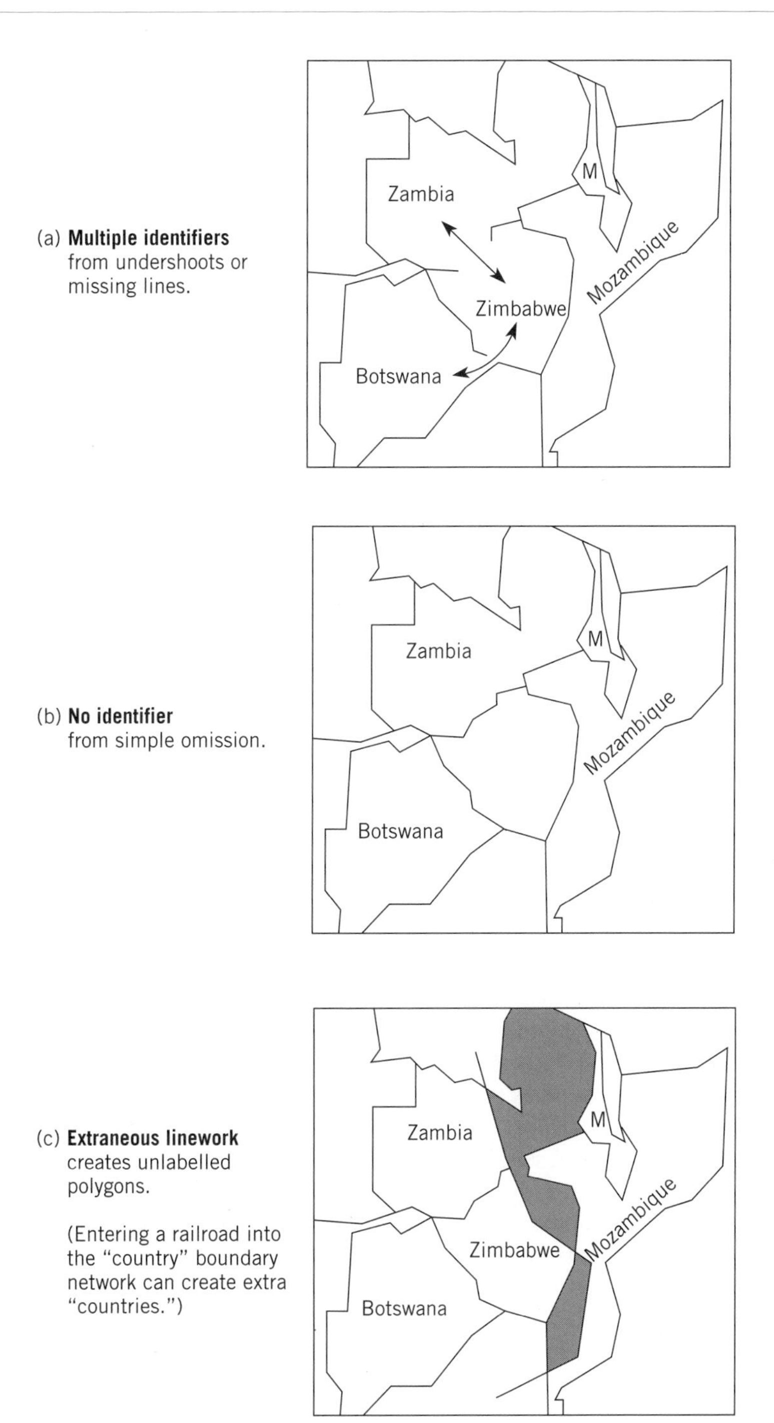

Figure 3-11: Digitizing errors detected while attaching identifiers (meatballs) to polygons: (a) Two (or more) label points may mean missing line or undershoot; (b) no label point may occur from omission, or (c) from extraneous linework.

Digitizing, at first glance, is limited by the source material. The results seem no better than the quality of the original maps. Certainly, the difficulty of digitizing is increased by errors and inconsistencies in the maps. However, the process of conversion is more than the digital version of a copy machine. By imposing a tougher set of relationships and integrity constraints, the quality of the maps can be improved. The digital representations force certain relationships to be more obvious and require greater consistency. The computer age has popularized the saying "Garbage in, garbage out." So long as the computer remains a passive repository, this will remain true. If the processing imposes a tougher set of integrity constraints, however, the digitizing process may upgrade the incoming map representation so that it is somewhat improved. It is hard to hope for more in a practical situation.

After Digitizing

After digitizing the source material, the result may not match your requirements exactly. Other procedures are frequently required to complete database construction. For example, one task involves building a larger coverage from units like sheets digitized separately (Beard and Chrisman 1988). Most of these tasks use the operations and transformations that will be presented in Part 2. This section will consider scale changing, geocoding, and temporal transformations that relate directly to the digitizing process.

Generalization and Scale Changing One of the most serious differences between a data resource and its intended purpose can involve the issues of scale. The data digitized may contain much more information than required by having higher resolution in attributes, geometry, or both.

Despite the protests of research cartographers, data sources are still described in traditional terms by the representative fraction, *cartographic scale*. The **Digital Line Graph** products of the US Geological Survey (USGS) are described as 1:24,000 or 1:100,000 or 1:2,000,000. When these are produced by digitizing the graphical products, this makes a certain sense, but only when considering the rules used by that particular agency in producing those series. For example, much of the 1:100,000 series was produced by photoreduction of the 1:24,000 series; this lineage is not always carried along. The DEM products of the USGS are described as 1:24,000. A few are produced from the contour plate of the graphic product, but most are produced independently by 30 m sampling on aerial photographs. The sampling density provides a much clearer description of these products. Scale will become even less indicative for new resources, such as municipal databases and the French BD-Topo, which are created directly by photogrammetric compilation without any traditional graphic product.

Scale has been a cartographic shorthand for a large number of decisions. When constrained by the graphic medium, more detailed scales typically did imply higher accuracy. However, this relationship is not obligatory. Detailed maps maintained for

Digital Line Graph (DLG): A data format developed by US Geological Survey National Mapping Division that uses a topological vector model.

local property tax purposes by a county in the United States may be at a scale of 1:4800 or 1:2400, but the quality may not follow the scale. Unless there has been a recent subdivision, these maps may be 50 years or more out of date. The original surveys may be dubious, and new, more reliable information may be distorted to fit the old framework (Kjerne and Dueker 1986). Information from scales of 1:24,000 may be more current and more accurate in some respects. The mapping rules subsumed by scale include many issues of attribute measurement and completeness along with the geometric relationship.

Many cartographic techniques may be applied to reduce the quantity of data within some constraints on the information required (McMaster and Shea 1992). It is not the purpose of this book to provide a complete review of the tools of automated map **generalization**, but one tool is used frequently during map digitizing and illustrates some important principles about the role of axioms for geographic information processing. This line reduction algorithm has become universal (Douglas and Peucker 1973). It works line by line, assuming that the endpoints are required. At each step, a trend line is constructed between the last two points known to be required (Figure 3-12). The points between are examined to find the maximum deviation from the trend. If this deviation exceeds some tolerance, the point of greatest deviation is selected, and two more trend lines are created for **recursive** treatment.

This algorithm of reducing the number of points along a line is certainly not the fastest. Since it must find the maximum deviation from each trend line, it cannot work in one sequential pass. There are much simpler filters that can weed out nearly collinear points with much less computation. However, these weeders can be tricked. For example, on a curve of large radius, they may decide that each point is essentially collinear, forgetting that the overall trend is moving away from the earlier point. Often it is worth paying the price of computation to obtain a better result. Applying a series of doublings to the tolerance can reduce an 85-point line to 5 points (Figure 3-13).

Any line reduction technique will fail if extended beyond its intended purpose. The Douglas method works line by line and thus does not try to alter the fundamental topology. Some situations may require removing whole objects to perform radical changes in scale. Line reduction cannot be expected to remove whole objects or merge them with their surroundings. The thin point in western Maryland is a common example of a spatial neighborhood that may not be properly treated by line reduction (Figure 3-13). The extended Voronoi network data structure introduced in Chapter 6 would provide the neighborhood information to avoid these problems, but few software packages take advantage of these possibilities.

Attaching Attributes by Geocoding Another method to generate new representations attaches attributes to existing objects through indirect measurement. For example, a hospital has records of patients with a particular disease. To examine

Generalization: In cartography, conversion of a geographic representation to one with less resolution and less information content; traditionally associated with a change in scale.

Recursive: A programming procedure that invokes itself to subdivide a problem.

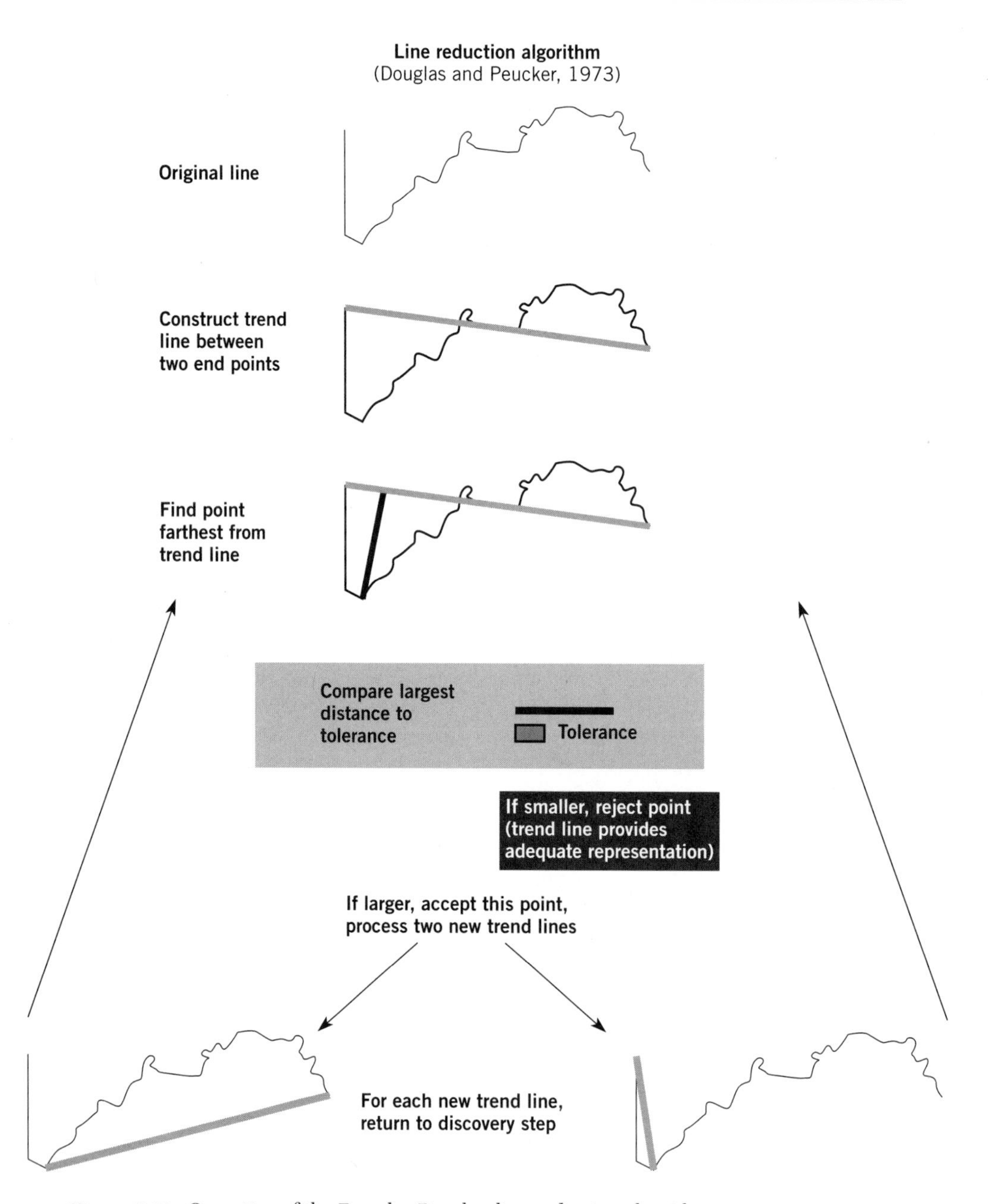

Figure 3-12: Operation of the Douglas-Peucker line reduction algorithm.

questions of environmental causes, each patient's home could be located on a map and digitized. Frequently, however, the location is assigned indirectly. Using simple tabular data processing, the patient records could be aggregated by city or postal delivery code, applying a choropleth framework.

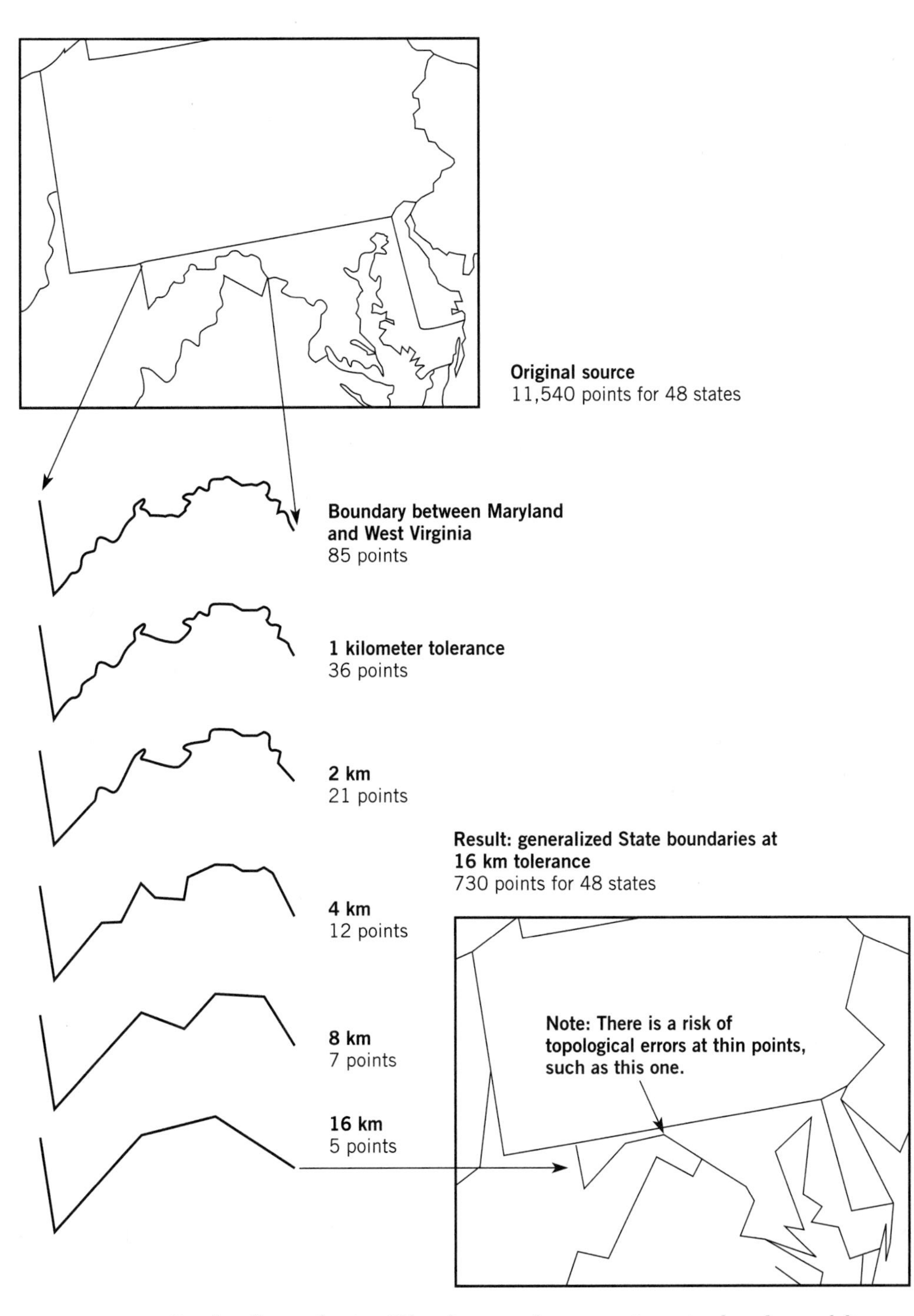

Figure 3-13: Results of line reduction. When lines are close, removing points based on each line separately may cause topological errors.

At a more detailed level, the street address provides a spatial referencing system for each structure in a city. Address matching, one of the original applications of the **DIME** files, converts from postal addresses to coordinates. The matching method varies depending on the rules that apply to addresses in various regions. Many cities in North America and Europe number houses sequentially with odd and even numbers on opposite sides of the street. The DIME files (and their TIGER replacement) record the name of the street and the range of addresses on each side. A house number can be interpolated in this range. The interpolation works better where the address range is more realistic than the grid referenced house numbers used in midwestern and western cities. Addressing rules are not universal, however. There are exceptions, as well as places with totally different systems. Some countries, like Japan, number houses by date of construction, not position.

Beyond the details of the specific addressing system, address matching represents a general set of network-oriented transformations. Dynamic segmentation (mentioned in Chapter 2) attaches highway information to a measured position along the route, and similarly, hydrological information may be attached to a river reference system. Each of these procedures attaches new attributes to existing objects in a database. These techniques can be used as an alternative to digitizing. Of course, the procedures involve assumptions that must be justified in each particular case.

Temporal Updates and Changes Perhaps the weakest component in geographic data handling involves time. If an existing resource of information fails to suit the requirements because it is out-of-date, then a process of updating may be more effective than a total reconstruction. Some updating is done from aerial photographs or other imagery, accompanied by visual interpretation of differences. There are many techniques for detection of change that use the analytical methods described in Part 2. If change is defined as a difference in two snapshots, then it is difficult to separate it from error in the sources. A more direct approach seeks information resources that record the changes directly. For example, an analysis of urban growth can use building permits as well as imagery to locate new residential construction.

GIS techniques offer many possibilities for improved treatment of time and change. The methods of representation are no longer tied to the mass distribution of standardized printed products. A computer database can be designed to maintain the historical record as the information evolves. Scientific applications require temporal information for process studies, not to mention the importance in legal proceedings. Still, the current state of the technology also presents a huge risk. Due to the limited frameworks for representation, time is not managed easily. The history of changes to the database may disappear with each `delete` command, leaving much less trace than in the era of paper maps.

DIME: Dual Independent Map Encoding, a digital database of streets and other census boundaries developed to conduct the 1970 US decennial census; an early and prominent implementation of a topological data structure.

SUMMARY

Representation places measurements in a structure for further processing. The data models for representation parallel the frameworks for measurement, though mixtures can be chosen in various ways. Vector models can be implemented by complex structures to represent topological and attribute relationships. Raster models appear simpler, but the simplicity can hide decisions made in generating the measurements.

At the practical level, constructing a new representation depends on hardware to measure and software to structure the results. All geographic data requires registration into a global reference system, and verification of data quality provides the basis for all further processing.

Creating the database is just the beginning of the exploration. The decisions concerning measurement and representation have their own logic, but they must also match the requirements of the analytical steps that form the topic of the next part of this book.

PART 2

TRANSFORMATIONS AND OPERATIONS

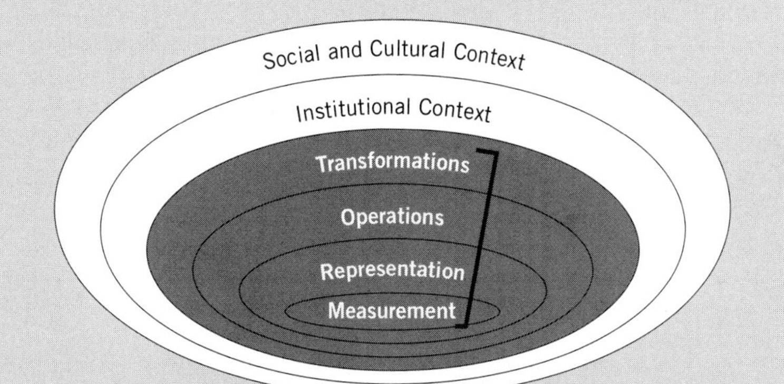

Part 2 moves outward on the ring diagram (Figure P-1) to the pair labeled "Operations and Transformations." Unlike measurement and representation, which occur in a specific order, the distinction between an operation and a transformation may be rather difficult to draw. Operations are the procedures that manipulate the information to construct new relationships or to make new measurements. Some special operations convert the information into a new measurement framework, so they will be termed transformations.

Each chapter in Part 2 is organized around a particular group of tools—operations and transformations. Operations will be presented roughly in order of complexity, from those that use no geometry to those that are most comprehensive. The concept of measurement framework creates a particular role for transformations that was not a part of earlier taxonomies. Chapter 4 begins with the simplest tools, those oriented toward single attributes. Chapter 5 introduces the key operation of overlay and builds a complete framework for combining attribute values using overlay. Chapter 6 adds spatial relationships such as buffers. Chapter 7 introduces surfaces and operations on neighborhoods. Chapter 8 considers a range of spatial models that require much more comprehensive processing. Chapter 9 presents the transformations between measurement frameworks.

Taxonomies of GIS operations

There are many alternative taxonomies of the operations in a GIS. Dana Tomlin (1990) developed one sequence to present map operations, ranging from the simple to the complex. It makes good sense to consider the simple operations that work on a single map, then those that work locally on two maps, and so on. However, Tomlin's scheme fails to include all possibilities (and thus provide the "algebra" promised), because it forces all measurements into a single raster representation and does not distinguish between a representation scheme and a measurement framework. Furthermore, Tomlin's terminology for the operations becomes a bit obscure for the more complex operations. Goodchild (1987) followed the flow of Tomlin's basic operation, adding some elements, such as information attached to pairs of objects—introduced as relationship control in Chapter 2. Burrough (1992) argued for "intelligent GIS" essentially by recognizing more spatial relationships. This book will develop a composite approach to operations developed on the basis of the measurement frameworks described in Part 1.

CHAPTER 4

ATTRIBUTE–BASED OPERATIONS

CHAPTER OVERVIEW

- Present operations that modify attributes without involving the spatial component.
- Describe interactions of spatial and attribute components caused by attribute operations.

The spatial aspects of GIS distinguish it from other kinds of information handling. The bulk of Part 2 will deal with operations rooted in geometry or spatial relationships, but this concentration on the spatial component should not ignore tools based on attributes. Despite their simplicity, operations based on attributes are important because they play a constant role in the applications world. Almost every operation of a more complex nature is preceded or followed by some form of housekeeping requiring this set of tools.

Attribute operations will be treated in two stages in this chapter. First, simple mathematical or set operators convert existing values into new ones, operating within one measurement framework. Some operations reduce information content, while others attempt to increase it, using some source of external information. As discussed in the second section, attribute operations for geographic information interact with the spatial component. These simple operations can also produce apparent changes in the measurement framework.

MANIPULATING ATTRIBUTES

Attribute values often encapsulate the final objective of GIS analysis. A forest products company wants to know the expected volume of timber to be harvested over a series of years. A transit authority wants to estimate changes in ridership on their route system. A tax assessor must determine a fair market value for every parcel in the county. The attributes intended at the end of the analytical process are rarely the items originally measured; they must be estimated from other information. The procedures available to manipulate attributes depend on the numerical properties of the measurement scale and external assumptions provided.

Reducing the Information Content

It may sound wasteful to throw information away, but many circumstances involve reducing a detailed source into a simpler form. These operations often occur as preparation for other more complex steps discussed in later chapters. There are many possible techniques, but a few common examples will serve to describe the general character. Each of these can be organized based on the level of measurement input and output (Figure 4-1).

Group A grouping procedure (at the top of Figure 4-1) takes a detailed classification and produces a cruder classification by merging some of the classes. An external source must specify which categories belong together in the resulting classification. For example, a detailed land use classification could be simplified into urban/rural groupings or into groups based on degree of human disturbance of ecological processes.

The syntax for grouping can vary. Some software provides complicated text commands like `recode land_use assigning 3 to 4 to 5 to 9 assign-`

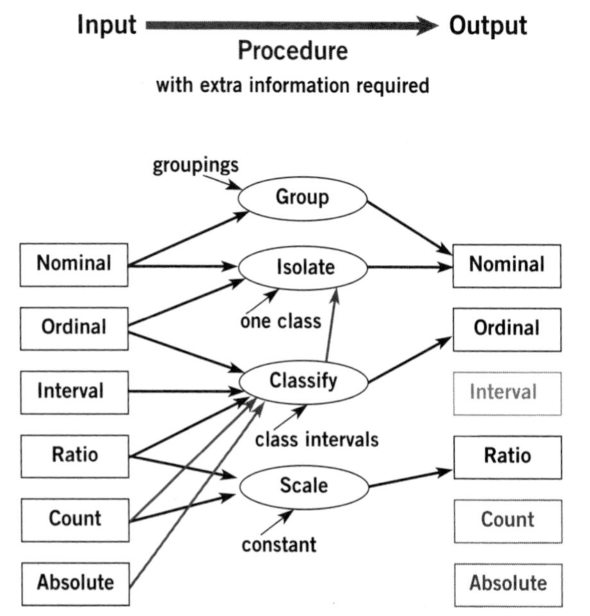

Figure 4-1: Procedures with one attribute input that reduce information content.

`ing 1 to 14 to 6`. Other systems may work on the basis of conversion tables that specify the output value for each value input. No matter how the command is entered, the mathematics remains simple. Nominal data should be treated by enumeration of the possibilities. For convenience, some languages might provide a range of values to recode to a new value. This shortcut may save much time in entering the commands when the categories are ordered based on some hierarchy. For instance, all the forest categories of a land use coding scheme might be given numeric values in one block without implying any ordinal scale. While a nominal classification simply states that each category is different, some differences may be larger than others.

Isolate An isolation procedure (second from the top of Figure 4-1) produces a simple nominal classification (selected and not selected) from a broader range of input. It can be applied to identify a single category from a larger nominal classification or a range of values from an ordinal classification. For example, an industrial category in a land use coverage could be isolated to depict industry as an isolated object. Similarly, using an ordinal class, an attribute of expected soil loss could be simplified to isolate those areas over a particular threshold.

Since an ordinal class can be constructed using a range in any higher measurement, isolation operators commonly apply to any kind of measurement. Isolation is one of a broader range of selection operations. Some selection works by region, a spatial component. When selection works through the attribute, the result isolates certain geographic information from its context. Database query languages such as Structured Query Language (**SQL**) give primary attention to selection with commands like `select parcels where owner = 'Smith'` or `select soils where permeability > 0.785 and texture = 'silt_loam'`.

Classify A classification procedure (third from top of Figure 4-1) produces ordinal categories from a higher level of measurement, such as ratio. A set of class intervals establish the breakpoints between adjacent categories of the ordered output. Classes can be constructed for different objectives as commonly applied in thematic cartography. An **equal interval** classification preserves some of the numerical properties of an interval or higher scale. A **quantile** classification chooses the breakpoints between classes based on a count of the objects in each class. Frequently the breakpoints come from a regulation or a technical specification. For example, the Conservation Reserve Program targeted lands higher than a particular amount of soil loss compared to the "tolerable" quantity. Land over the threshold is eligible; land under it is not.

SQL: Structured Query Language; a standard interface for access to a relational database through queries that select records matching logical expressions.

Equal interval: A classification procedure that divides the total range of attribute values by the number of classes; breakpoints are spaced at equal intervals, whether the class has any members or not.

Quantile: A classification procedure that assigns an equal number of objects into each class. The interval of each class will vary unless the distribution is completely uniform.

Scale A scaling procedure (bottom of Figure 4-1) simply changes the units of measure for a ratio or a count. For example, elevation can be converted from feet to meters or money from dollars to yen. If the representation of numbers were ideal, multiplying a ratio measurement by a constant should retain all the information content. However, it is common to round off measurements at some level of resolution. For example, the population of a city may be given in millions with one decimal point for hundred thousands. The integer nature of the original count has been lost, but the result retains the most significant portion of the measurement.

These four procedures show how an attribute can be changed toward a lower information content. Each involves some external decisions, such as the categories to isolate or group, or the ranges to isolate or classify.

Increasing the Information Content

The levels of measurement set up a rough hierarchy of information content. While it is possible to reduce the information content easily, increases in measurement level require some external source of information. Three procedures: rank, evaluate, and rescale, provide examples (Figure 4-2).

Rank A set of nominal categories, such as vegetation types, should be considered unordered. Using additional information about preferences, the categories can be converted into an ordinal scale (top of Figure 4-2). For example, one ordering could rank the habitat potential of each vegetation class for a species of reptiles; another ordering of the same vegetation classes would rank preferences for a bird species. Sometimes rankings are very simple, with fewer categories than the original nominal system, but the new categories are ordered from high to low. The soil capability information pre-

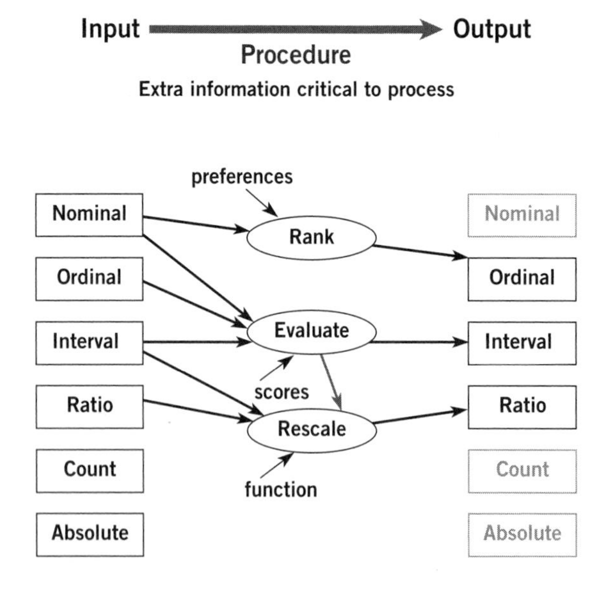

Figure 4-2: Procedures with one attribute input that increase information content.

sented in Table 2-5 ranks limitations from slight to moderate to severe. These values are assigned using indirect measurement through the soil series categories.

Evaluate The process of indirect measurement can also upgrade lower levels into interval and ratio scales (middle of Figure 4-2). An evaluation process takes a category and assigns it an interval (or higher) measurement. For example, vegetation classes might be assigned the expected density of a species of reptile. In some GIS applications, ordinal scales are given interval values more out of convenience. For instance, slight limitations could be scored "0," moderate gets a "1," and severe gets a "2." For this assignment to mean anything, there must be some justification that severe limitations are just as much worse than moderate as moderate are from slight. Without this kind of information, an evaluation procedure is likely to obscure more than it reveals.

Rescale A continuous measurement can be rescaled using a mathematical function to represent some other property (bottom of Figure 4-2). In some cases, the rescaling can upgrade an interval measurement into a ratio scale, for example, by converting temperature data through a nonlinear equation into the biomass produced by a tree. Such a function would involve substantial understanding of the plant and its dynamics. More commonly, a ratio scaled measure is rescaled to another ratio scale through a nonlinear funtion. For instance, Galileo's physics of an inclined plane demonstrated that the potential energy of soil particles (and hence the potential for erosion) does not scale directly with the angle of the gradient, but rather with the sine of the angle.

All these operations that increase information content can be performed on a single attribute, with the additional information provided in the form of external tables or mathematical functions. These are the simplest operations in any GIS, usually no different from similar operations in other forms of software.

Combining Pairs of Input Values

Attribute-based operations are not restricted to a single attribute. Most arithmetic functions (addition, subtraction, multiplication, and division) combine two values. If the data structure provides two values for the same entity, then a number of attribute-based operations become available (Figure 4-3). This section presents four examples, and then considers how these operations are performed.

Cross-tabulate A pair of categories (for either nominal or ordinal scales) can create a cross-tabulation, thereby coding the combination of the two values (top of Figure 4-3). In the past, the codes for the categories from each source were simply concatenated. For example, the "fractional codes" used in the 1930s built up complex numeric codes by assigning a digit each to represent dozens of attributes. The US National Wetlands Inventory maintains this tradition with a multicharacter code where each position can carry another attribute. Some software creates new categories for each distinct combination, giving a compact list of the actual pairs rather

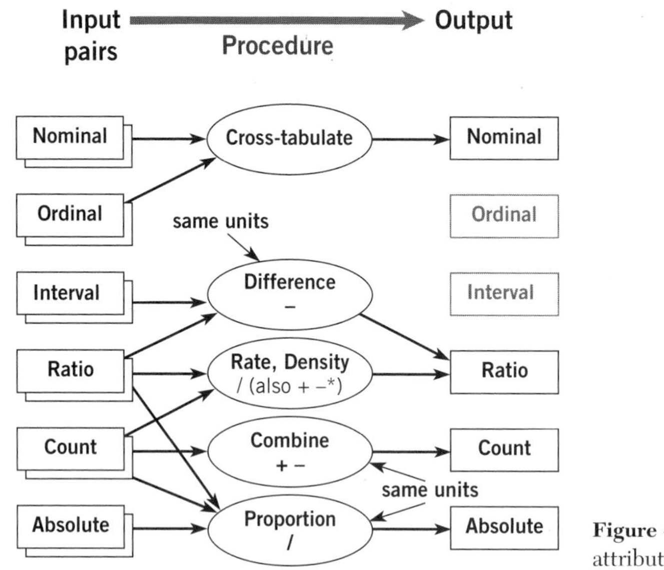

Figure 4-3: Procedures with two attributes input.

than the possible permutations. In any case, the new category combines information from both sources.

Sum and Difference The basic operations of arithmetic combine two sources at the interval or higher levels (second from top of Figure 4-3). Differences of two interval measures on the same scale produce a ratio measure. Similarly, two measures counting the same kind of objects can be combined by addition or subtraction. These operations depend on comparable units of measure. Adding ordinal categories, for example, presumes a whole series of assumptions.

Rate and Density Ratio measures get their name from the central role of division in their construction (third from top of Figure 4-3). A distance in meters is a ratio between the standard rod and the measurement. All kinds of rates and ratios can be constructed as derived measures from a variety of input sources. The ratio of a quantity to area is usually called a density.

Proportion One particular form of division gives a higher level of measurement (bottom of Figure 4-3). If the two values are measured on the same scale, say population, then division produces a proportion (on an absolute scale). For example, dividing the population over 65 by the total population gives a proportion over 65.

Performing Combinations All these operations require that two attributes attach to the same geographic object, using the same measurement framework. Consequently, the implementation differs between raster and vector representations. In the most common implementations of raster GIS software, each attribute is stored as a separate array of values, called *image planes* or *layers*. Each procedure generates a whole new raster file. Thus, raster operations on pairs of attributes are effectively indistinguishable from overlay operations (described in the next chapter).

In the vector representation, the attribute values are more distinct from the geometric structure. The **relational database** model provides a conceptual structure—tables of records, whatever the implementation. This model borrows the format of the geographical matrix, with one record for each object in a certain class, but with more rigor. A new attribute value, derived from any of the procedures discussed in this chapter, becomes another column in this table (Figure 4-4). Procedures that operate on pairs of values do not require overlay if the two input values come from the same table. In the vector form, most of these operations can be performed using regular commercial databases or spreadsheets, not any particular geographic software.

In some cases, an attribute value does not contain a measurement, but represents a relationship between different objects. In the example of corn yield estimates (Figure 2-17), each soil mapping unit (the polygon) is connected to a soil series (the class in the soil taxonomy). Initially, the identifier for the soil series may be the only attribute for the polygon. The soil series has dozens of possible attributes for different applications (see Table 2-5). The most compact (and least redundant) structure maintains the attributes of the soil series in a table by soil series, rather than duplicating those attributes onto the thousands of soil polygons or onto even more pixels. In the terminology of relational database management, the soil series identifier attached to the soil mapping unit is a **foreign key**, linking the two attribute tables. When needed for some purpose, the value of each soil series attribute can be associated with the polygon (Figure 4-5). This process is called a **join** in a relational database and a

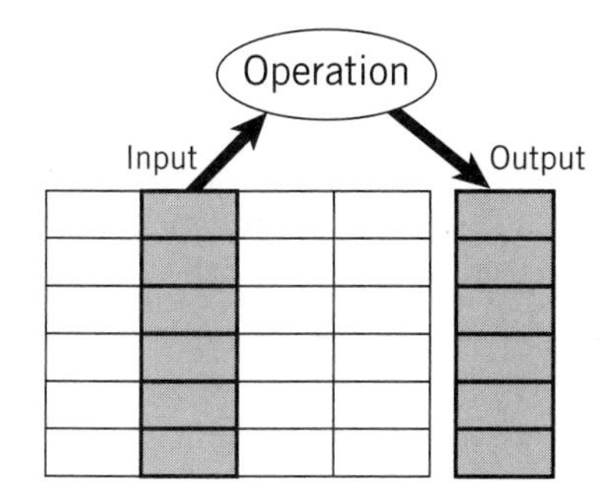

Figure 4-4: Attribute operations take values from a number of columns and produce a new column.

Relational database: A data model based on set theory. Each set has elements that can be uniquely defined by a primary key. A table (relation) stores all records for a set. Each record in a table has the same columns for attribute values. Relationships between tables are constructed by storing the key to a record in the other table.

Foreign key: Item in a relational table that contains a value identifying rows in another table; represents a relationship between two elements of a relational database.

Join: Procedure that attaches values from a database table to another table based on matching a foreign key to its primary instance.

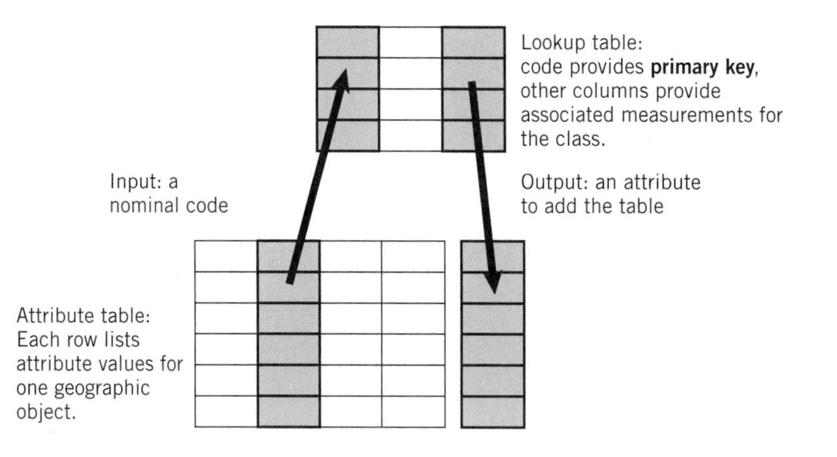

Figure 4-5: A join operation associates a geographic object to another table using a unique key. Attributes of that table can be added to the geographic objects as a new column.

table–lookup in other software, such as image processing packages. If the size of the table is manageable, the lookup mechanism can be quite quick, but a general purpose join requires massive data processing. The indirect measurement techniques introduced in Chapter 2 are implemented using an evaluation or ranking operation implemented through a join or a table–lookup.

INTERACTION OF ATTRIBUTE AND SPATIAL COMPONENTS

The operations described in this chapter are all designed to manipulate the attribute. However, these attributes are attached to geometric descriptions. Changing the attributes can also change the spatial component of the database. These relationships depend on the measurement framework and the system of representation. This section will consider aggregation (the spatial consequence of grouping categories) and isolation and then will connect these operations to broader issues of cartographic generalization.

Spatial Consequences of Aggregation and Isolation

The grouping and isolation operations reduce the richness of detail in categorical attributes. Although these operations seem limited to the attribute component of geographic information, they also can influence the spatial component and thus the measurement framework.

As described in Chapter 2, a categorical coverage involves measuring the boundaries between a specified set of classes. The refinement of categories can vary enormously. For example, a land use inventory can be interpreted into the 37 classes of the Anderson Level II land use/land cover system (Anderson and others 1976). This widely used system has one or two categories for commercial land uses and many more for natural resources. For comparison, the South East Wisconsin Regional Planning Commission (SEWRPC) used 79 classes for its 1975 inventory. This coding system approximately doubles the number of classes compared to the Anderson system, but the refinement is far from evenly distributed. There are 12 categories that match the Anderson category "Commercial and Services" (Table 4-1). Yet, in the less urban parts of the landscape, such as wetlands and forests, the Anderson system is more refined. Any survey must choose the system of classes with a balance of uses in mind.

TABLE 4-1: Example of different detail in codes for land use inventories

Level II Anderson codes (selected examples)	*SEWRPC land use code (under closest match)*	
12 Commercial and Services		
	210	Retail Sales and Service—Intensive
	220	Retail Sales and Service—Nonintensive
	432	Retail and Service—Related Parking
	436	Government and Institution—Related Parking
	437	Recreation—Related Parking
	611	Local Government
	612	Regional Government
	641	Education—Local
	642	Education—Regional
	661	Group Quarters—Local
	711	Cultural Public
	712	Cultural Nonpublic
41 Deciduous Forest		
42 Evergreen Forest		
43 Mixed Forest		
	940	Woodlands
61 Forested Wetlands		
62 Nonforested Wetlands		
	910	Wetlands

Source: Anderson and others (1976) USGS Professional Paper 964; South East Wisconsin Regional Planning Commission.

For different applications of land use, certain distinctions between categories would not be important. For each purpose, the classes of a detailed inventory can be grouped together to assemble more general categories. Regrouping categories has consequences when that attribute served as control for the coverage. When adjacent polygons in the detailed version abut, they may now need to merge into a large, simpler region, a process termed *aggregation*. This operation is the spatial consequence of the attribute operation of grouping a category used as control.

Few GIS packages perform aggregation with direct and elegant expressions. In the vector world, a classification system can be simplified in the attribute tables. Then a simpler geometry is produced by deleting all boundaries that have the same new category on each side. This operation is called *dissolve* or "dropline" aggregation, though it will only drop lines when merged categories are contiguous. In a raster package, grouping changes the values of the pixels, and there is no further geometric process to recognize contiguous objects.

In terms of information content, an aggregation operation seems quite similar to an isolation operation. Both reduce the content of the attributes, but the *spatial* consequences differ. If everything else is aggregated into a background class, then a specific category is isolated from all others. The isolation operation must be recognized as a transformation between two measurement frameworks. Starting with a categorical coverage, it produces an isolated object view in which the specific category is surrounded by the void. This differs from other aggregations that maintain an exhaustive classification of the whole region, thus continuing the categorical coverage framework.

Cartographic Generalization

Aggregation and isolation operations are simple components in the tool kit of cartographic *generalization*. These two operations are driven by the categories, and the geometry of the input coverage simply appears or disappears. If the map had been constructed for the simpler set of categories (or for the single object surrounded by the void), there is no guarantee that the same boundary lines would have been drawn. A less refined set of categories might change the rules of line sinuosity, minimum polygon size, and minimum width. If the aggregation is relatively mild, perhaps there is little difference. Converting the detailed SEWRPC categories (Table 4-1) into some highly generalized urban/rural split may not work simply by grouping the categories. Urban and rural at a generalized scale may not depend only on the detailed category, but also on the surrounding context. Figure 4-6 shows the SEWRPC and GIRAS land cover maps for the same region of Walworth County, Wisconsin. The SEWRPC land use maps include the pavement of roads in a transportation category, grouped with utilities at the urban end of the list. Are they urban or rural? Roads occupy areas that continue from towns into rural areas. In town, they certainly belong to the town, but in the rural area, the road might vanish into the background of fields and forests. A geometric filter or a neighborhood recognition is far beyond the capability of this sim-

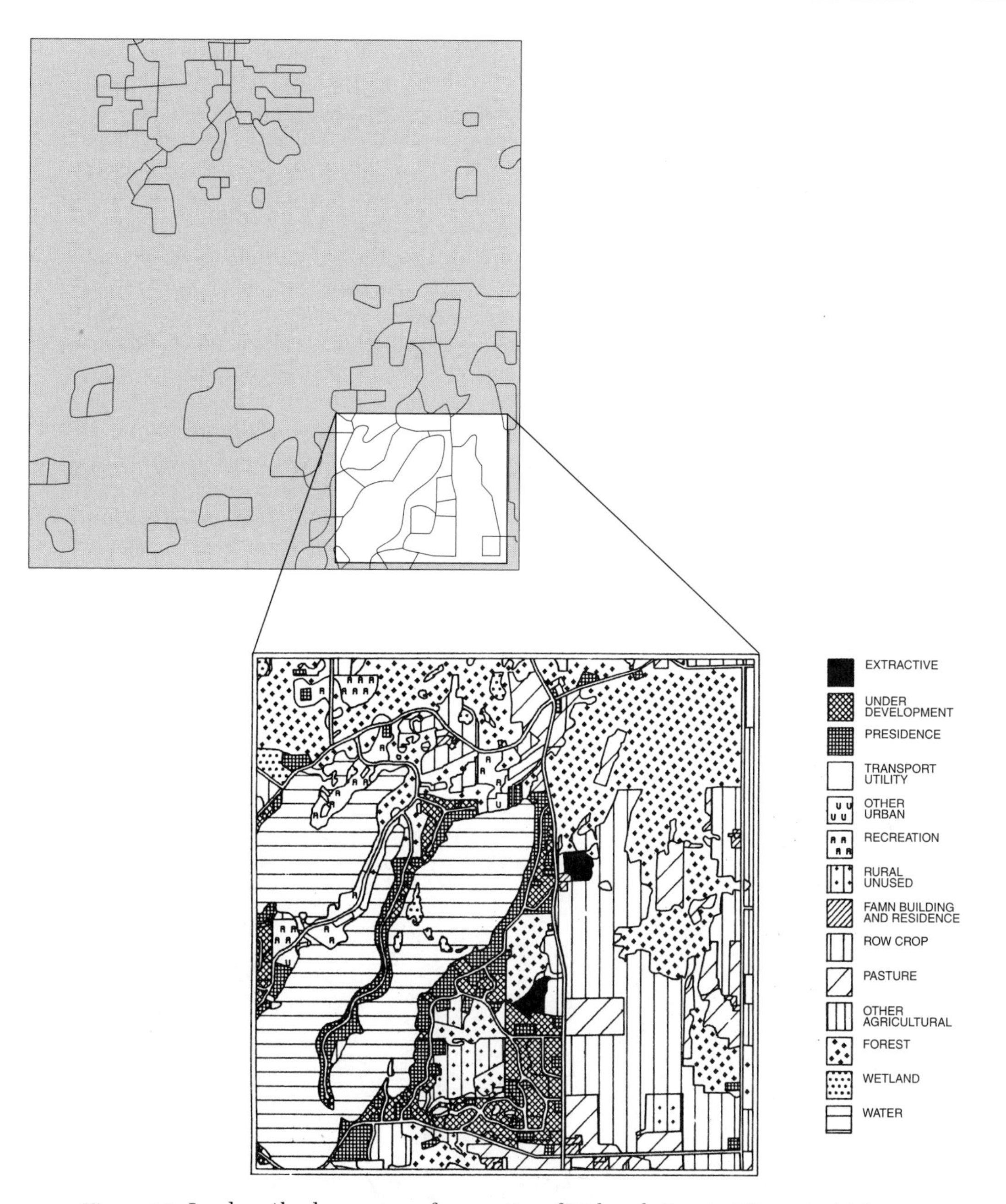

Figure 4-6: Land use/ land cover maps for a portion of Walworth County, Wisconsin: (a) from GIRAS digital data, 1976 photos, produced by USGS for display at 1:250,000; (b) from SEWRPC, 1975 photos, produced at a scale of 1:4,800. Both sources classified using respective systems described in Table 4-1. Illustration from Beard, 1987, p. 61.

ple aggregation operation. Recent research on cartographic generalization emphasizes that there are a number of distinct tools that must be used in combination (Beard 1987; Brassel and Weibel 1988; McMaster and Shea 1992).

The isolation operation creates the same generalization difficulties as aggregation, but it also includes a transformation from an exhaustive framework to an isolated object view. Exhaustive measurement requires compromise between the various classes, and the selected class may not be represented as it would have been if it had been mapped alone. A simple example arises from land uses that coexist in three dimensions. A road may pass underneath a building, or even through a mountain. In downtown Seattle, Interstate 5 passes under the Convention Center for a few blocks (Figure 4-7). From the land *cover* perspective, the road is not visible. If the road is then selected out of a land cover map, it may have disturbing gaps. Hence, just because there is a category "Road" in an exhaustive coverage, one cannot assume that it is represented as one would construct an isolated object "Road." Differences in measurement framework are particularly apparent in the conversion between raster representations. It makes sense to code an isolated object into a raster representation using a rule like presence/absence, where a road is shown in all cells it crosses. By contrast, an exhaustive classification is usually controlled by some area rule such as dominant type in the cell. Thus, a selection of one class from an exhaustive coverage may not match the desired object view, particularly for linear features narrower than a pixel.

An analyst must know the limits of the source materials before assuming that any particular mathematics apply directly. On first impression, these problems appear to be inaccuracies in the data, but they arise as much from the assumptions of the model and the measurement framework.

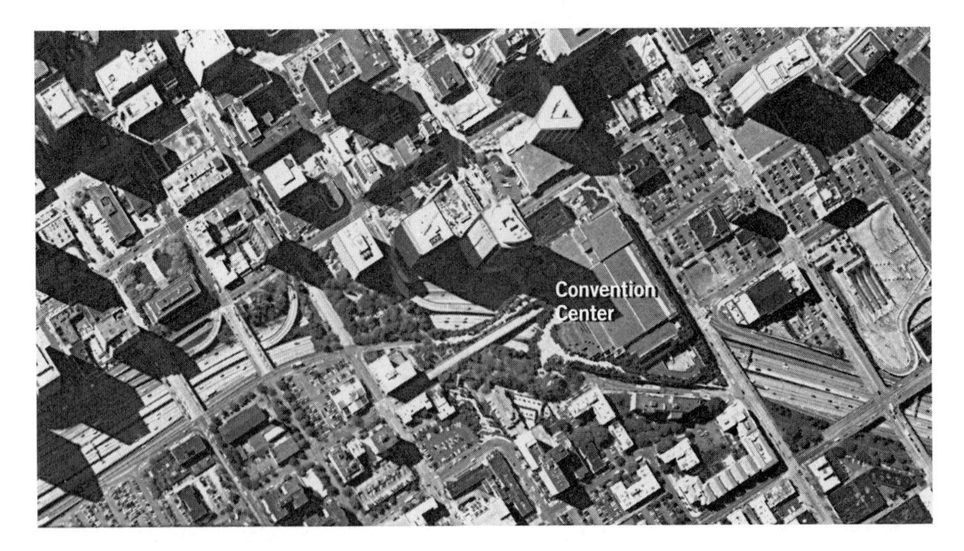

Figure 4-7: Interstate 5 disappears under Washington State Convention Center. Air photograph taken April 25, 1995 by Washington Department of Natural Resources.

SUMMARY

The simple attribute-based operations introduced in this chapter play a critical, though far from flashy, role in GIS applications. As a general limitation, a derivative representation cannot contain higher resolution (or information content) without having some external source (and the assumptions that go along with incorporating that external information). Although they manipulate the attributes alone, these operations influence the spatial component. In some cases, they act as transformations between measurement frameworks.

CHAPTER 5

OVERLAY: INTEGRATION OF DISPARATE SOURCES

CHAPTER OVERVIEW

- Review origins of map overlay analysis for site suitability.
- Describe geometric operations that establish connections between diverse sources.
- Introduce direct analysis of overlay results for change detection and error analysis.
- Present taxonomy of rules for combining attributes.

The tools presented in Chapter 4 treat attributes related to a single set of geometric objects. Many problems in geographical analysis require integration from a number of sources. The process of overlay discovers the basic spatial relationship between objects using geometric measurements. Then, attributes from the sources can be analyzed or combined. Direct analysis of overlay results serves many purposes, but change detection and error investigations provide the simplest examples. Combination also applies to any circumstance with multiple attributes, even if the attributes attach to the same geometric objects. A discussion of this combination process, including examples, forms the major part of this chapter.

DEVELOPMENT OF MAP OVERLAY

There are many applications that integrate different sources of geographic information. One of these applications, generically termed *site suitability*, played a lead role in developing the overlay technique. Site suitability examines social, economic, physical, biological, and other criteria to locate potential sites for some purpose. Site suitability motivated McHarg's (1969) *Design with Nature*, the book often seen as a harbinger of GIS development. McHarg championed sensitivity to the landscape and attention to multiple factors that constitute the environment. He implemented his vision by combining gray-scaled maps. Each component of the environmental system resulted in a separate "map overlay" in a literal sense—each map was made on a separate transparency. The cartographer used dark shading in the areas considered sensitive and left the rest transparent. All these layers were combined by placing the transparencies on top of each other (in registration) on a light table (Figure 5-1). The operation is called overlay after this physical procedure and its linkage to techniques in the photomechanical graphic arts (see Box). Once all the transparent overlays were assembled, visual interpretation could distinguish the areas of least sensitivity. McHarg (1969, p. 34) considered and rejected most of the mathematical solutions that his followers later adopted.

Connection to Photomechanical Reproduction

The development of the overlay method depended in part on the availability of geometrically stable materials. Before the middle of the twentieth century, a transparent material (like tracing paper) was likely to be the least stable material available for use. Stable material, like treated linen, was barely translucent on the best of light tables. Any graphic to be printed in multiple colors required separate impressions for each ink and thus separate originals for each color. The availability of stable photographic material contributed directly to the development of cartographic methods based on photographic exposure using multiple negatives. Sherman and Tobler (1957) described a "multipurpose cartography" that broke away from the allocation of a single negative for each color plate. This technique uses dozens of overlays, each one consisting of a group of features selected for having a common attribute that might require distinct graphic representation. Photographic methods served as the model for overlay analysis before the computer database (Alexander and Manheim 1962; McHarg 1969). To be able to adapt to all potential uses, the photographic process requires one overlay per category (Steinitz and others 1976), certainly a cumbersome pile of transparencies.

McHarg's book was enormously popular and very quickly influenced many aspects of North American and European politics. In particular, the concept of screening through map overlay was absorbed as the fundamental logic for a whole generation of regulations, beginning with the US National Environmental Policy Act. Of course, McHarg was hardly the first person to use map overlay. Steinitz and his students (1976) recount a rich history of the map overlay method dating back to the late

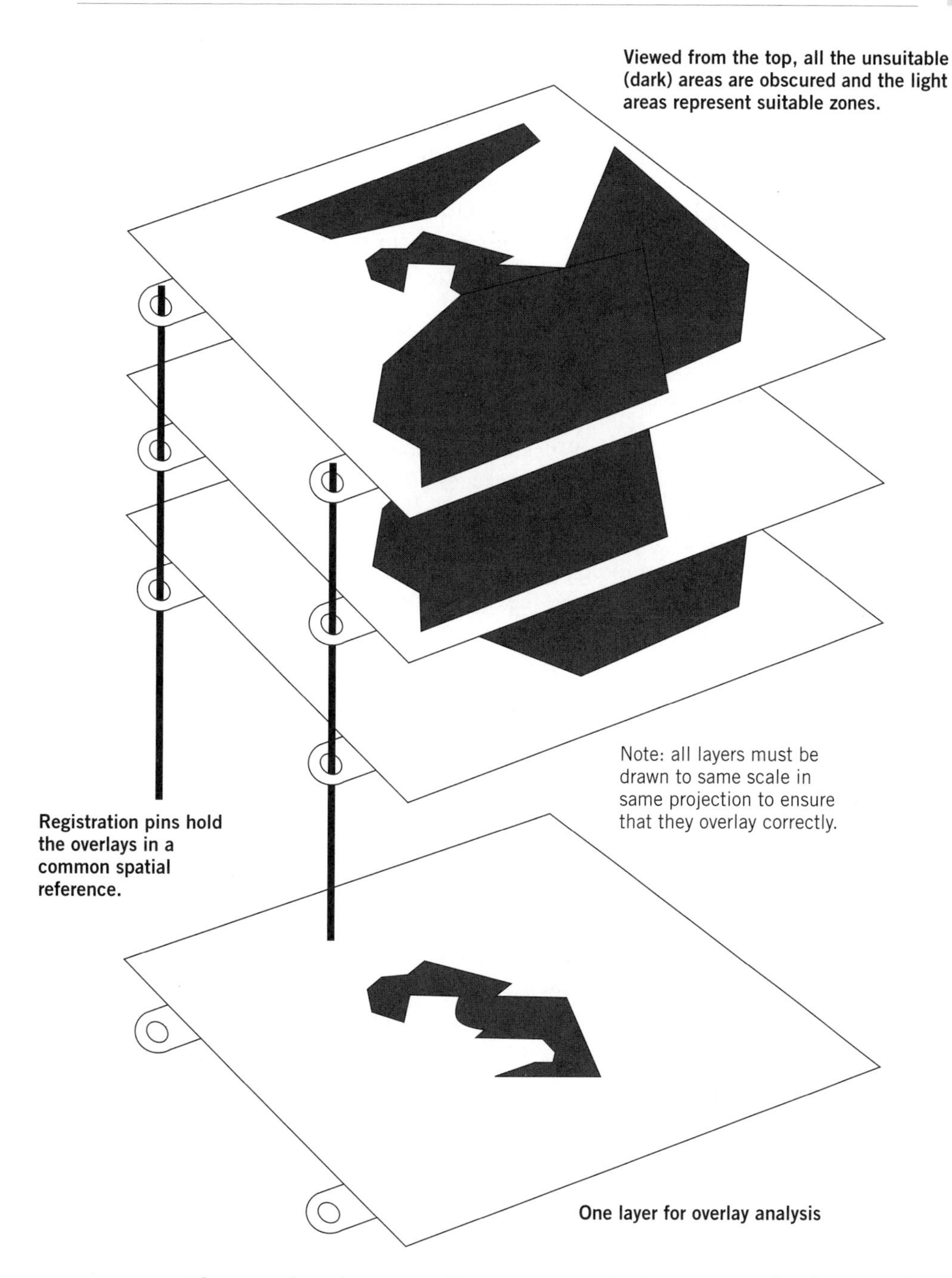

Figure 5-1: The manual overlay process: Transparent "overlays" are registered and combined to make a composite suitability mask.

nineteenth century. For example, Warren Manning (1913), a landscape architect connected to Frederick L. Olmsted (whose office designed New York's Central Park and similar projects), produced a plan for Billerica, Massachusetts, in 1912 that combined four maps: soils, vegetation, topography, and land use. By 1950, a textbook for town and country planning in Great Britain described the combination of attributes from transparent overlays (Tyrwhitt 1950). Philip Lewis (1963) produced recreation plans for Illinois and Wisconsin as the composite of various environmental factors. Thus, map composites were well established, though quite labor intensive and error prone, prior to the analytical solutions possible using a GIS.

The overlay problem became one of the early objectives for GIS software development. Programs for performing grid-based overlay analysis appeared in the 1960s and came quickly into professional use for the early wave of environmental impact analysis (on nuclear power stations and the like). The Map Analysis Package (developed by Dana Tomlin in the mid-1970s) became a standard for instruction of GIS, and it has been the inspiration for similar packages used in many classrooms (IDRISI, MAP II, ARC/Grid, and others.). There were some early algorithms to overlay vector representations (such as MAP/MODEL in 1969, and the original Canada GIS), but not until the late 1970s did prototype systems (like MOSS and Harvard's ODYSSEY) deliver practical overlay solutions. By the 1980s, commercial software packages built their basic functions around polygon overlay processors (such as ARC/INFO, MGE, DeltaMap, and others). Thus, overlay played a role in developing both raster and vector software strategies.

THE OVERLAY OPERATION

The reason overlay played a key role is that most applications of geographic information must integrate information from different sources. In the terminology of relational databases, map overlay serves as a kind of *join*, the procedure that links two tables based upon a common key (see Figure 4-5), but the use of the database term may make it seem easier than it really is. Normal joins operate through a *foreign key*, a value stored in one table that creates a relationship to a record in another table. Map overlay starts with no such direct correspondence between the layers. Instead it uses the geometric description to discover the connections. The digital map overlay procedure, whether raster or vector, depends on the absolute location of each feature. The spatial reference system provides the geometric basis to connect the two sources. Thus, the first issue in describing the overlay operation involves the spatial reference system, followed by the geometric intersection processing, finally leading to the attribute combination procedures.

Registration: A Universal Requirement

In the manual graphic form, overlay depends on physical registration of the transparencies. The graphic arts industry uses a range of devices to maintain registration, such as holes punched in the overlay that match a set of pins. These physical techniques work well for multiple overlays in photographic cartography as long as they remain at a consistent scale, projection, and area of coverage. In contrast, digital overlay is not restricted to such consistent sources. Most applications involve sources collected using different methods and at different scales. In the era of manual methods, map overlay implied recompiling each overlay onto a common base. The redrafting process, being a visual interpretation, could adjust the layers to become more consistent. Instead of an exterior framework of pin bars, registration came from the content of the map.

Manual integration consumes time, and redrafting introduces additional error. As digital methods replaced manual overlay, the computer offered the chance to escape many of the limitations of the physical media. The external framework of registration moved from the pin bar to the abstract mathematics of a coordinate system. The registration transformations described in Chapter 3 apply to the preparations for overlay. The least-squares procedures to compute a transformation can be used as a diagnostic tool to determine if points on two layers do actually correspond before performing an overlay.

Raster Implementations of Overlay

The overlay procedure is actually an inherent feature of the space-controlled measurement frameworks. Once two maps are rendered into the same grid system, the pixel becomes the base object for both (Figure 5-2). Thus, the methods that operate on pairs of attributes (Figure 4-3) become overlay operations with very little difficulty. The computation of overlay results simply translates into **Boolean** or arithmetic operations on a cell-by-cell basis.

Of course, the trick is to get the geographic information into this common grid reference. Tomlin (1990) begins with an abstract model that all maps consist of a large number of points with a common set of attributes. Conceptually, the points are infinitely small, but the pragmatics of resolution make these points discretely sampled, not continuous. Tomlin's reductionist model does permit some powerful developments, but there are hidden assumptions. First, the operation requires the two grid systems to be identical. A different projection, sampling interval, or even rotation will lead to a mismatch. The process of *resampling* an image (or a cellular database) may be a necessary prerequisite for overlay. In addition, not all geographic information converts from its measurement framework into the grid measurement framework with equal ease. This is particularly true for collection zone tabulations. Due to their complexity, these transformations between measurement frameworks will be discussed in Chapter 9.

Once a raster database organizes a number of layers into the same geometric framework (implementing Tomlin's model), overlay analysis requires only some relatively simple calculations on the attributes of each cell. Most raster software packages begin with the operators described in Chapter 4 that combine pairs of attribute values. Despite the number of possible operators, they are all rather primitive, so it usually takes a few steps to get from the source layers to the desired result.

Boolean algebra: System of operations applied to sets (and logical propositions); Boolean variables are zero or one, hence strongly connected to modern computing; originated by George Boole in 1847.

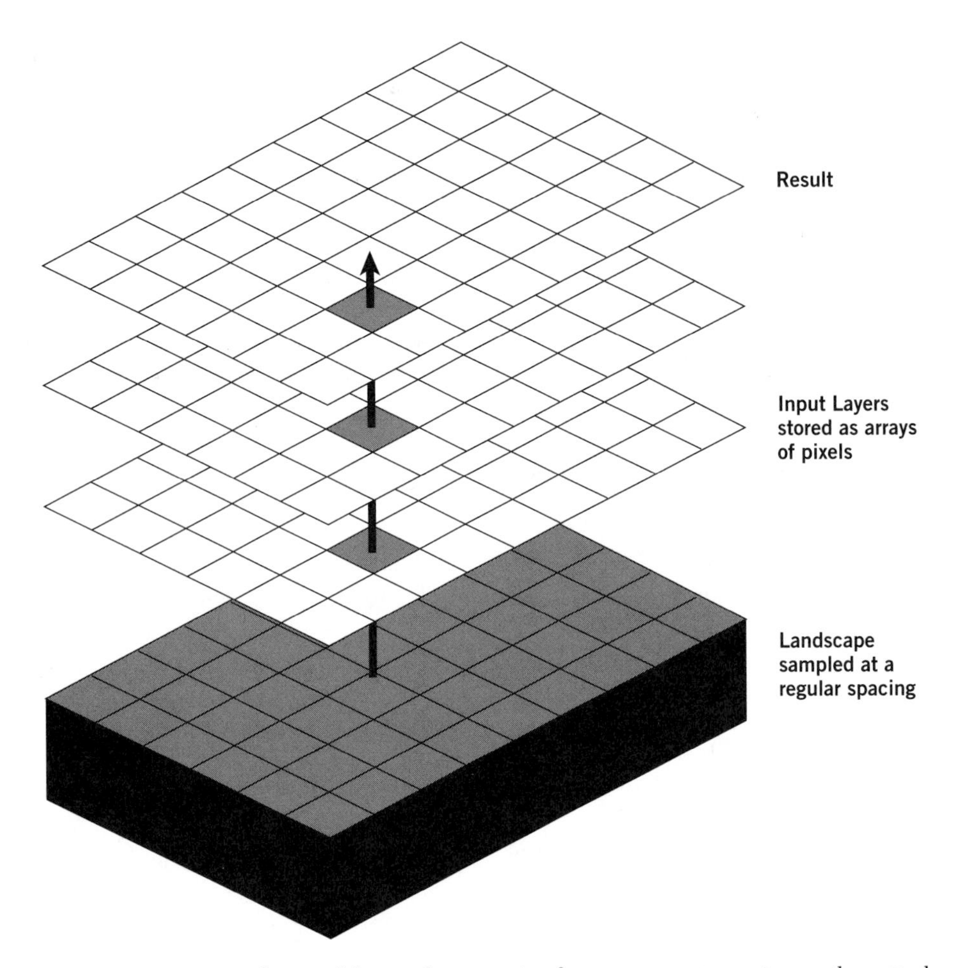

Figure 5-2: Conceptual view of the overlay operation for raster structures. Any mathematical operator can be applied to a pair of values, if justified by the measurements. Results become a new map layer, permitting arbitrarily complex results from simple operations.

For example, Figure 5-3 shows a hypothetical overlay analysis. Three source maps are represented in a common grid by applying one of the rules for spatial control. Attribute processing might include a threshold that classifies the original measurements into two categories, such as suitable and unsuitable. The binary maps can then be combined with Boolean operators, such as AND (set intersection) and OR (set union). These two operations are frequently required by the logic of environmental regulations and suitability analyses. Disqualifying factors often combine with the others using OR, because an area is excluded if disqualified in any way. In constructing an exclusion, certain conditions must occur simultaneously; these combinations can be constructed with AND.

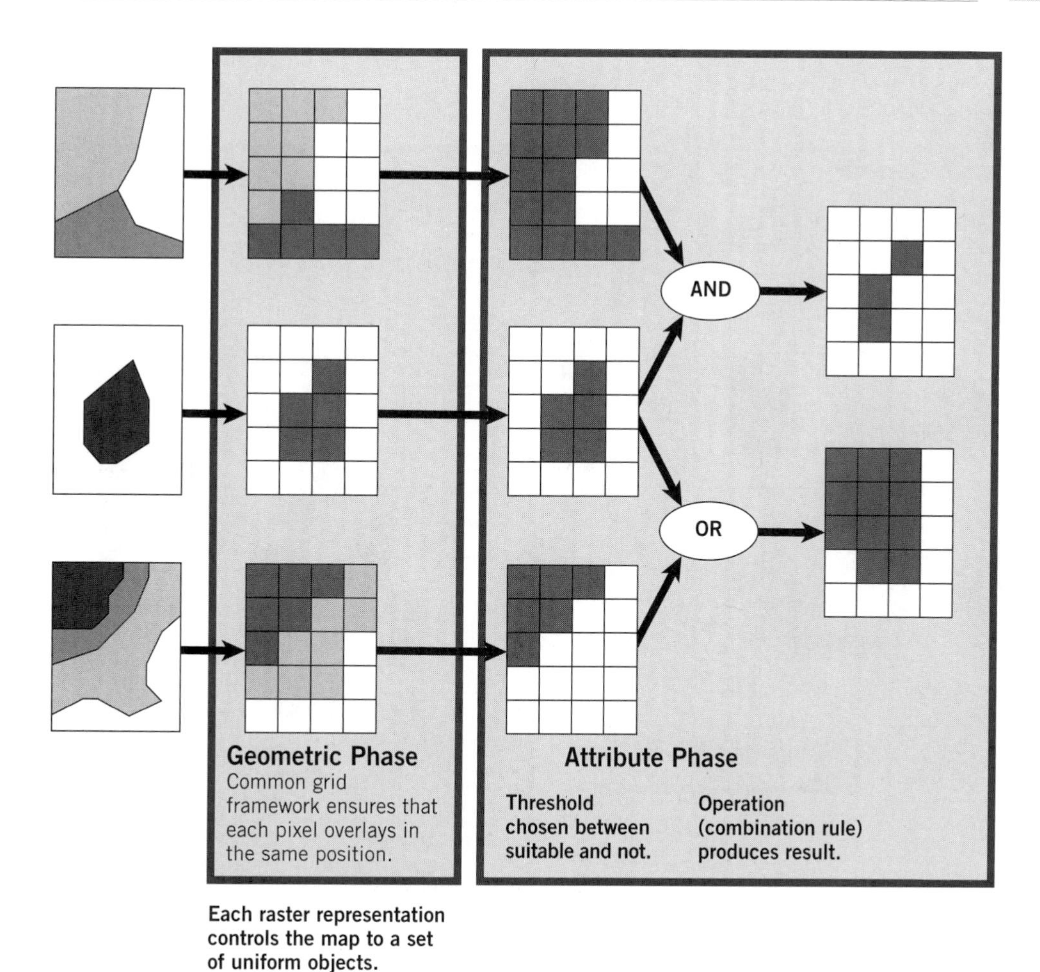

Figure 5-3: Operations for Boolean combination of raster categories.

An alternative approach creates a new category for every distinct combination in the pair of maps (LocalCombination in Tomlin (1990); `COMBINE` in Map II; see Figure 5-4). This method uses the cross-tabulation operation introduced in Chapter 4. The resultant categories can be grouped into any combinations desired. Finding all the possible combinations involves extra work, and some software is limited in the number of categories it can handle. However, as an exploratory measure, the complete combination forces the analyst to consider all factors. The key shown in Figure 5-4 links the new categories back to the categories on the original. Without a relational database to manage these critical linkages, the user often has to perform the analytical extractions without any assistance from the software.

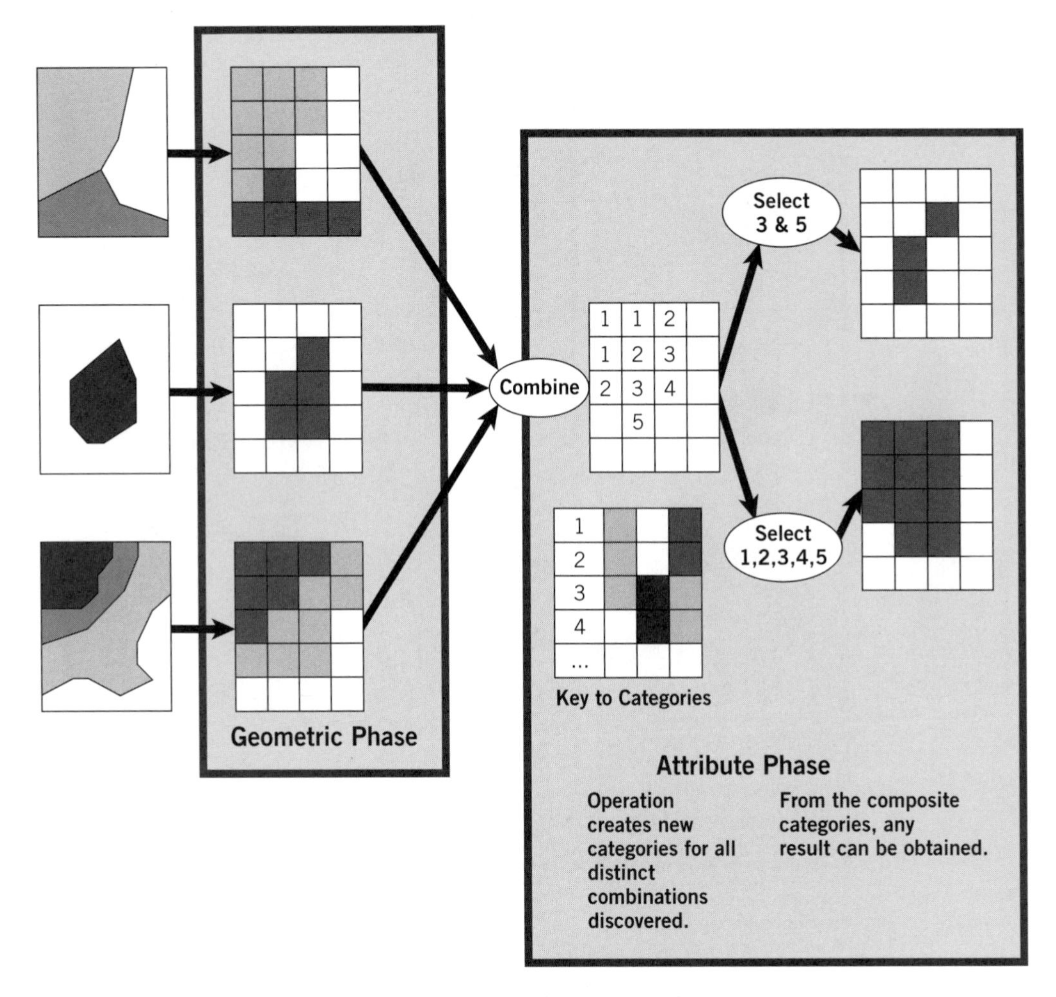

Figure 5-4: Composite combination approach for raster overlay.

These alternative approaches work reasonably well inside the constraints of the raster system. Even a simple analysis can be performed using many different combinations of the basic operations.

Vector Implementations of Overlay

The vector solution to the overlay problem is not as easy as the raster approach. The representation does not guarantee a set of common objects; they must be created geometrically. "Polygon" overlay produces a composite geometric representation where each area has a key to the attribute tables for the two source layers (Figure 5-5). Similar procedures can also combine points and lines with polygons. Following the geometric steps, the attributes of the objects may be combined.

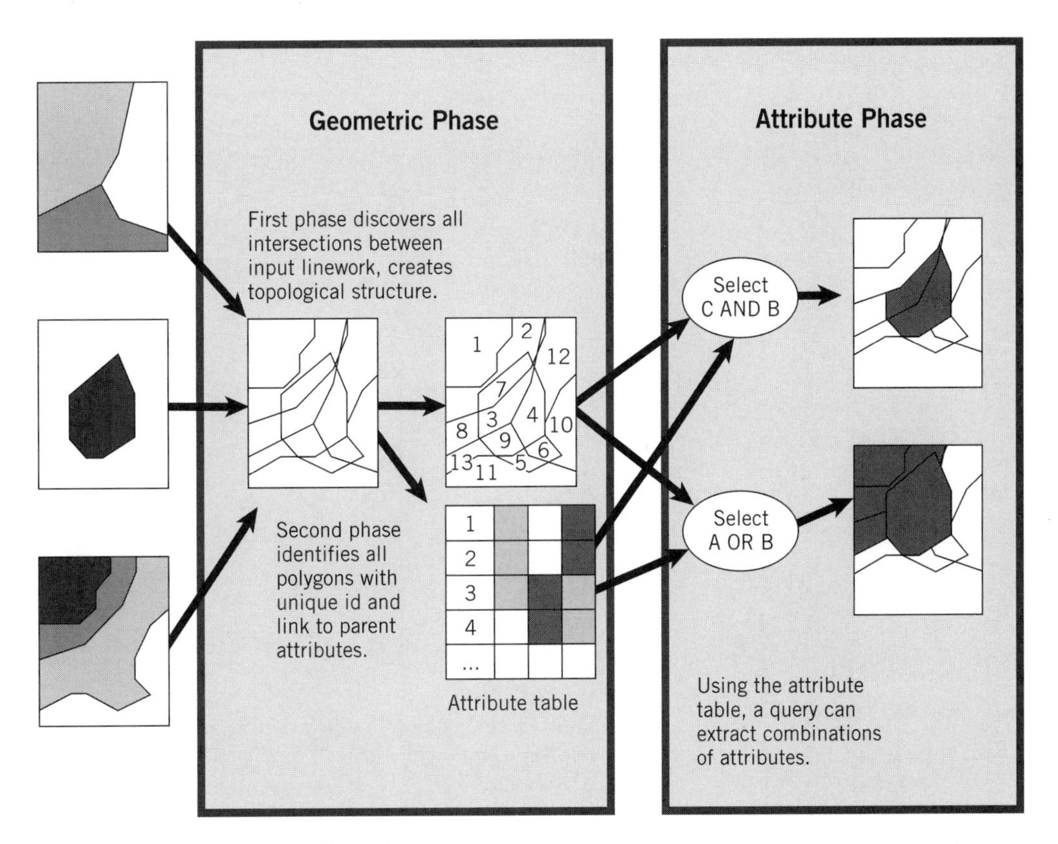

Figure 5-5: Vector overlay: schematic view. A geometric processor constructs a composite topological structure first by finding all intersections, then by labeling all the polygons with a unique identifier and linkage to the source attribute tables. This creates a single coverage with links to all the attributes. Attribute-based operations can produce results from the table and apply to the coverage.

Geometric Intersection Processing Intuitively, the polygon overlay operation involves a pairwise comparison between the objects in the two sources. Polygons that lie outside each other clearly do not require any effort, and polygons totally enclosed by another are not very complicated. Complicated processing is required when they overlap partially (Figure 5-6). The process first finds all intersections between the boundary lines of the polygons. These intersections become new nodes in the composite topology. For this stage, a topological database built of chains eliminates tedious calculations involving shared boundaries of neighboring polygons. With the newly constructed topological network, the procedure must then label all the new objects with a unique polygon identifier. A new attribute table provides access to the attribute tables from the sources. Efficient overlay processing requires careful use of geometric neighborhoods to limit calculations.

If the overlay calculation had infinite resolution, a single feature represented as a boundary on two independent sources would be unlikely to coincide perfectly. The

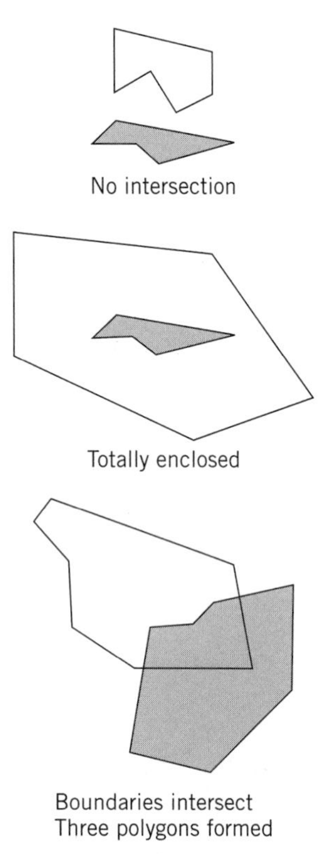

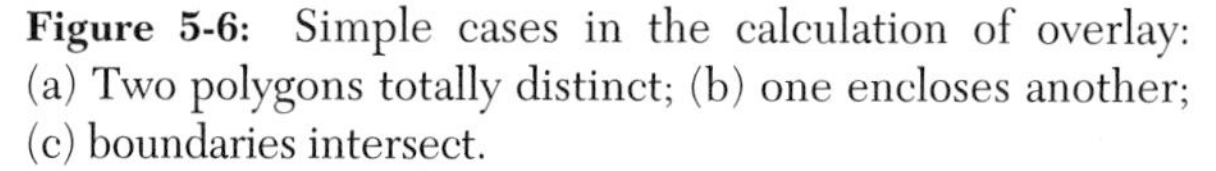

Figure 5-6: Simple cases in the calculation of overlay: (a) Two polygons totally distinct; (b) one encloses another; (c) boundaries intersect.

result is a flurry of small objects, often called *slivers*, created by slight differences in the representation of boundaries that should have been the same (Figure 5-7a). Slivers have been known to be a major problem ever since the earliest days of the Canada Geographical Information System (CGIS); the vast majority of polygons in the CGIS database were trivial objects created through overlay (Goodchild 1978). The word "sliver" implies thin shapes, though some slivers are quite compact (Lester and Chrisman 1991). Slivers can occur for several reasons, including misregistration of sources, inaccuracy of one source compared to the other, temporal differences between the sources, and more. Most current software offers at least one mechanism to combat slivers caused by overlay. One method was originally termed an *epsilon filter* (Dougenik 1980) to acknowledge the work of Julian Perkal (1956), who recognized the basic mathematics, but it is now more commonly called the **fuzzy tolerance**. This technique can remove slivers (Figure 5-7b), but it can also filter away most of the detail in the boundaries. The fuzzy tolerance is a fairly blunt instrument, and current research promises to provide some more sophisticated capabilities, such as multiple

Fuzzy tolerance: A distance within which intersections and points will be treated as coincident. To be processed correctly, the fuzzy tolerance cannot be handled immediately (otherwise a point might be moved twice and beyond its original tolerance). A "cluster" of points must be grouped so that no point is moved more than the tolerance.

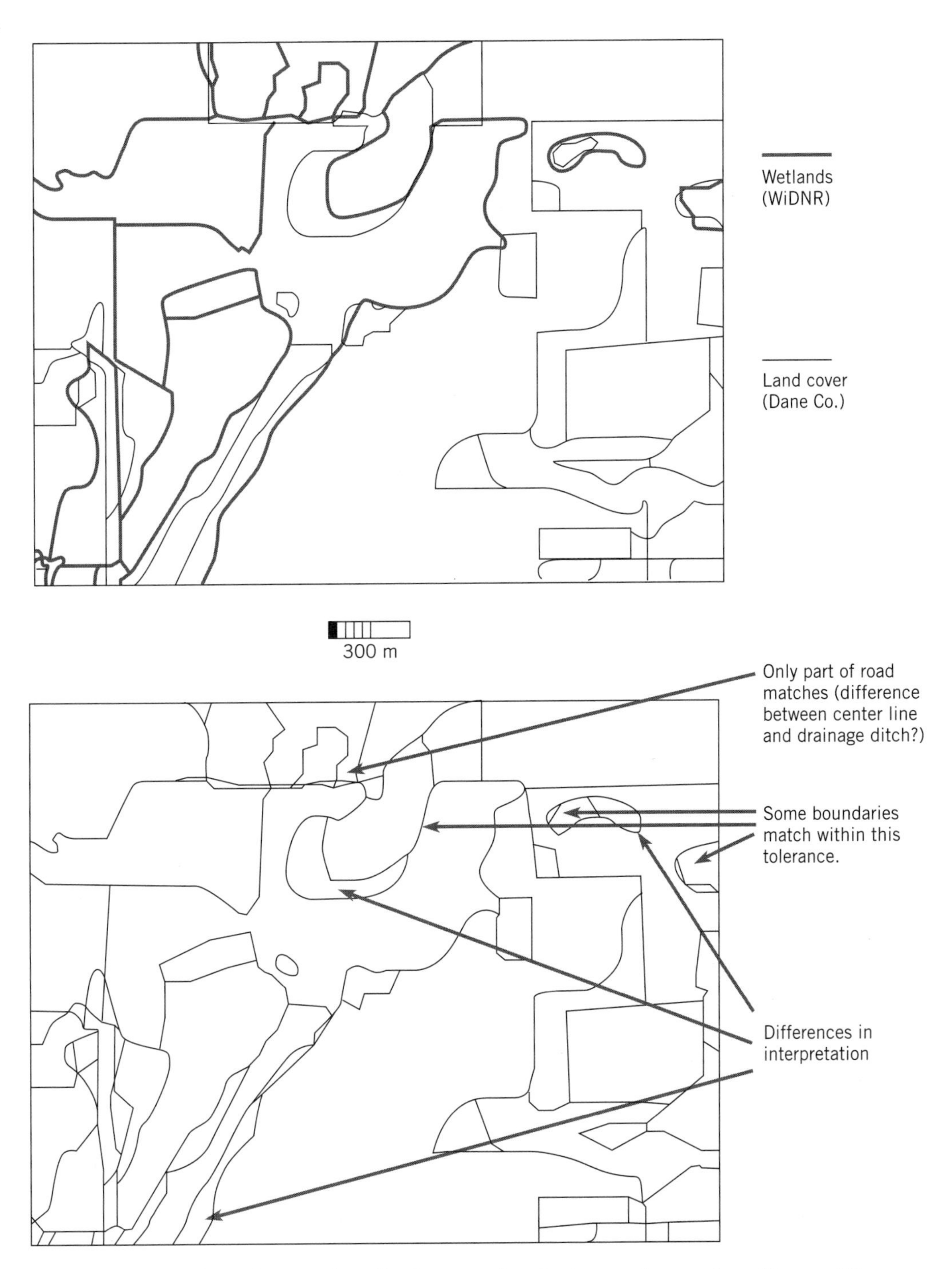

Figure 5-7: Slivers from overlay of land cover and wetlands maps for a portion of Oregon, Wisconsin: (a) Original source layers, both derived from photointerpretation (Wetlands interpreted at 1:24,000 scale; land cover interpreted at 1:15,840; in both classifications, farm fields and wetlands should have similar boundaries); (b) overlay with 30 meter tolerance. (Shaded box in bar scale shows tolerance distance.)

tolerances tied to different error estimates (Pullar 1991; 1993; Harvey 1994). The epsilon model may survive in software implementations for many years, because the single parameter is simpler to comprehend.

The geometric integration of polygons overlaid on polygons represents the most complex case. Software built for this job can easily determine *point-in-polygon* or line-in-polygon as well. In the case of points, no new topology is required, but the processor must still use the geometry to determine which points belong in which polygons. If the boundary lines are treated as imprecise using the epsilon model, then some points may fall not inside a polygon, but on a boundary line. For example, Blakemore (1984) found that only 55% of industrial locations in Britain could be safely assigned to political jurisdictions, considering the probable error in the polygon boundaries. The other 45% were too close to the boundaries (0.7 km in this case) to be certain.

Attribute Handling Using Results of Overlay Once the geometry has been used to construct a new common framework, attributes are attached to the objects created by the overlay. The attribute handling following point-in-polygon provides a clear example of the role of measurement frameworks in this process. After the geometric process assigns each point to a polygon, the attributes can be interpreted in different ways. The simplest assigns polygon attributes to the points, as Blakemore assigned the political code (area) to each industrial plant (point). Alternatively, as described in the digitizing procedures (Chapter 3), a single label point located inside each polygon can be used to identify the polygon. The process transfers point attributes to the polygon. However, if multiple points occur inside the polygons, some other rule must apply. Adopting a choropleth framework, a polygon attribute can be obtained by counting the number of points in the polygon or the sum of their attribute values. Using Blakemore's data, the number of plants or the total industrial employment could be tabulated for each political unit.

The fragmentation of polygon overlay makes polygon-on-polygon more complex. Areal attributes must be examined closely to ensure that their measurement framework has not been misinterpreted. Overlay can split areas into many pieces. Typically, the attribute values are simply copied from the parent or referenced indirectly using the parent identifier as a foreign key into the parent table. Either method assumes that the attribute is uniformly distributed in the parent object and that the attribute applies equally to every piece split from the parent. This assumption is defensible, within limits, for most categories and some derived ratios, but it does not apply to extensive measures, such as population. The appropriate procedure for extensive measures requires areal interpolation, discussed in Chapter 9.

Once the attributes are assigned to the composite, the tools used for attributes of a single map (as described in Chapter 4) apply. The new attribute table allows a query to reference characteristics of all the source layers. A new attribute can be calculated using measurements from different layers, choosing appropriate combination techniques from those presented later in this chapter.

The geometric operation is typically the time-consuming part of vector overlay and is done as a preprocessing stage without much operator interaction but at some cost in time and storage. Once the geometric overlay is constructed, many different attribute combinations can be investigated using simple tools and at low cost. This balance of costs would argue in favor of overlaying all layers into one integrated coverage as a part of constructing the database (Chrisman 1975; Frank and Kuhn 1986; Herring 1987). For certain kinds of analytical projects, like a site selection, this may be the appropriate strategy. One integrated coverage does force the recognition of common features to combat slivers. While it may sound efficient, an integrated database makes it much more difficult to maintain each layer's internal consistency as changes occur. In a less centralized institutional arrangement, each custodial organization should maintain its own layer, then overlay them with the current version of other layers as needed for analytical purposes.

If the overlay serves a specific analytical query, the sequence of operations should be reconsidered. Comprehensive overlay produces a complete composite similar to the raster combination operator described in the previous section. Instead of processing all these boundaries in order to ignore them later, it makes sense to aggregate each coverage into the categories of the query before performing the overlay. Simplifying each layer to the boundaries necessary for the specific purpose also has the advantage of reducing the chance that slivers and fuzzy tolerance effects will be caused by the nonessential boundaries. Advanced spatial database systems offer complex intersection processing through the query language, not as distinct processing steps.

Comparisons of Performance and Capabilities

Despite the trenchant rhetoric supporting vector or raster approaches, the analytical method for overlay does not differ dramatically. Both representations use geometric relationships to create a cross-reference between attributes. In the raster case, the integration occurs in the creation of the raster, while in the vector case, it has to be discovered for each combination. Folk wisdom contends that "raster is faster, but vector is correcter." This sounds catchy, but it is not necessarily true. This adage came into common usage at a conference in the early 1980s. A raster plotter manufacturer handed out buttons reading "Raster is Faster," probably a true statement when confined to the printing of graphic images compared to a pen plotter. The second part of the phrase was added to homemade buttons by various conference attendees. When applied to analytical operations, raster may seem faster, in that the only step called "overlay" is done after the integrated database is constructed. If the process of creating the raster is included and the resolution is strictly comparable (causing rather large increases in the number of pixels), however, the speed advantage might be diminished. Similarly, the vector method may preserve the cartographic crenulations of boundaries at a level of resolution far beyond the reliability of the boundaries. All the effort and expense may be driven by cartographic convention, not a careful analysis of the accuracy of the information. Thus, a vector overlay may not be any more correct. Both raster and vector analysis can be used properly or improperly.

OVERLAY FOR DETECTING DIFFERENCES

The simplest applications of polygon overlay operate on the presumption that the attributes of the two sources should match. This assumption applies (for different reasons) to change detection and error testing. In both cases, some methods of analysis work directly from the results of overlay.

Change Detection

If two categorical coverages of the same region are available for two different times, an overlay can detect the differences. If the coverages used identical categories, the analysis is particularly easy. Those areas with the same category did not change, while those with different classifications appear to have changed. For example, a land use map for Cwmbran (on the edge of Wales in the United Kingdom) was created in 1967 as part of the Second Land Use Survey (Coleman 1961). I created another land use map in 1981 using similar categories and on the same base (Figure 2-9). After overlay, the composite can be classified into areas that changed and those that did not (Figure 5-8). The specific transitions from category to category can be summarized as a transition matrix (Table 5-1).

This form of analysis simply reports the results of the overlay; it does not need produce another attribute other than the pairs of values from the two sources. Of course, neither map is perfectly error-free, so the apparent change includes some amount of error mixed in with the actual change. The overlay in Figure 5-8 applied a 20 meter tolerance to avoid slivers, but this does not cover all the sources of error. Despite the potential for error, change detection remains a major application for overlay processing.

Accuracy Testing

Testing with Points

The well-established methods for testing *positional accuracy* depend on direct pairing of observations, much in the tradition of statistics in other fields. The criterion embedded in the US National Map Accuracy Standard (US Bureau of the Budget 1947) is that 90% of points tested should fall within some fixed distance (0.02 inch or 0.5 mm) of their correct position on the map product. The particular threshold in this test, and the revisions to this test adopted by the American Society for Photogrammetry and Remote Sensing (1989), are less important

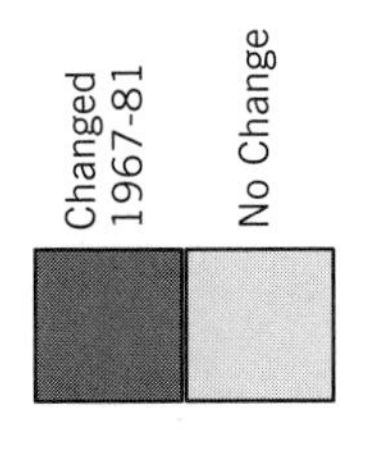

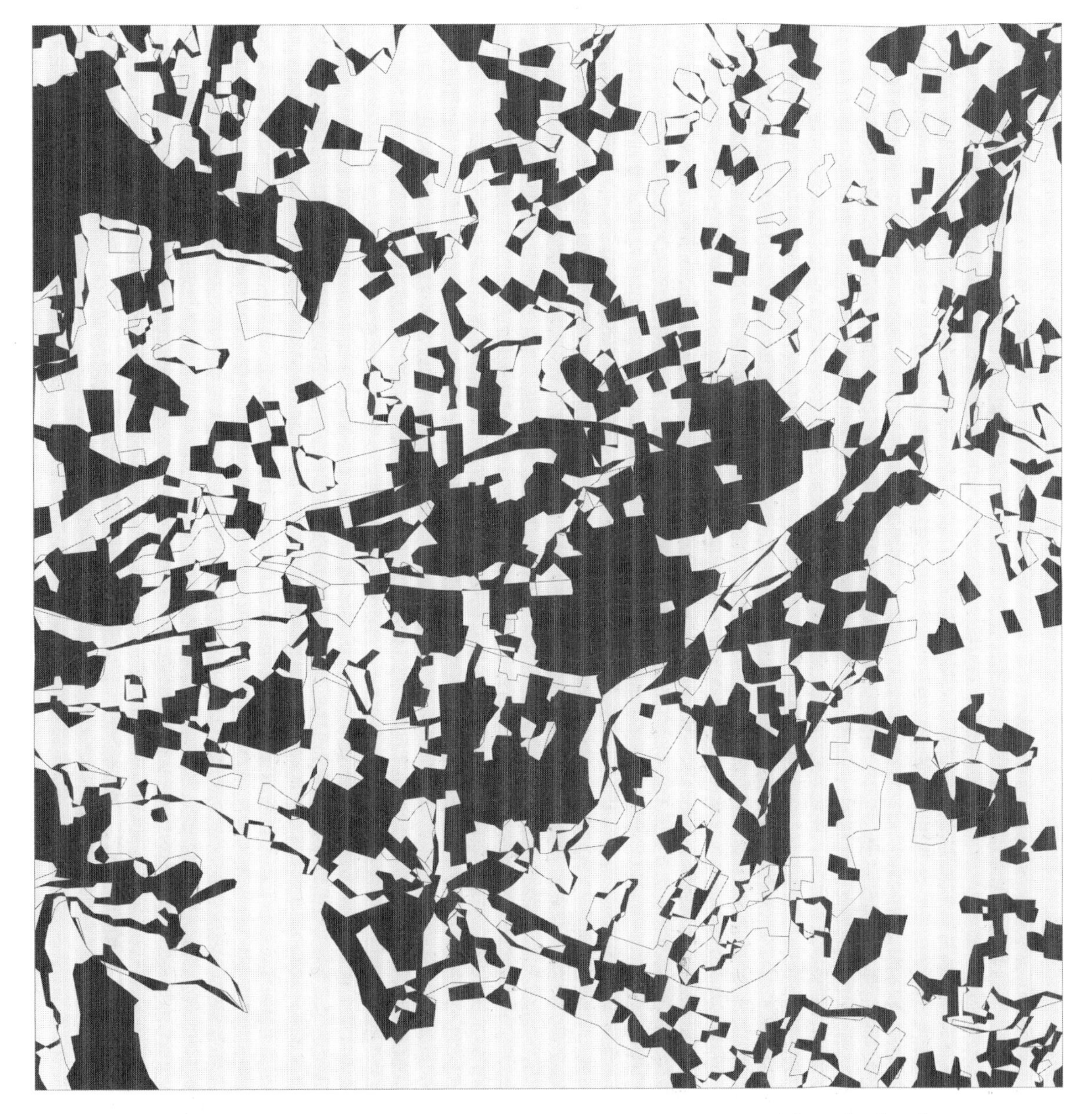

Figure 5-8: Change detection for Cwmbran, Gwent (UK). Area covered is 8 km by 8 km. Areas of change between 1967 and 1981 maps.

TABLE 5-1: Transition matrix: land use change 1967-81 Cwmbran, Wales

1967/81	*Resid.*	*Indust.*	*Open*	*Pasture*	*Crops*	*Woods*	*Health*	*Water*	*Trans.*
Residential	689.2	23.0	20.8	7.7	0.0	1.2	0.8	0.0	0.7
Industrial	30.3	215.0	3.1	5.7	0.0	10.2	9.1	0.0	32.4
Open Urban	17.6	13.9	157.9	9.4	0.0	2.4	1.6	0.0	0.0
Pasture	324.2	98.6	167.4	2715.	150.1	101.6	23.3	0.1	75.9
Crops	2.1	2.7	1.5	197.7	20.3	7.3	0.0	0.0	1.6
Woods	13.9	7.0	19.3	46.5	1.5	337.8	12.8	0.5	3.6
Health	17.0	2.9	1.0	50.4	0.0	7.9	490.3	0.1	0.0
Water	0.0	0.0	0.0	0.0	0.0	0.6	0.1	134.5	0.0
Transition	28.9	38.8	26.1	0.3	0.0	0.0	1.6	0.0	1.4

Figures in hectares tabulated from results of polygon overlay of land use maps compiled at 1:25,000 for an 8 km by 8 km around Cwmbran, Gwent, Wales. Categories grouped from classes used in Coleman survey (1967) and Anderson codes (1981). Residential includes commercial and retail, Industrial includes transportation and mining (Cwmbran had steel mills and coal for the steel facilities). Health includes gorse, bracken, and tundra vegetation not of great value for pasture.

Source: Chrisman (1982a).

than the implied pairing of observations. These tests are restricted to *well-defined points*, points that can be determined to be the same on two products, mostly based on their identity, for instance, a particular named lighthouse or a particular highway intersection. These features, of course, occupy space on the ground, but they can be conceptualized as points. The difference in position for each pair of points provides a direct measure of positional accuracy.

To test for attribute accuracy, some kind of pairing is required. Commonly, a point sample is drawn to determine the actual classification from **ground truth** (Fitzpatrick-Lins 1981). Through a point-in-polygon procedure, the classification at the point can be compared to the classification at that point in the material tested. While this seems to be similar to the positional accuracy test, the definition of "same" is now that it has the same coordinate (space-controlled), not the same object identifier (attribute-controlled).

Map overlay provides an alternative tool for testing attribute accuracy, but this tool differs from the traditional techniques applied to testing maps (see Box). Overlay can verify that the two sources agree on some basic elements of the landscape, testing logical consistency between the two sources. Some relationships are so uncommon as to signal errors in the data. Classic examples would be a navigational buoy on dry land or a floodplain that does not cover the river itself. An overlay analysis can also be used to test the attributes of one source against another. For instance, the Wetlands Subcommittee of the Federal Geographic Data Committee (Shapiro 1995) used polygon overlay to assess the consistency between six different inventories of wetlands for Wicomico County, Maryland. No one source was judged to be more accurate; the analysis located areas of agreement and disagreement. Taken at the broadest level, wetland/not wetland, the four polygon inventories unanimously agree for only 4% of the areas classified as wetland by one source. The two sources that were most comparable still show significant differences in the summary table (Table 5-2).

Differences found in an overlay test do not mean that the sources must be totally rejected. After all, the compilation process involves weighing evidence from a variety of sources. Some differences found in Wicomico County come from different purposes in the original inventories. Using the tools of logical combination, a consistent product can be constructed using the most reliable elements from the sources available. For instance, one source can be judged to have the more accurate (or more current) shoreline, so disagreements can be resolved by trusting that source on the land-water distinction. Overlay is not an entirely automated procedure. It requires judgments about the validity and compatibility of the sources.

Interpreting Overlay Results

In both change detection and overlay accuracy testing, the procedure creates a cross-tabulation as the primary raw result (for example Tables 5-1 and 5-2). Shown in tabular form, these matrices summarize the area in all possible combinations of the categories. A number of techniques can analyze these tables. The **diagonal** of the matrix is the focus for some of the simpler techniques. The percentage correct (applied to error matrices) is simply the sum of the diagonal divided by the total fre-

Ground truth: A determination of geographic attributes judged to be of higher accuracy; usually applied to a point classified into a land use/land cover category.

Diagonal: Cells in a square matrix whose row and column indices are the same. In a transition matrix, they represent no change; in an error matrix, they represent no error.

TABLE 5-2: Wetland classification comparison

		FWS-NWI					
		Pal	*Lac*	*Riv*	*Est*	*Upl*	*Total*
MD-WRA	*Pal*	7,214	33	8	15	7,311	14,581
	Lac	32	458	0	0	58	548
	Riv	178	5	567	41	61	852
	Est	24	0	1	1,044	48	1,117
	Upl	3,193	49	29	74	136,705	140,050
	Total	10,641	545	605	1,174	144,183	157,148

Pal=Palustrine; Lac=Lacustrine; Riv=Riverine; Est=Estuarine; Upl=Upland; shaded areas represent acreage for each system upon which both data sets agree.
Results from overlay of Fish and Wildlife Service National Wetland Inventory (photo interpreted from 1981-82 color infrared photography) with Maryland Water Resource Administration (photo interpreted from 1988-89 photography). Figures in acres. Aggregated at the "system" level of the Cowardin coding system.
Source: FGDC Wetlands Subcommittee, (Shapiro 1995) p. 49.

quency (in the overlay case, the area of the region). This percentage is easy to understand, but it may not represent the only (or even the best) measure of the matrix. Differences in the number of categories and in the distribution among these categories can change the meaning of this single figure. As a remedy, some suggest the use of **Cohen's kappa**, a measure that deflates the percentage correct by the amount that would be expected by chance (Rosenfield and Fitzpatrick-Lins 1986; Congalton 1991). This measure represents a simple case of applying statistical methods, such as discrete multivariate analysis, to the matrices that summarize the results of overlay. Unfortunately, there are some large differences between regular statistical measure-

Cohen's kappa: A measure of agreement between two classifications. Defined as (observed accuracy − change agreement) / (1 − chance agreement) where the chance agreement is estimated by the cross-product of marginal frequencies (statistical independence model).

ment frameworks and the circumstances that produce these matrices. The model for chance agreement is statistical independence, which is not really the baseline expected either within or between coverages. Both change detection and error analysis begin with a presumption of no change or no error. In addition, the quantities in the matrix represent areas, a ratio measure with an arbitrary unit of measure, not counts of discrete independent objects. By definition, categorical coverages expect extremely strong correlation between adjacent points, so the model behind kappa (and other discrete multivariate techniques) does not fully apply.

While the previous group uses a single index, other techniques produce two measures to describe dependencies. Each entry in a change or error matrix can be compared to its row and column total, giving a proportion. In the remote sensing accuracy literature, the proportion comparing a cell to the total ground truth for that class is called the *producer's accuracy*, while the figure divided by the total from the satellite source is called the *user's accuracy*. In a change matrix, these have to be interpreted respectively as the proportion unchanged at the earlier time period (row) and the proportion unchanged at the later time period (column). Zaslavsky (1995) has adopted similar measures inside a technique called *Determinacy Analysis* based on simple logic. For instance we might say in Table 5-1 that Industry in 1967 implies Industry in 1981. This statement is "accurate" according to the row proportion (70.6% of the 1967 Industry remained Industry in 1981), but only "complete" according to the column proportion (53.1% of 1981 Industry is covered by this statement). The statement that the area classified as Transition in 1967 converted to an "urban" use (Residential, Industrial, or Open Space) is 96.6% accurate, but it is only 5.0% complete in accounting for all urban uses.

Determinacy statements can also describe the relationships in accuracy tests. In the comparison between the National Wetland Inventory and the Maryland inventory (Table 5-2), the statement that Estuarine on the Maryland map predicts Estuarine on the NWI is 93.5% accurate and 88.9% complete. By contrast, the connection between Palustrine classification is 49.5% accurate and 67.8% complete. By adding explanatory categories, determinacy analysis tries to obtain the most accurate and complete statements to describe the results in a matrix. Since it does not use the condition of independence as a benchmark, it can be applied to error matrices and change matrices without requiring statistical assumptions.

A TAXONOMY OF OVERLAY COMBINATIONS

The overlay process, whether performed in a raster or vector representation, provides access to all the attribute values that occur at one location. Direct analysis of these results give some insight, but often the overlay serves as an intermediary to obtain other results from the combination of attributes. A number of different rules can be applied to combine the attribute values placed in contact by overlay.

At a generalized level, the possible forms of map combination can be organized into three groups by the rules applied to the attributes assembled by overlay (Figure

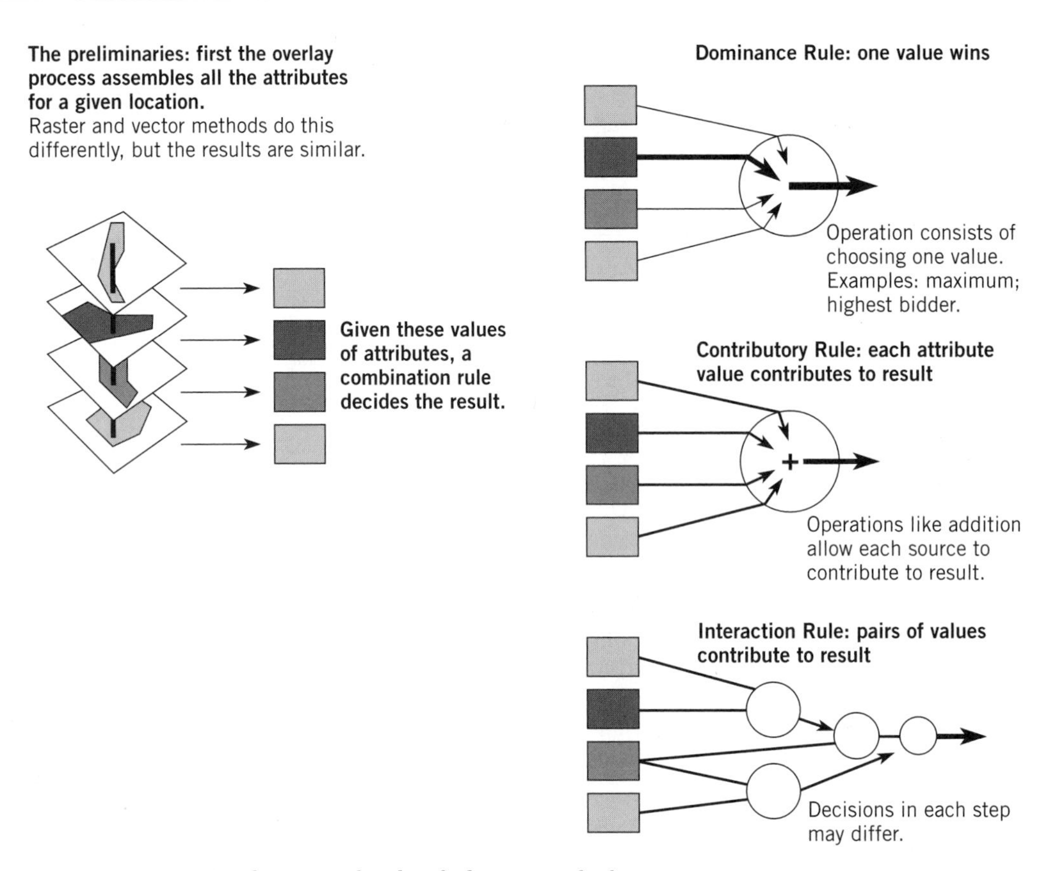

Figure 5-9: Combination rules classified in terms of information use.

5-9). The simplest method selects one value from those available. This will be termed a *dominance* rule, since other attributes are ignored. The next group uses each layer's attribute value to create a composite result, often using a mathematical operation like addition. Because each value contributes to the combination without regard for the others, the rule will be termed *contributory*. The third group goes beyond independent contribution to exploit the interaction between values. These will be termed *interaction* rules. Each of these three groups can have many variants; those listed in Table 5-3 are intended as illustrations, not an exhaustive set.

Previous Taxonomies of Overlay Operations

In the landscape planning literature, various procedures to combine suitability information are listed without an organizing principle. In the early period of GIS development, Hopkins (1977) presented a rather complete consideration of the possibilities for map combination, guided by Stevens' levels of measurement. At the time, this helped organize the possibilities, but the terminology developed in Hopkins' paper is rather confusing. Furthermore, the understanding of measurement frameworks was only partially evident at that time. Stevens' scheme focuses on the arithmetic of attributes, leaving out the other axioms involved in geographic measurement. The taxonomy presented here is developed to remedy some of the difficulties in Hopkins' presentation.

Another, more recent stream of research has relied on the "multicriteria decision making" literature. This interdisciplinary field has worked on problems of rendering a decision for environmental problems that involve more complex criteria than traditional optimization methods from operations research (Voogd 1983; Hobbs 1985). There have been some examples of applications of these multicriteria methods in GIS (Carver 1991; McCartney and Thrall 1991). The complex measurement basis of geographic information requires some care in adapting multicriteria techniques to the map combination problem.

Dominance Rules

A dominance rule determines the result of combination by selecting a single value from those found at the same place. The selection is governed by a set of external rules, not the combination of values available. The one value selected then dominates all the others. Various possibilities arise based on the rules that choose the single value from those available.

TABLE 5-3: Attribute combination methods following overlay

Dominance Rules	
Exclusionary screening	One strike and you're out
Exclusionary ranking	Extreme value from rankings
Highest bid	Extreme value from continuous data
Highest bidder	Records identity of extreme value
Contributory Rules	
Voting tabulation	Sum of binary exclusions
Weighted voting	Weighted sum of binary factors
Linear combination	Sum of "ratings" (mean, etc.)
Weighted linear combination	Weighting and rating game
Product	Multiplication of factors
Interaction Rules	
Integrated survey	Informal judgement, "gestalt" interpretation
Factor combination	Rank with conjoint measurement
Rules of combination	Formal interaction tables

Exclusionary Screening The simplest form of dominance is *exclusionary screening*, a rule that could be summarized: "One strike and you're out." Screening imposes a binary vision of the world, assigning priority to any detrimental attribute. This rule may be appropriate for factors that by themselves are so damaging that there is no need for further study. If a factor could be mitigated by some technology or could be neutralized by some other factor, then the method may exclude some areas falsely. This dominance rule also gives the same result for those areas with a single detrimental factor and for those areas with all the possible negative factors.

The Pennsylvania project to site a disposal site for low–level radioactive waste (**LLRW**) provides one example of a siting project centered around a GIS and the overlay tool. The first phase of this project disqualifies any portion of the state that exhibits one of 18 characteristics (Table 5-4). Some criteria are simple, and some are much more complex. Any one disqualifying factor excludes the area as a potential site. Knowing this logic, the contractor could expend its efforts somewhat strategically. The 18 factors were not equally well established or equally expensive to study. There was no need to spend time and money creating a digital coverage to exclude an area that was already excluded. The screening phase was conducted in three stages of increasing spatial resolution. Stage One excluded about 23% of Pennsylvania using components of seven of the factors. Stage Two refined the layers used in Stage One and added three more to exclude some 46% of the state. Stage Three (Chem-Nuclear Systems, Inc. 1994) included all 18 factors and excludes 75% of the state (Figure 5-10).

The disqualification stages relied on many sources. Some disqualifying criteria came from a single source, as slope gradient was derived from digital elevation data generated by the US Geological Survey. Other criteria, such as protected areas, required information from all levels of government. Some sources, like the National Wetland Inventory, did not use quite the same set of categories, requiring some interpretive reclassification. Other disqualifying factors required compilation and digitizing efforts by the contractors. For example, coal mines have been recorded individually on surveys of diverse scale, some dating back to the Works Progress Administration in the 1930s. This project paid for the conversion of these records, which are seemingly unrelated to their mission. Such a requirement creates a large barrier to the use of GIS for less well-funded concerns. It is ironic that a search for a 500-acre site requires such massive information processing.

Under the best conditions, a screening analysis will remove the bulk of the study area, leaving just a few alternatives for further examination. However, there is no guarantee that it will work out that well. One risk is that the criteria exclude only a small portion, leaving a huge area for future study (and the methods developed below).

LLRW: Low-Level Radioactive Waste; low-level waste is generated (on the order of 160,000 m^3 for the US per year) by nuclear power plants, hospitals, and various other industries; excludes waste from weapons construction and the spent fuel from nuclear reactors (high-level waste).

TABLE 5-4: Disqualifying criteria for a low-level radioactive waste disposal facility

Criteria	*Stage Implemented*		
	One	*Two*	*Three*
Masking facilities			X
Active faults		X	x
Geologic stability			X
Slope			X
Carbonate lithology			
Outcrop at surface	X	x	x
Within 50 feet of surface, >5 feet thick		X	x
Potential for subsidence		X	x
Evidence of subsidence			X
River floodplains			X
Coastal floodplains	X	x	x
Important wetland			X
Dam inundation			X
Public water supply			X
Surface water intake			X
Wildlife area boundaries (many sources)	X	X	x
State forests and gamelands	X	x	x
Watersheds	X	x	x
Oil and gas areas	X	X	X
Agricultural land			X
Mines		X	X
Protected area boundaries (various)	X	X	X

X first stage identified
x previous criteria updated by more detailed material
Source: simplified from Chem-Nuclear Systems Inc. (1994) Table 4-1 and 4-2 pp. 7–10.

Another more troubling risk is that the whole study area might be ruled out. If the criteria cover a wide set of environmental conditions, they may exclude each element in the landscape for some reason or another. External constraints (such as a decision that the facility *must* be accommodated) may require some relaxation of the criteria to find a possible site. The dominance approach provides no hint regarding the most prudent way to relax the restrictions.

Figure 5-10: Portion of disqualifying coverage from Stage Three for northeast portion of Pennsylvania. Scale 1:1,000,000; most elements derived from 1:24,000. This region included large regions not disqualified in previous stages. Source: Chem-Nuclear Systems, Inc. (1994), Appendix A, Plate 3.

Exclusionary Ranking A somewhat more complex dominance rule relies on ordinal measurement. If the attributes in the various layers are ranked according to a common scheme, then a rule for combination would select the most extreme (highest or lowest) value at each location. This *exclusionary ranking* is simply an evolution of exclusionary screening. The concept of limitations used by US Department of Agriculture and the UN Food and Agriculture Organization (FAO 1976) ranks soils and other resources according to limitations from severe to moderate to slight. These rankings were mentioned in Chapter 1 as a prototype for ordinal measurement, but the concept is not limited to a single measure. It also provides a rule to combine multiple factors. These rankings imply a dominance rule for the various ingredients (slope, drainage, etc.) that contribute to each evaluation. When combining various factors (such as availability of water, oxygen, or nutrients), the FAO method selects the most limiting ranking. This logic is in common use for land resource assessment (Kiefer 1967; Burrough 1986, p. 95). This method makes sense as long as there is no interaction between factors, for example, no way for a "severe" limitation on one factor to be mitigated by some other factor's value.

Highest Bid/ Highest Bidder Dominance rules can also apply to interval or ratio attributes, as long as each is expressed in the same units. The *highest bid* rule differs from exclusionary ranking only in allowing numerical results in place of the ranked categories. When the highest value overrides the others, it simulates bidding, a basic rule of a market economy, and many forms of rational decision making. For example, in von Thünen's location theory, purchasers offer a price at a market center for agricultural produce, but progressively higher transportation costs reduce the net profit to producers that are located farther from the market. Thus, a producer grows the crop that offers the highest profit at that distance from the market.

If highest bid records the maximum value, then another related dominance method could be devised that assigns the location to the *highest bidder,* the source that contributed that highest value. This rule processes ordinal, interval, or ratio measures, producing an identifier for the extreme value. For example, Young and Goldsmith (1977) prescribed land use alternatives for an area in Malawi based on the most suitable of six alternative crops and competing uses.

Dominance rules are perhaps the lowest form of combination, since they are all-or-nothing. Yet, this logic does play an important role in many applications. It is particularly important to use a dominance rule for absolute exclusions that precede other steps. Some of the GIS literature belittles these kinds of operations as being too simple, but they are useful exactly because they are simple and understood by all parties. Environmental regulations are often simplified into binary criteria for simplicity and clarity. The concept of *highest and best use* is basic to land economics and property appraisal. These culturally sanctioned decision rules provide the basic rationale for many GIS applications.

Contributory Rules

The next group of operations uses all the information from the various sources. The values from one layer contribute to the result without regard for attribute values from other layers. The combination process can use any arithmetic operation, but the most common example is addition. Addition is, after all, considered the basic rule for extensive measurement scales. As long as the measurements extend along the same axis and in the same units, addition models the appropriate combination. Each contributory rule depends upon certain assumptions about the source materials.

Voting Tabulation The simplest form of voting applies to nominal categories. By examining all the values present at one location, the procedure can select the category that is most frequent (the mode of the distribution) or least frequent or other variations. This differs from the dominance rules because the choice of the value comes from the set of values available, not from some external rule. This procedure can produce either the value selected or the number of votes cast for the winner.

Voting can also be applied to the layers used in an exclusionary screening. If the positive factors are scored as one and the negative as zero, then the sum of the values can tell how many positive factors occur at the location. Or the scores could be turned around to tabulate the number of negative factors. This method, which might be best termed a *voting tabulation,* provides more information for site suitability or environmental regulation. Compared to exclusionary screening, it helps to know that an area has been excluded for multiple reasons. This result can be obtained by using addition on the binary layers instead of Boolean operations.

The process of counting exclusions seems quite reasonable, but it leads directly to a troubling question. Is an area excluded on two criteria twice as excluded as an area only excluded for one reason? For example, following the Washington State (1990) critical area designations created by the Growth Management Act, are wetlands and geologically hazardous areas of equal sensitivity? Should the components of the geological hazards (seismic risk, slope stability, erosion) be given separate status? Should each source of landslide information be counted separately or together? These questions simply cannot be answered in the general case. There is simply no universal calculus that tells that one floodplain equals two aquifer recharge areas. In the colloquial, this is the problem of *apples and oranges*. At one level it is possible to generalize and to call it all "fruit," but the unit of measure is not as stable or as comparable. In some rough sense, the count of critical areas may act as an ordinal scale, though even that requires a stretch of measurement rules.

Weighted Voting Another form of voting tabulation weights each factor by a number. The weight is supposed to express the relative importance of the factor. The attribute values are still scored zero or one. Carver (1991, p. 324) adopts a seven point ordinal scale (from 7 = very important to 1 = unimportant) as the weight. While this allows finer gradations of preference, there is little logic to suggest that "very important" is exactly seven times larger than "unimportant" or that "important" belongs at three or four. The wording of such ordinal scales conflicts with their use as numeric values on an interval scale.

For the Pennsylvania LLRW facility, Chem-Nuclear Systems (1992) proposed a variant of a weighted voting called rank-sum for the phase designed to follow the exclusionary screening (Table 5-5). The votes consist of zero–one thresholds of nine factors. These nine factors are ordered according to importance, from nine down to one. This rank is used as the weight. The score for a site is the sum of the weighted votes, so that the factor chosen as most important contributes a value of nine if present, zero if not, and the lowest factor contributes one or zero.

These forms of voting tabulation approximate popularity of each alternative based on presence or absence of certain factors. Voting makes sense when the preponderance of the evidence rules and there is no interaction between the factors. Each factor contributes its weight to the result democratically, but the result depends on how votes are counted.

TABLE 5-5: Rank–sum developed for Phase II of Pennsylvania LLRW project

Major Factor Group	*Technical Assessment (0 or 1)*	*Public Rank (1-9)*	*Combined Technical and Public Rank*
Land Use	T	P	T × P
Transportation Effects	T	P	T × P
Weather	T	P	T × P
Geology	T	P	T × P
Surface Water Features	T	P	T × P
Water Supplies	T	P	T × P
Wildlife	T	P	T × P
Natural Resources	T	P	T × P
Cultural Resources	T	P	T × P
		TOTAL SCORE: *summation of T × P*	

Major Factor Groups are scored for their presence/absence, and then the ranking of each factor determines the score for each site.

Source: Chem–Nuclear Systems Inc. (1992) pp. 4–10.

Linear Combination The more common form of a contributory rule involves a more general use of addition. A linear combination adds up the values for a particular location. Compared to the voting rules, the measurement scale becomes even more important.

A simple example of linear combination shows many of the hidden assumptions. In 1982, the Dane County (Wisconsin) Regional Planning Commission (1980) was charged with selecting a site for solid waste disposal. Six site factors ranked on the scale of slight, moderate, and severe were represented by the numbers one, two, and three (Figure 5-11). Once assembled, the ratings were added up for each grid cell. The minimum possible value was six, from a ranking of "slight" on all factors. If the landfill site could have been found with this score, the method would have replicated exclusionary screening. As noted above, it is not always possible to find a site that will pass all the exclusionary filters. The proposed landfill had some cells with composite ratings of eight and nine, values that can each occur with two possible combinations. For example, a value of eight implies either two values of "moderate" (1+1+1+1+2+2=8), or one "severe" (1+1+1+1+1+3=8). Does one "severe" (and one "slight") equal two "moderates"? In this case, the factors may not hold equal importance to a landfill. Some factors can be remedied with engineering efforts, while others cannot. The relative environmental risk is not strictly equivalent. A result of all "moderate" scores (2+2+2+2+2+2=12) may rank more favorably than

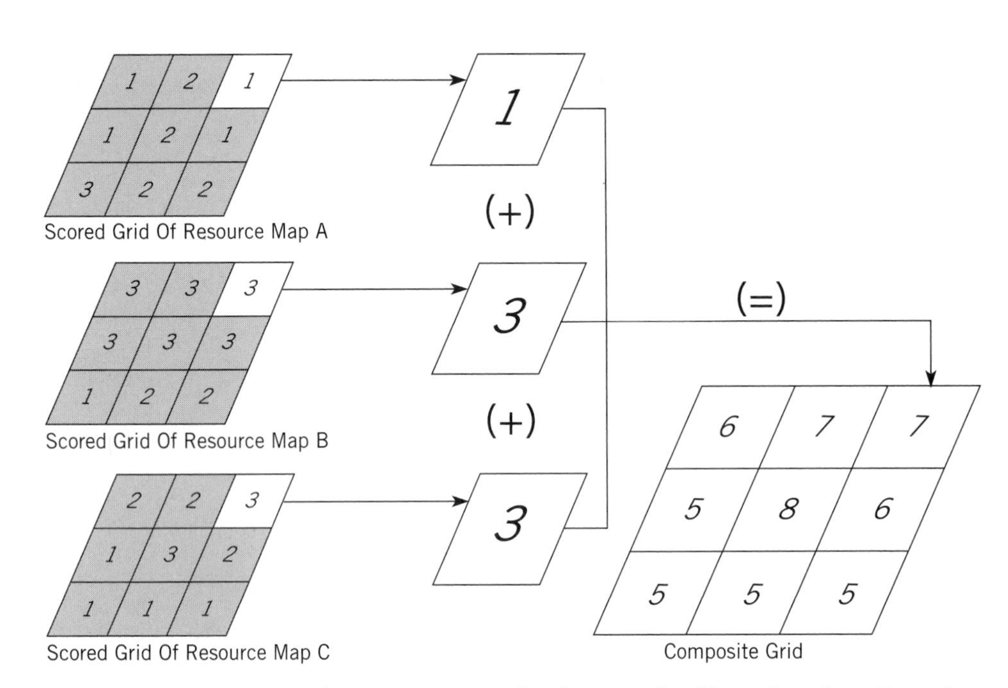

Figure 5-11: "Linear combination" in its simplest form simply adds up the values. Example shows the ratings assigned to categories for the Dane County Solid Waste Plan (Dane County Regional Planning Commission, 1982). Ratings for grid cells were added to obtain a composite ranking.

(1+1+1+1+1+3=8), if the one "severe" rating totally prohibits the landfill's economics. In conclusion, simple addition does not provide an unbiased site selection method for this landfill siting database. In its defense, the Regional Planning Commission did this study using manual methods in a predigital era, but the technology is not really the issue. Rules for combining scores must be justified, whatever technology is applied. Using addition imports a set of axioms that may not apply to the measurements available.

Despite years of criticism from sceptics like Hopkins, many practitioners of GIS still use linear combination for ordinal scale data. For example, Cornwell and Rohardt (1983) report a site selection study for a steel plant in Nigeria. The study rated seventeen factors on a common ordinal scale of suitability. The overall suitability was then calculated by the sum of all the rankings. In their paper, the authors recognize some drawbacks in adding ordinal scores. They seek a solution in rescaling the numbers assigned to the ordinal scores, not in some other method of combination.

Various methods have been proposed to convert ordinal rankings to a continuous scale. Carver (1991) suggests a proportion of the range for each variable between minimum and maximum. Eastman and others (1993) describe standard deviations from the mean and histogram equalization. Each of these techniques depends on the distribution within the study area. The mean and standard deviation make sense only for interval scales. Histogram equalization converts all distributions to rankings, which removes any sense of an extensive measurement. The median will always be given the midpoint value, no matter what kind of distribution. Rankings remain orderings, not interval or ratio measures.

As presented in Chapter 4, new information can evaluate each ordinal category on a true ratio scale. For example, categories from a number of factors can be rated in terms of dollar cost to a developer (Table 5-6) (Fabos and Caswell 1977). As a scale, money has a strong attraction, though the 1977 dollars in Table 5-6 certainly would need adjustment to be taken seriously by current developers. The particular values shown here would certainly not apply in all surroundings. In earthquake coun-

TABLE 5-6: Assigning a ratio-scaled measure to ordinal categories

Factors and dimensions	*Estimated added costs (in dollars)*
Depth to Bedrock:	
0–2 feet	$20,000
2–5 feet	5,000
5 feet +	200
Depth to Water Table:	
0–3 feet	$5,000
3–5 feet	1,400
5 feet +	0
Drainage:	
Poorly to very poorly drained	$5,000
Moderately poorly drained with hardpan	1,400
Other	0
Slope:	
15% +	$*****
8–15%	1,300
0–8%	0
Topsoil:	
Poor (0–4 inches)	$1,500
Fair (4–6 inches)	600
Good (6 inches +)	0
Bearing capacity:	
Plastic and nonplastic; silts and clays; peat; muck	$1,500
Other	0

Source: Fabos and Caswell (1977) *METLAND Landscape Planning Research*, Research Bulletin 637, Massachusetts Experiment Station.

try, the cost of building on peat or muck should be substantially higher. In Hong Kong and other places, a 15% slope gradient is certainly not an absolute restriction on construction. Each marketplace and physical environment values these categories differently.

As much as one can argue with the specific values assigned, there are also problems that arise simply from trying to monetize all the factors. Why should the cost to the developer be the only cost considered? How can we include benefits to society that are not traded in a market of any kind (like scenic beauty)? On the more technical level, the sharp breaks between categories point to the weakness of assigning ratio values to measurements that are just categories. The METLAND project expected to obtain most of those physical factors from soils map and similar sources, using indirect procedures. Consequently, there will be very sharp differences in development cost along boundaries that may represent rather gradual transitions between soil properties.

Weighting and Rating The full form of the linear combination method requires a slightly more general form (Equation 5-1) to include a weighting factor for each attribute rating. The insertion of a weight simply recognizes that the simple sum assigns an implicit weight of one to each factor.

Equation 5-1

$$V_j = \frac{\sum_i w_i r_{ij}}{\sum_i w_i}$$

where

V_j refers to the resultant value for each object j

r_{ij} refers to the rating of each object j for each attribute i

w_i refers to the weight assigned to attribute i

Weighted addition is essentially identical to the procedure termed *ideal point analysis* in the multicriteria literature (Hobbs 1985). A more neutral label for this method might be *weighted linear combination,* but in many applications it turns into "The Weighting and Rating Game." The selection of weights and ratings is often established iteratively based on their results, without sufficient attention to the assumptions involved.

The US Department of Agriculture's Land Evaluation and Site Assessment (LESA) methodology uses weighted sums of many factors (Table 5-7). As applied in a county in Kansas (Williams 1985), the land evaluation component of the score is based on a single index crop, grain sorghum. This figure is similar to the corn grain yield shown in Figure 2-17, an indirect attribute of eight soil productivity classes (in turn a grouping of soil mapping units). Thus, the largest contributor to the score has only eight possible values, not a continuous measure from 0 to 100. On this scale, an increment of one represents 0.6 bushels of sorghum per acre per year. The 21 site assess-

TABLE 5-7: Implementation of weighted linear combination: Land evaluation and site assessment for Douglas County, Kansas

Land Evaluation

Grain Sorghum yield *rescaled to set highest value at 100*

Site assessment		*Source*	*Weight*
1	% of area agricultural within 1.5 miles	Land use, buffered	10
2	Adjacent agricultural land	Land use, buffered	7
3	Farm size	Parcel area	2
4	Average parcel size within 1 mile	3, neighbor sums	4
5	Agrivestment in area		3
6	% of area zoned ag within 1.5 miles	Zoning, buffered	8
7	Zoning of site (for agriculture)	Zoning, rescored	6
8	Zoning for alternative (residential use)	Zoning, rescored	6
9	Nonfarm AND poor soil	Zoning, soil overlay	6
10	Need for urban land	Land use/City limits	8
11	Compatability of adjacent uses	Land use, neighbor	7
12	Unique features		3
13	Adjacent to unique features	12, buffered	2
14	Floodplain or drainage way	Hydrology	8
15	Suitability for on-site septic	Soils, rated	5
16	Compatibility with Comp. Plan	Plan	5
17	Designated Growth Area	Plan	5
18	Distance from city limits	City, buffered	6
19	Distance from transportation	Roads, buffered	5
20	Distance from water service	Water, buffered	4
21	Distance from sewer lines	Sewers, buffered	4
	(Reweighted to sum to 200 points)	Total	114

Overall score = Land evaluation (100) + Site assessment (200)
Source: Williams (1985), *p. 1928.*

ment factors are scored on a scale from 0 to 10, then weighted. A few of these factors are based on distance measurements from lines or areas. The tools to produce these buffers are described in the next chapter. The distance measures in this study were rated on a logarithmic scale, for some reason. For example, the distance to water lines (factor 20) was scored 0 in the same 100 m cell as the water line, 2 for less than 1/8 mile, 4 for 1/8 to 1/4 mile, 6 for 1/4 to 1/2 mile, 8 for 1/2 to 1 mile, and 10 for more than 1 mile. These ratings do not predict the costs of installing pipes very closely.

This study targets the conversion of agricultural land to residential use. Another set of factors and weights would be required for another application. As the author points out (Williams 1985, p. 1930), the LESA linear combination assumes independence of the factors, but suitability for septic fields (factor 15) is irrelevant if the new residential development is attached to the sewer system (factor 21). Similarly, the highly weighted factors for percent agricultural (1) and percent zoned for agriculture (6) may simply provide double weight for strongly related maps.

Specialists in multicriteria methods react with suspicion to the choice of weights using a "magic number" method. A system of rating scores and weights can be justified if the measurements turn into a common ratio scale of measurement. The magnitude of a weight must express the amount of one quantity that one is willing to trade for another. In the LESA case, the weight of 4 assigned to distance from water lines means a specific relationship between bushels of sorghum and each increment away from water lines. Moving 1/8 mile away from the water lines converts to 0.8 bushels of sorghum; beyond 1 mile to 4.2 bushels. It is doubtful if the value of 4.2 bushels of crop productivity can compensate for the cost of installing 1 mile of water line over any reasonable budgeting period.

One method to prepare a defensible system moves beyond the technical level to recognize the institutional and social context. Any procedure, no matter how boneheaded, is adequate if all participants accept it. Many environmental decisions become heated public issues, and the technical components can become the overt battlefield for disagreements with other origins. Faced with this situation, there have been many attempts to use conflict resolution methods in the technical process. Methods of compromise and consensus building, such as Delphi panels, have been used in various site selection processes over the past 20 years (see, for instance, Dames and Moore 1975; McCartney and Thrall 1991; Eastman and others 1993). Bringing the group to agreement on a set of weights is a fine goal, but it will not remove the mathematical objections to weighting as a technique. The process of social agreement may often find a compromise by taking a middle value, thus importing the same set of measurement assumptions.

Nonlinear Combinations Beyond the linear combinations, there are many alternative mathematical functions to combine variables. As with addition-based methods, all the available values contribute to the result according to the rules of algebra. Most of the criticism applied to the linear methods becomes even stronger as the functions become more powerful.

One common method involves a multiplicative rule in the place of addition. Sometimes the unsuspecting user is not aware of the switch to multiplication. If the values for the variables are transformed by a logarithmic functions (one of the suggestions in Cornwell and Rowhardt 1983), then the addition produces a result that is essentially multiplicative. Other adoptions of the multiplicative form are more conscious, such as Storie's (1933) index of crop productivity. One of the most common examples of an overtly multiplicative model applied in GIS is the Universal Soil Loss

Equation (**USLE**) (Wischmeier and Smith 1978). USLE is an empirical model, calibrated by a series of experimental plots. The amount of soil eroded by a given rainfall event is directly measured in a trough built at the bottom of a plot of a particular soil on a 9% slope. Using a set of calibration studies, the empirical relationships have been summarized in the parameters of the equation. For example, the soil factor adjusts for the difference between the soil tested at the University of Missouri and the results for any other soil class. These factors derive mostly from inference about soil properties rather than testing. USLE combines all these influences as multipliers that increase or reduce the estimate of tons of soil lost over a given time period.

A multiplicative rule is particularly sensitive to errors in the original sources, since small changes in one factor can influence the whole result. Multiplication is particularly sensitive to zero values, which can dictate the result despite the contributions from other values. With a zero value, multiplication acts as a dominance rule, though with nonzero values, each value contributes to the result.

Interaction Rules

The contributory rules assume that each attribute should contribute to the result without regard for the specific level of some other attribute. Effectively, this rules out interactions between factors. However, interactions between attributes should be considered to be the norm, not the exception. The research community has struggled with interrelated factors in developing techniques for a number of years. Few of these methods are in common use, perhaps due to the complexities in implementation.

Integrated Survey One interaction rule involves **integrated survey**, the landscape evaluation approach that makes multicriteria evaluations of suitability a part of the fieldwork process (Mabbutt 1968). The key to the integrated survey method is that the fieldwork uncovers associations by discovering the *process* that causes factors to combine to make the landscape suitable for a given purpose. Examining the interrelationships in the field is a demanding process that is not easy to replicate, even with the same team. The ultimate weakness is that the field team must be sent out with the appropriate mission. Evaluation is conditional on the purpose, precluding a multipurpose database for unknown future requirements. Still, this method remains one of the best designed to create scientific consensus about the landscape.

Factor Combination While the integrated method relies on a survey that handles all the interactions, there are other methods better connected to current technology. The method that Hopkins (1977) called *factor combination* involves creating an overlay of all possible factors, then rating the combinations for their suitability. It differs

USLE: Universal Soil Loss Equation; predicts average soil loss (in tons) for a period of time (a storm event or a year) as the product of six factors: Rainfall intensity, Erodability of the soil, Length of slope, Slope gradient, Crop, and Practices.

Integrated survey: An approach to land evaluation that combines the opinions of many disciplines in producing a common representation of the processes that form a landscape.

from integrated survey in that various factors are mapped separately and then combined by overlay. *Integrated Terrain Unit Mapping* (Dangermond 1979) is a term for an automated form of factor combination.

Most of the literature on map combination throws up its hands at the prospect of ranking all the combinations of categories that arise from an overlay. Yet, in the multicriteria decision making literature, the procedures of *conjoint measurement* (Keeney and Raifa 1976) have been developed for exactly this purpose. Instead of trusting some linear combination of individual ratings, a conjoint method evaluates the result in the full multidimensional space including the interactions.

The number of factor combinations does increase rather rapidly with more factors. In the worst case, uncorrelated environment, the number of overlaid categories will be the product of the numbers of categories in the source layers. After a few overlays, such a product is frighteningly large. Due to associations in the environment, the number of categories is much less likely to rise that quickly. For example, in a four- factor inventory of forest characteristics in northern California, the product of the categories was $12 \times 5 \times 6 \times 3 = 1080$ potential categories. The 5 acre minimum mapping unit coverage only had 105 of these 1080 possibilities. Smoothed to 40 acres, only 64 survived.

Even if there are many combinations of categories, there may be no substitute for an explicit recognition of their interactions. Vector systems produce the factor combination as the result of the overlay process (see Figure 5-5), even if other combination rules process the attribute tables. Factor combination delays the assignment of a result, but it does not avoid it. Some scheme must specify the value for each combination of input. At least the factor combination method can accommodate all interactions.

Rules of Combination The factor combination approach challenges an analyst to consider all the possible interactions between factors, while the contributory methods offer the simplicity of performing the combinations automatically using an algebraic analogue. There is a compromise between these extremes, involving some of the simple rules of the contributory methods, modified by interaction tables. An analyst will decompose the problem into subsets on which simpler rules apply, then combine the subsets according to some other rule.

At an early stage in the development of geographic information processing techniques, the Honey Hill project (Murray and others 1971) developed a simplified method to accommodate a certain amount of interaction between the factors of an overlay. In this somewhat artificial prototype, each factor had three categories. A three-by-three table combined two factors into a three-category result, which was combined with the third factor in another three-by-three table. Certainly, the restriction to a three-level ranking limits this particular method, but the underlying logic can recognize interrelationships not treated by an arithmetical procedure.

Methods that include interactions between factors can be developed from any of the contributory methods. One example involves the Wisconsin Groundwater Contamination Susceptibility Map (Riggle and Schmidt 1991). Five statewide source maps were overlaid to create a composite (factor combination) of some 54,000 polygons. The categories on four maps were scored in the typical linear combination approach in which ordinal categories are given status as continuous measures (Table 5-8). The high values in this scoring system denote less chance of groundwater contamination.

Thus, carbonate rocks have greater risk than shale; shallow groundwater is easier to contaminate; and so on. The treatment of interaction comes from the table that gives a weight for these four factors based on the value of a fifth factor, depth to bedrock.

TABLE 5-8: Formula for groundwater contamination susceptibility score

Attributes	*Value assigned*
Type bedrock (TBV):	
Carbonate	1
Sandstone	5
Igneous / metamorphic	6
Shale	10
Depth to water table (DWV):	
0-20 feet	1
20-50 feet	5
> 50 feet	10
Surficial deposits (SDV):	
No material	0
Sand and gravel	1
Sandy	2
Peat	5
Loamy	6
Clayey	10
Soil characteristics (SCV):	
Coarse texture / high permeability	1
Medium course texture / high–medium permeability	3
Medium texture / medium permeability	6
Fine texture / low permeability	10

Weights-Based on value of Depth to Bedrock:

Depth to Bedrock	*TBW*	*DWW*	*SDW*	*SCW*
0-5 feet (> 70%)	13	1	0	1
0-5 feet (< 70%)	11	1	1	2
5-50 feet	6	2	4	3
50-100 feet	0	3	8	4
> 100 feet	0	3	8	4

GCSS = (TBW × TBV) + (DWW × DWV) + (SDW × SDV) + (SCW × SCV)

Note: High scores are less susceptible.

Source: Riggle and Schmidt (1990).

The weight given to the type of bedrock declines as the depth to bedrock increases. Thus, a shallow situation will be more strongly influenced by the type of bedrock. The zero value for depths over 50 feet mean that the factor has no influence on the result. Here the weighting serves to remove a factor from the contributory equation. The other three factors have weights that increase with depth to bedrock. Thus the importance of soil characteristics is greater for deep soils.

The values and weights in the Wisconsin groundwater study are difficult to justify as pure extensive measures. The source materials available statewide were not sufficient for anything but the crudest determination of depth to bedrock (50-foot contours). The use of a linear equation may not be properly justified for the ordinal categories on the original maps. Nevertheless, this study did wrestle with the interaction between some of its factors.

Summary of Rules

This taxonomy of map combination distinguishes methods based on the amount of information used. In all cases, a number of map layers have been combined using overlay. Each location on the map has an attribute on each of the input layers. The dominance rules pick one of these values as the result; the exclusionary screening rule is used quite frequently in applying regulations because of its simplicity. The contributory rules use all the attributes available to create some kind of composite attribute value. The limitation is that each layer contributes its value without regard for the others. These methods are relatively easy to implement but full of assumptions. Interaction rules try to provide a method to treat the complexities of the environment. The result depends on the specific combination of attribute values for some layers taken together. It is not surprising that these methods are underdeveloped in practical cases. Truly defensible applications of GIS will require continued development of the procedures for map combination. Many models of environmental interactions exist, but they are not always implemented in their full complexity.

SUMMARY

Map overlay had its origins in a simpler world of physical maps drafted on transparent material and examined on a light table. In a GIS, overlay constructs the linkage between different sources of information using the geometry as a key. Direct analysis of overlay can quantify change and other kinds of differences. Attributes attached to the sources can then be combined using three general classes of rules: dominance, contributory, and interaction. Chosen with care, the composite information forms a strong base for geographic analysis because it can integrate diverse sources.

CHAPTER 6

DISTANCE RELATIONSHIPS

CHAPTER OVERVIEW

- Construct vector buffers around isolated objects.
- Perform distance measurements and construct distance fields for rasters.
- Introduce extended Voronoi representations of distance relationships.

Chapter 5 concentrated on the overlay method, a technique to integrate different sources. As with any tool kit, there is a risk of seeing the world through the viewpoint of that one method. To the hammer salesperson, the whole world is full of nails (that need hammers). With geographic information, there is a similar risk. Regulations and specifications are usually written to reflect what seems technically feasible. Scientists, naturally, tend to use the techniques with which they are most familiar, sometimes missing important insights. There are many tools beyond overlay. This chapter will introduce one such tool in some depth that will lead to a more general set of neighborhood tools in the next chapter.

The next step in expanding the range of tools from the strict locality of overlay is adding tools that discover distance relationships implicit in a spatial representation. These operations can construct a new representation using distance measurements from some existing representation, thus qualifying as a limited form of transformation. The nature of these transformations depends on the assumptions implicit in the measurement framework. Chapter 7 will continue in the same direction with more generalized methods that operate on neighborhoods.

EXAMPLES OF DISTANCE SPECIFICATIONS

Before considering the procedures used to implement distance measurement, it is useful to consider some of the motivations for using distance in geographical analysis. Distance can be a positive or a negative factor in location. This section presents examples of distance zones that occur in many application fields under somewhat different labels and with different scientific authority.

Exclusionary Zones Around Features: Buffers and Setbacks

Separation between objects implies a relationship that can be measured as a distance. Distance serves as the very prototype of a continuous measurement scale used by analogue in all other measurement situations. Yet, all forms of human society try to simplify the continuous variation of distance into categories of inside and outside, near and far, us and them.

The exclusionary logic introduced for map overlay applies to distance measurement under a number of terms. The word **buffer** is commonly used in environmental regulation, and it has been adopted by a number of GIS packages. Buffers are usually constructed outward to protect some element. In development regulations like zoning ordinances, the word "setback" appears. **Setbacks**, however, tend to move inward within an area by constructing lines parallel to the boundaries. Both buffer and setback imply a simple zone: inside and outside the critical distance (Figure 6-1).

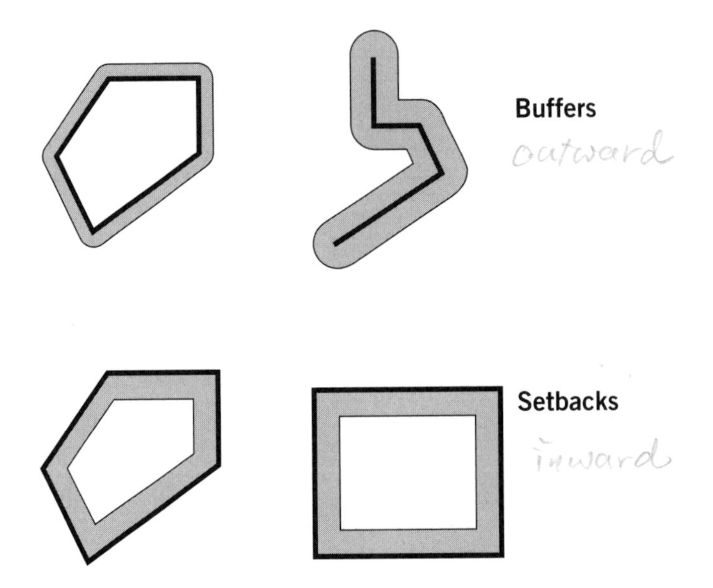

Figure 6-1: Diagram of simple buffer and setback.

Buffer: A zone constructed outward from an isolated object to a specific distance.

Setback: A zone inside a polygon constructed by a fixed distance from the edge of the polygon; typically used to restrict building or activities too close to the edge of a property parcel.

Land regulation often converts continuous distance into a category. Setbacks, for example, have a long history. In colonial America as early as 1703, there were regulations for a minimum front yard in Williamsburg, Virginia. Certainly the concept of such regulation was imported from European models. As town planning became more universal in the early twentieth century, the setback became an integral part of suburbia. Eventually, other environmental regulations imposed outward distance criteria (buffers). Changes in local ordinances require notice to adjoining property owners, usually specified as everyone within some given distance. Automated parcel records provide the basis for lists of property owners who must be notified concerning a land use action (Figure 6-2). This list can be constructed by a number of different procedures, but at the base, each one must construct a new geometric object (the buffer) and then decide what is inside that object.

Perhaps the most common application for buffers comes from the difficulties of environmental management for linear features within larger landscapes. Ecologists have long recognized the value of edges as habitat and as a locus for flows of all kinds. This scientific recognition spurred regulations based on buffers. For example, streams

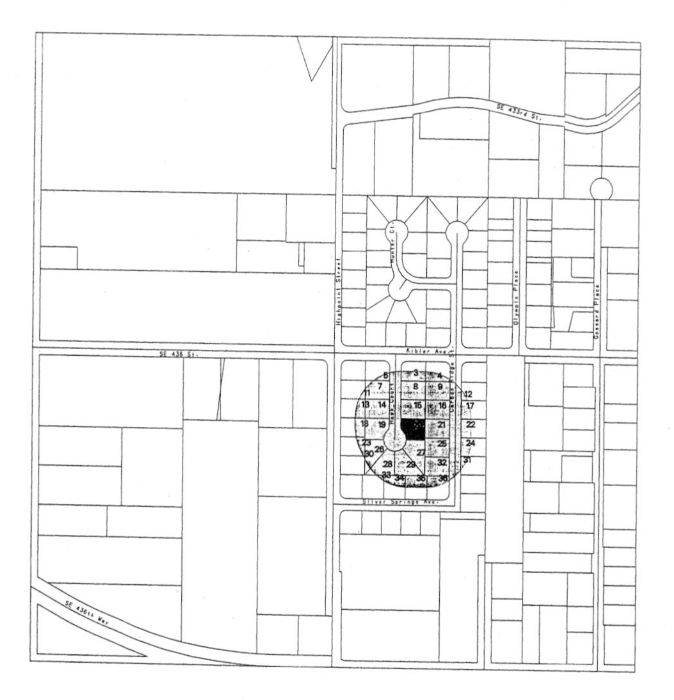

Figure 6-2: Buffer constructed around a parcel. Zoning regulations require mailing notices to all owners within 225 feet of the applicant's property when an owner applies for certain changes in land use. Source: Metro King County, Washington.

present a problem in forest management, and much of the current dispute about environmental regulation in the Pacific Northwest of the United States involves arguments about buffer definitions and widths. Thus, the State of Washington Forest Practice Rules and Regulations (1988) will serve to introduce the concept of buffers.

At the societal level, the Pacific Northwest region of the United States has a conflict between two environmental property regimes. On the one hand, land can be purchased, and the property owner can extract the economic benefits of the land by cutting trees. On the other hand, fish swim freely in the waters of the state and belong to whoever catches them, subject to state regulations and treaty obligations to Native Americans. The terrestrial and aquatic arenas create quite different attitudes about public and private rights, but they would not conflict if the two regimes remained distinct. However, the environment is not divided into two neatly separated zones. Many of the fish in question return from the ocean to freshwater to spawn. The reproduction of fish depends on cold water temperatures and gravel stream bottoms. Both characteristics can be influenced by logging. Removing trees can open the stream to direct sunlight, raising the water temperature to a level where the fish fry cannot thrive. Furthermore, removing vegetation can increase sediment flow into the stream, covering the gravel.

Thus, there is an environmental interaction where the activities of one group can influence another. Under the simplified rules of exclusive ownership (pure capitalism), the timber owner would have the right to remove any trees, thus causing a reduction in fish survival. As with many such interactions, the consequences would not be immediately detected, particularly when the fish resource is diffuse and under collective ownership. The economic benefit of logging the trees can be seen immediately, but the reduction in salmon stocks is much harder to attach to a specific logging operation. This situation characterized the logging practices during the first century of European development in the region. Eventually, the rights of the fisheries became more apparent. Through the procedures of state regulation of forest practices, forest landowners were required to change their practices. The nature of these regulations is strongly contested between the various parties, and the political power of the various interests can be seen in the width of buffers.

The Timber/Fish/Wildlife Agreement (Washington State 1987) represents one recent compromise, created before the federal interest in protection of the spotted owl habitat created a further set of negotiations. Timber owners are restricted in their activities near streams, in a carefully arranged geometric framework. The goal is to ensure that the fish habitat remains shaded and free of sediment, while permitting some economic benefit from the timber on these lands. Trees within a certain distance of the stream contribute to the shade that maintains the cool water habitat. In addition, logging triggers erosion. Undisturbed vegetation can trap sediment before it reaches a stream channel. The clean geometry of exclusive landownership thus has been modified by a new geometric construct—a buffer called the Riparian Management Zone (RMZ). The regulation (Table 6-1) changes streams from essentially linear features into areas. Different buffer widths protect

TABLE 6-1: Riparian Management Zone (RMZ) Requirements

Water type and width (feet)	*Maximum RMZ width (feet)*
Type 1 and 2 >75	100
Type 1 and 2 <75	75
Type 3 > 5	50
Type 3 < 5	25

Note: a minimum width of buffers is set to 25 feet for all water types and widths. Types of water are defined in WAC 222-16-030. Note that Type 3 channels must exceed 5 feet unless classified by residential use.

Source: Washington Administrative Code (WAC) 222-30-020 (4c).

Definitions of water classes:

Type 1: All water inventoried as "shorelines of the state"

Type 2: Water diverted for use by 100 residences
Within a campground of 30 units and within 100 feet
Used by substantial numbers of significant fish
(defined as channel > 20 feet and gradient < 4% OR lakes > 1 acre)

Type 3: Water diverted for use by 10 residences
Used by significant numbers of anadromous fish
(channel > 5 feet and gradient < 12% and not upstream waterfall > 10 feet
OR lakes < 1 acre connected to anadromous streams)
used by significant numbers of resident game fish
(channel > 10 feet and gradient < 12% and summer flow > .3 cfs
OR lakes > .5 acre)
Protect downstream water quality (>20% of flow in Type 1 or 2)

Type 4: Not Type 1, 2, 3 and channel > 2 feet

Type 5: Not Type 1, 2, 3, 4; intermittent or seasonal flow

Source: WAC 222-16-030

each different class of streams, based on expected fish habitat and political compromise. Unfortunately, the widths of the RMZ buffers may not be large enough to reduce sedimentation sufficiently (Castelle and others 1992). The more recent federal forest plan adopted riparian buffers of 300 feet as a further attempt to preserve habitat.

Protection of fisheries was not motivated by a cartographic model of the world. The shade on the streams clearly comes from nearby trees. The Riparian Management Zone concept includes many field-based determinations about the temperature sensitivity of the water, the streambed materials, and other concerns that influence the regulation. But then, to ease implementation, the regulation is converted into the simple cartographic format of an exclusive buffer zone, encapsulating the political tension between competing forces. The buffer concept can become a major regulatory element as a surrogate for understanding the actual interactions.

Beyond Buffers and Setbacks

Buffers can be used to protect some features, to push activities away. Other situations may use proximity as a positive factor and may seek to minimize distance. For these purposes, it makes more sense to use the continuous measure of distance, not a buffer.

A substantial body of location theory addresses the use of distance as a factor in various kinds of decisions. Often the simpler location theories are based on an analogue to a model from physics. Various physical laws specify different relationships over distance: sound pressure diminishes with the inverse cube of distance, and gravity diminishes with the inverse square of distance. Simple transportation cost would seem to increase as a linear function, but usually terminal costs and economies of scale make it rise more slowly than a linear function. Each of these numerical relationships must be considered in converting from strict distance to the purposes of a given project. Geographic reality is often much more complicated than the simplified worlds of a physical model. Careful construction of geographical relationships can model much greater realism. The tools for such models will be introduced in later chapters.

DISTANCE MEASUREMENT

Distance is a relationship between two points in space. Because measurement frameworks constrain relationships, each data model treats distance differently. Given these differences, the methods connected to a system of representation adopt different tactics. Yet, there is some unity of purpose hidden behind the differences.

Distance Relationships

Although distance is a relationship between two points, it is often simplified and considered to be an attribute. Distance is an appropriate attribute for a line segment because the segment embodies the same relationship. Distance can be totaled for aggregations of line segments as long as this aggregate distance is understood to be distance along some network, and not the distance as the crow flies. These distance measurements do not require complex logic, because the distances can be computed from the coordinates of end points. In practical application, however, the concept of *river mile* or *highway distance markers* runs into difficulties as the networks change. If you shorten the road near the beginning, do you go move all the markers and change everything in the database? Mark Twain's *Life on the Mississippi* measures the passage up the river in a flexible sequence of old mile markers, cut-offs, and new channels before the Corps of Engineers tamed the fluvial processes.

In much of the site selection literature, distance is mentioned as a criterion measured from a linear or areal object, but the geometry is far from simple. The distance required is usually the minimum distance to any point or line along the description of the object—essentially a dominance operation on a set of simple distance calculations.

As with operations such as exclusionary screening, the use of distances often begins with an isolation operation that extracts some element from a more complex coverage (see Chapter 4). Internal complexities are simplified into a binary distinction, and distances are measured from these isolated objects outward into the void. Starting with this simplified conception, elaborations like variable widths (as required for the RMZ regulation) can be added. In any case, the process still starts by creating an isolated object view.

Construction of distances varies depending on the underlying representation model of the GIS software. In the vector environment, there are so many possible point pairs for constructing distances that distance relationships are usually simplified into buffers. In the raster environment, the process is less tied to a specific threshold. It is simpler to construct a distinct distance measure to all the discrete points of the grid. Despite these capabilities, however, users tend to construct buffers as binary thresholds from a more continuous measure. This section will discuss both vector and raster separately and then discuss similarities. In both cases, the distance processors construct a map layer to be used with others in an overlay application.

Constructing Buffers with Vector Data

As the examples above demonstrate, the external context of the application often provides specific widths for a buffer. In some cases, these values come from clearly established science, though other distances are fairly arbitrary. Most vector software constructs buffers of a specified width. Some software can vary the distance according to other attributes, so that the RMZ buffers in Table 6-1 could be computed in a single operation.

To review the measurement process, the buffer is just another form of an isolated object, a contour line drawn at a specified distance from the selected feature. The location of the contour line can be calculated from the coordinates of the relevant portions of the original feature. There are a number of methods to obtain this measurement. The simple method treats each chain in isolation; the generalized Voronoi network, discussed below, offers an alternative.

Buffer construction in its vector form is usually performed in a series of steps hidden from the user. Each boundary chain is converted into a sausage, a locus of parallel lines and circles that stake out the contour of the chosen distance away from that particular line (Figure 6-3). Then all these sausages are passed through an overlay processor or some other geometric process to discard zones of overlap. The process

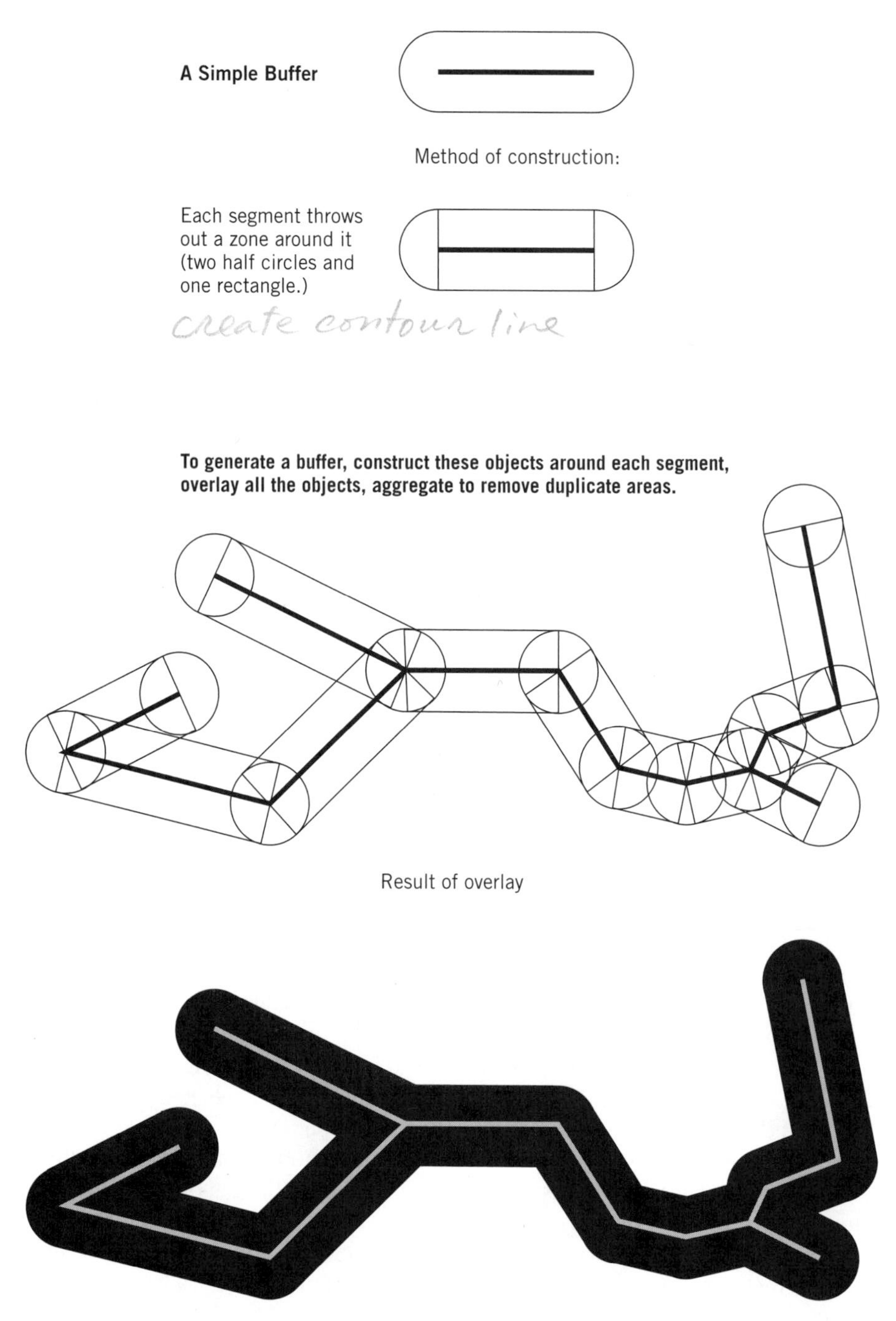

Figure 6-3: Construction method for buffers in a vector representation.

creates an isolated object, a contour of the chosen distance, that can be used in any other operation.

Measuring Distance in a Raster

The measurement process changes in the raster environment. Because space is controlled, the distance from the selected cells to other cells can be measured directly without slicing the distance into a contour form. The process usually begins with a selection operation to isolate the starting cells and to define the void in which distances will be measured. By definition, cells in the selected object have a distance of zero. The algorithm for distance construction operates iteratively. Each cell with a value propagates that value plus the cell width to its neighbors. Hence it has been called `spread` in some implementations, an apt sense of how it works. Figure 6-4 shows the sequence

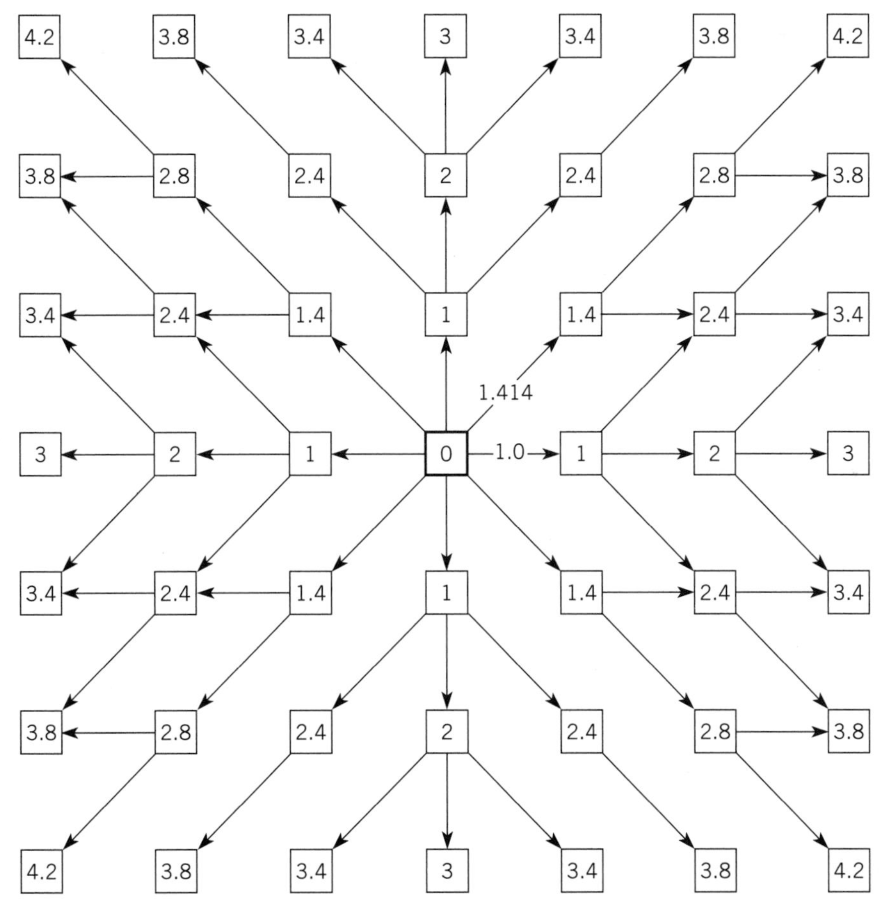

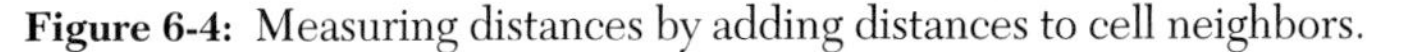

Figure 6-4: Measuring distances by adding distances to cell neighbors.

of operations for a few cells. The result is a continuous field surrounding the original selected objects. Each sample point can be assigned its distance measurement.

At short distances, the granularity of the pixels only permits certain integer multiples of the pixel width. This can be difficult if a particular threshold cannot be represented in whole cell widths. In addition, diagonal distances can be a bit troublesome, depending on the implementation. Early software stored distances only as integer counts of pixels, which did not permit the diagonal neighbors to have the correct distance of $\sqrt{2}$ times cell width. Now, most raster software provides a method to represent diagonal distances in a close approximation of the proper proportion. For a given buffer distance, finer pixels lead to better results at a cost in greater storage and computation.

This iterative approach can be propagated as far as required. It takes longer to fill the whole matrix than it does to detect the nearest neighbors, but the process is not beyond reason for databases of practical size. The spreading algorithm produces a continuous distance surface outward from the starting cells. A threshold can convert these measurements into a simple category inside or outside a tolerance (a classification operation). This simply approximates the result of the vector buffer operation. More interestingly, the distance from the selected feature becomes a value for the pixel that can be compared to the distances computed from some other feature. By subtracting one value from the other, the zone closer to one than the other will be the positive or negative values in the result. For example, the archaeological sites in a region of New Mexico include two periods called Mesilla and El Paso. The relationship between these two periods can be studied by constructing a distance surface around each. The difference of these two surfaces shows those areas closer to one or the other (Figure 6-5). There are many potential applications for such an analysis of nearest facility including central place theory, industrial location, and allocation of service areas for all kinds of services.

The iterative approach to distance measurement permits some additional controls not found in the vector case. The search process can be constrained to respect a barrier or to move only downhill on a topographic surface. More important, the iterative process, since it works cell by cell, could weight the distance based on some impedance to travel through the cell. Both enhancements move away from the pure Pythagorean distance to some interaction with a surface. These operations will be considered in Chapter 7, which deals with surfaces and more advanced neighborhood operations.

Comparison

To compare the two approaches, the vector method permits the analyst to select a buffer width and to construct that buffer contour directly. Current computation methods are fairly crude but effective. In the raster implementation, the representation is more approximate. The features have been simplified into the cellular representation, thus losing perhaps a half-cell width of resolution, particularly for linear networks. In

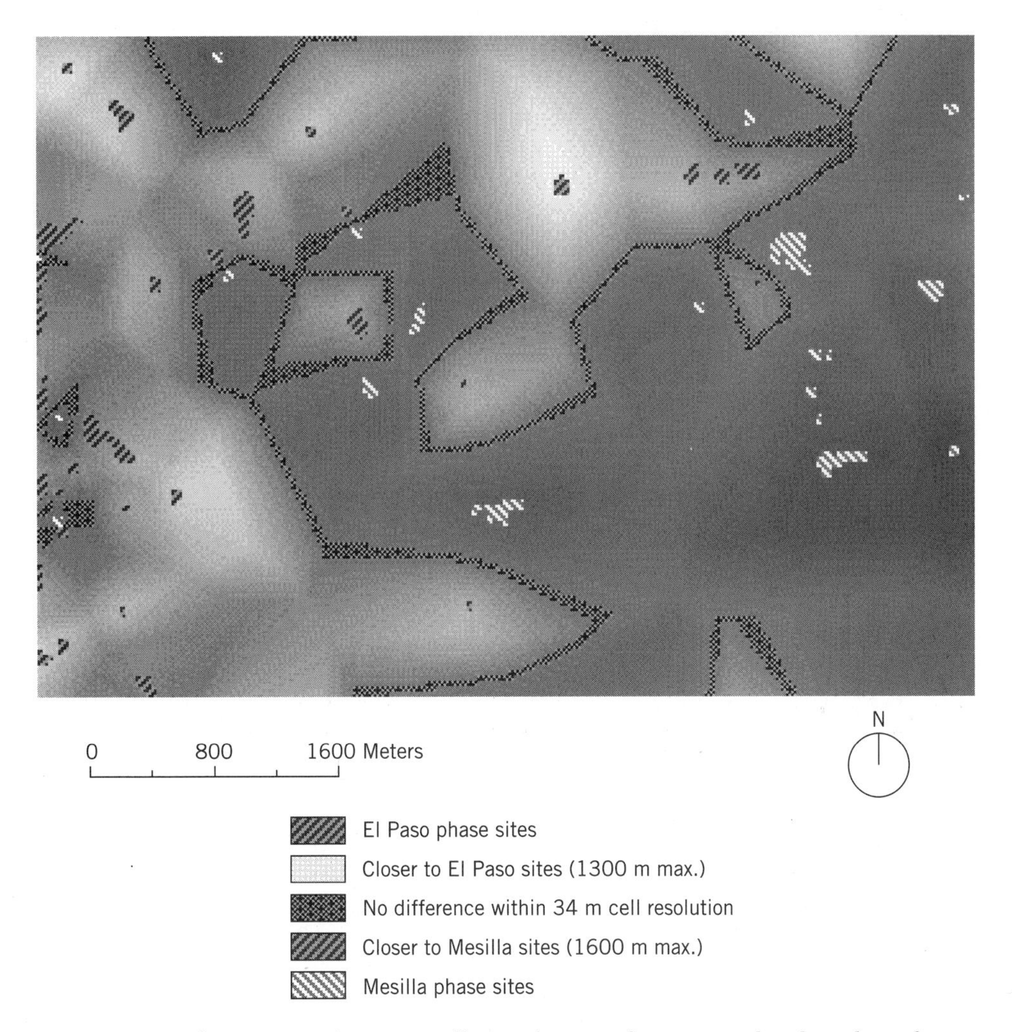

Figure 6-5: Subtracting two distances. Difference between distance to archaeological sites from two different periods, Mogollan Plateau, New Mexico. Source: Linse (1993).

addition, distances are constrained by the grid. For buffer widths of three cell widths or more, the raster method provides a reasonable approximation.

The raster method permits a more continuous approach to the distance measurement than the fixed-threshold calculation now performed by most vector software. A continuous approach allows numerical comparisons between distance relationships. This is not an inherent property of the raster method, however, because some experimental vector methods calculate a similar result. These experimental methods use the generalized Voronoi diagram, discussed below.

GENERALIZED VORONOI DIAGRAMS

Vector and raster methods are locked in the classical opposition between control by attribute and control by space. Buffers in the vector view must be isolated contours, because some form of attribute control seems necessary. Similarly, the raster can estimate distances but only inside the discrete steps of the grided space. In most GIS packages, these are the only alternatives provided.

Recently, a thread of research has taken a fresh approach. Much as the Triangulated Irregular Network (TIN) structure (introduced in Chapter 2) does not fit the opposition between raster and vector, the alternative works by a form of control through relationships. The roots of this approach run rather deep in a number of fields. Consider the set of polygons that assign a space that is closest to each of a cloud of points (Figure 6-6). In geography, these polygons have been called *Thiessen poly-*

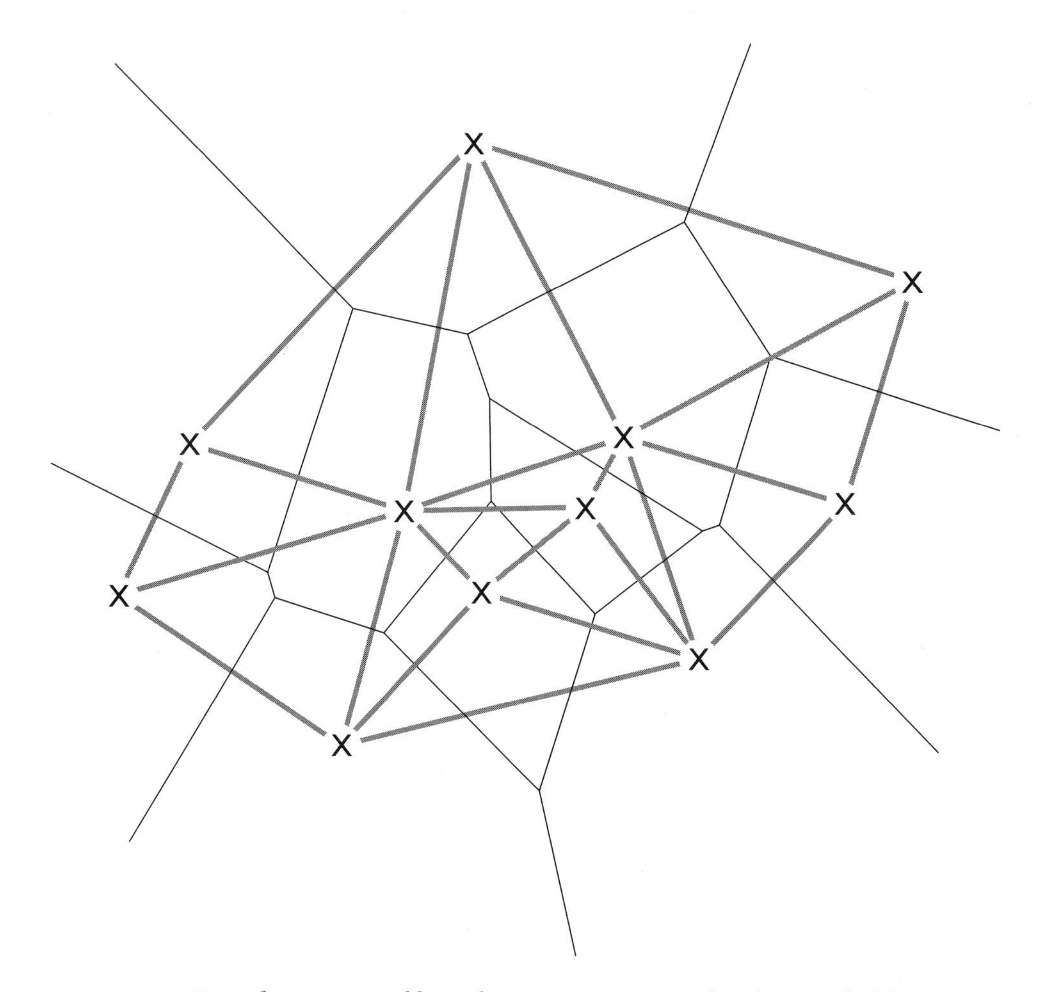

Figure 6-6: Zones for nearest neighbors of points: Voronoi network is shown in dark lines and Delaunay triangulation connecting neighboring points in gray.

gons after a climatologist who used them to perform a transformation from point climate stations to watersheds (see Chapter 9). In mathematics and computer science, the same construction is called the **Voronoi network** after earlier and more precise work by a Russian mathematician. This book will use the term Voronoi network for this construction. Whatever the name, the polygons control for the relationship of being nearest to a particular point. Each side of a polygon is a perpendicular bisector of the line between two neighboring points. Construction is easier when approached as a triangulation (named after Delaunay) connecting the neighboring points (gray lines in Figure 6-6). Construction of **Delaunay triangulations** and Voronoi networks is well established (Preparata and Shamos 1985).

A number of researchers have developed the extensions required to construct proximal zones around lines as well as points. The extended Voronoi diagram captures the *relationships* of nearest object that may be intuitive for the human vision system but not included in the established representation schemes. Each zone represents the part of the plane that is closer to one particular object than any other. The edges of this diagram include three geometric primitives (Figure 6-7): the perpendicular bisectors used by the basic Voronoi network for point–point situations, angle bisectors for line–line relationships, and parabolic sections for line–point relationships. The extended network thus encapsulates a model of distance. Any vector buffer can be produced from this diagram by contouring the network at the desired distance. There will be no overlay or geometric postprocessing required.

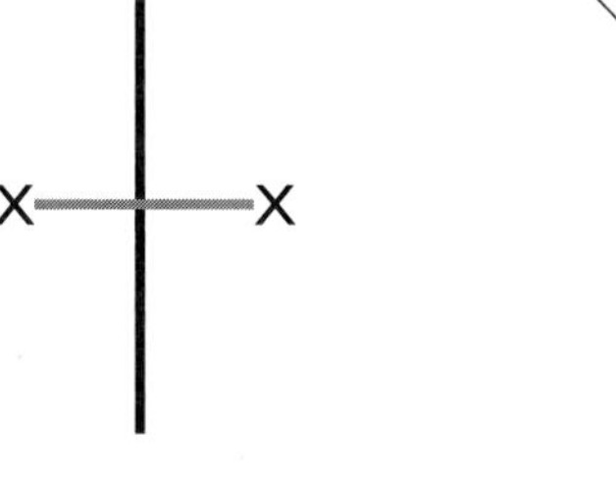

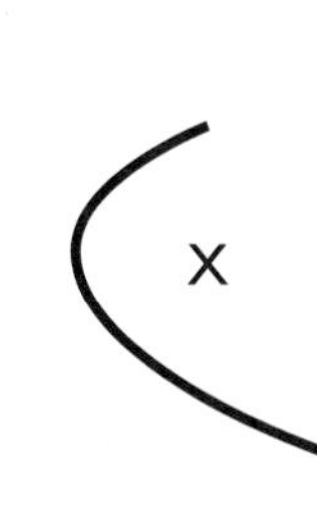

Figure 6-7: Basic distance relationships between points and lines in a plane. These three cases divide a space into the area closer to one or the other. Point–point neighbors are separated by perpendicular bisector; line–line by the angle bisector; point–line by a section of a parabola.

Voronoi network: A set of lines that divides a plane into the area closest to each of a set of points. The lines are perpendicular bisectors of the lines connecting nearest points (Delaunay triangulation).

Delaunay triangulation: A network that connects each point in a set of points to its nearest neighbors; topological "dual" of the Voronoi network.

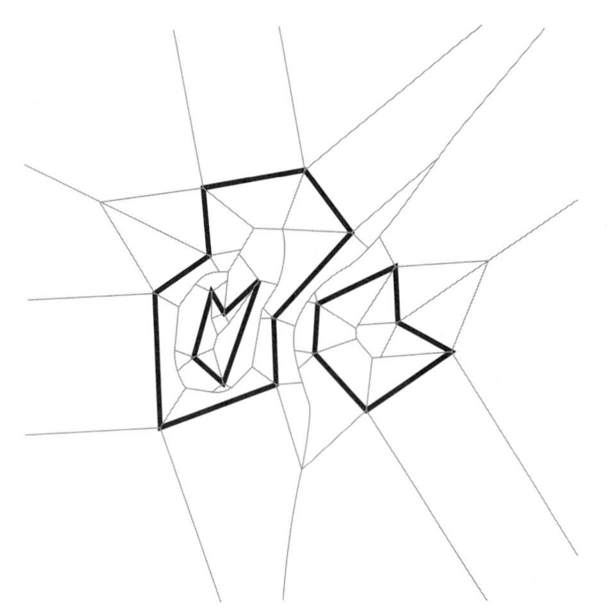

Figure 6-8: Extended Voronoi network creates zones nearest to a collection of points and lines.

Recently, at team at Laval University has implemented a practical solution to calculating these extended Voronoi networks (Gold 1992) (Figure 6-8). It may take some time for these innovations to move from the research community into the commercial packages.

DATA QUALITY ASPECTS OF DISTANCE RELATIONSHIPS

The distance operations discussed in this chapter work on a single representation, so they do not provide a method to test one source against another. However, by combining with the overlay testing method discussed in Chapter 5, buffers can serve a role in data quality tests. More important, the distance relationships can serve as a model of certain kinds of error.

The construction of buffers applies meticulous geometric rules to information that may be nowhere near as precise as it seems. A buffer tool can also be interpreted as a statement about uncertainty in the representation. In the place of the exact model, the measurement of boundary lines should be considered from a statistical point of view. The linework obtained could be modeled as a true location, perturbed by some amount of measurement error. Many cartographic processes introduce error in positioning a line, and each error may have a distinct form. A band around the line provides a first approximation of the likely position of the true line.

As a statistical model, this would imply a set of contours for the probability of finding the true line at a given distance from the measurement obtained. The width of this zone may be uncomfortably large. For example, in considering just the line width error, the digitizing process, and roundoff for the US Geological Survey **GIRAS** data, the standard deviation of the error might be 20 meters on the ground. Constructing a buffer of this width may indicate the potential variation in the area calculations due to measurement error in the linework. The area in a 20 meter buffer can range up to 10% of the area in a GIRAS coverage, depending on the density of linework (Chrisman 1982b).

The appropriate statistical procedures to handle boundary imprecision have not been developed, though it remains a key issue of research concern (Goodchild and Gopal 1989; Mark and Csillag 1989; Chrisman and Lester 1991; Goodchild and others 1992). It is particularly difficult to distinguish the various forms of error in boundaries of categorical coverages. Some models deal simply with measurement error, while others focus on the lack of sharp transitions between the categories (Burrough 1989; Wang and others 1990; Leung and others 1992). Each kind of error will occur in different amounts, and thus different models and tools must be used to approximate the behavior of the error distribution. A fixed width buffer is not the only model for error, but it can help visualize the potential influence of uncertain boundaries.

SUMMARY

Distance transformations construct a new geographic representation based on distance relationships implicit in some original coverage. Just as raster and vector offer two distinct approaches to controlling space or attribute, the construction of distance transformations varies between these basic approaches to spatial representation. Current vector methods permit the exact construction of a buffer (a contour for a specific distance from a selected object). Raster methods permit the storage of the distance measurements, sampled at discrete points. Raster results can treat distance in a more continuous manner, not just the sharp edges of a buffer. Despite these differences, regulations are usually written with specific thresholds. The output from a distance transformation can be considered a new coverage for use with the other tools, such as overlay. Alternatively, a new measurement framework, the extended Voronoi diagram, can represent the distance relationships without the compromises inherent in the vector or raster techniques.

GIRAS: Geographic Information Retrieval and Analysis System, a project conducted by US Geological Survey in the 1970s; produced vector interpretations of land use/land cover stored in an early topological data structure.

CHAPTER 7

SURFACES AND NEAR NEIGHBORS

CHAPTER OVERVIEW

- Describe properties of surfaces and how they are calculated.
- Review geometric component of neighborhood construction.
- Develop taxonomy for combining attributes discovered inside a neighborhood.

The previous chapter introduced a set of operations to handle distance relationships. These operations belong to a larger set that deal with neighborhoods—the spatial context around each value. To extend to a more general treatment of neighborhoods, the surface provides a clearer model than the discrete objects that formed the major focus of the previous chapters. Thus, this chapter begins with a review of surfaces and then presents a system to understand neighborhood operations for all kinds of models, not just surfaces. This chapter deals with near neighborhoods, leaving more complicated operations to the next chapter.

SURFACES

The concept of a surface implies a distribution of a continuous attribute over a two-dimensional region. A number of distinct measurement frameworks share this conceptual model. Surfaces have only one value at any point. Thus, from a topological perspective, a surface is simply a plane deformed into the third dimension. Commonly, surfaces are called "two–and–a–half" dimensional constructs. Surfaces are two-dimensional in topological form and are measured in three dimensions. There is no reason to assign the value of one half to this combination.

Multicomponent Surfaces

Strange as it may seem, the single *value* of a surface may require more than one number to represent it. If a surface has a single ratio measure, it can be called a **scalar** field—the single value can be represented by a single number. Some geographic quantities require more than one number to characterize them; yet, they remain a single phenomenon. Mathematics and physics use two classes that extend the concept. A **vector field** has a quantity and a direction at each point, while a **tensor** is a higher order multicomponent measurement (Figure 7-1). The term "vector" used for a system of geometric representation (presented in Chapter 3) comes from the same root. The vector representation permits a line segment to have a distance and direction from its point of origin. To avoid potential confusion, the term *multicomponent* will be used for surfaces with more than scalar values. Vector fields and tensor fields sound like terminology from science fiction, but dynamic models of the atmosphere, hydrology, or migration specify varying rates of movement that are properly modeled as multicomponent quantities.

Just as in the pure mathematics of fields, the key assumptions about geographic surfaces relate to the property of continuity. A continuous function in mathematics comes in degrees of smoothness. The lowest form is simply *piecewise continuous*, meaning that there is one value over the whole domain, but there may be some abrupt jumps. A continuous function has a smooth transition between all values, meaning that there is always a value in between any two points, no matter how close, and its value is intermediate between the values of its neighbors. The higher orders of continuity demand that the rate of change between points also varies continuously (Figure 7-2). These mathematical properties are not always discussed for geographical distributions, because the infinitely small displacements described by the theory seem irrelevant to geographic measurement.

The property of continuity describes behavior of neighboring values on the surface. The simplest relationship, and the focus of the basic operations of differential calculus, describes the rate of change in a surface. Rate of change on a topographic surface is usually called *slope*. While the topography is a scalar field, the rate of change is a multicomponent quantity, having a magnitude and an orientation. At each point,

Scalar field: A surface whose value can be represented by a single number.

Vector field: A multicomponent surface whose values have a quantity and a direction in space. Newtonian physics can be expressed as vectors. This mathematical term is the origin of the term "vector" applied to a geometric data structure, but the connection is indirect.

Tensor: A multicomponent surface of higher degree than a vector. The theory of relativity requires tensors to handle space-time and electromagnetic fields.

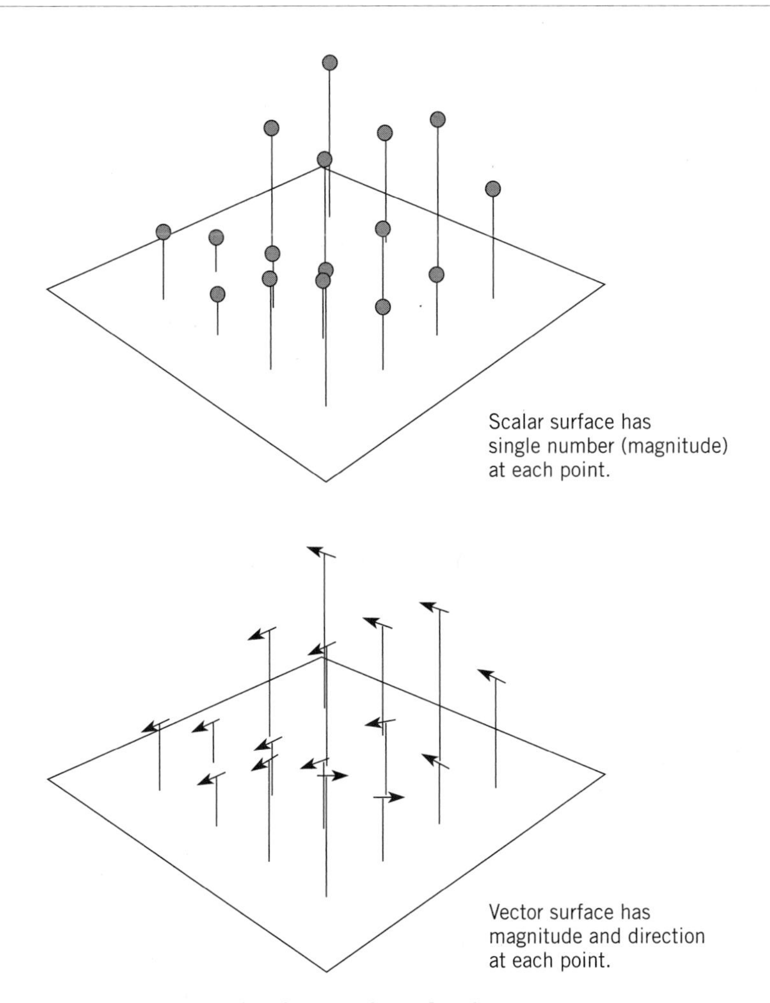

Figure 7-1: Diagram of types of surfaces: Scalar and multicomponent.

the slope measures the plane that is tangent to the original surface at each point (Figure 7-3). The term "slope" is often used ambiguously for the vertical component; properly, this should be called *gradient*. The direction of the slope is usually called *aspect*. These two components can be represented as two distinct numbers, but together they measure one property of the surface.

Topology of Surfaces

The combination of gradient and aspect is quite useful in the analysis of geographic surfaces. The topographic surface (earth–air interface) provides the most apparent

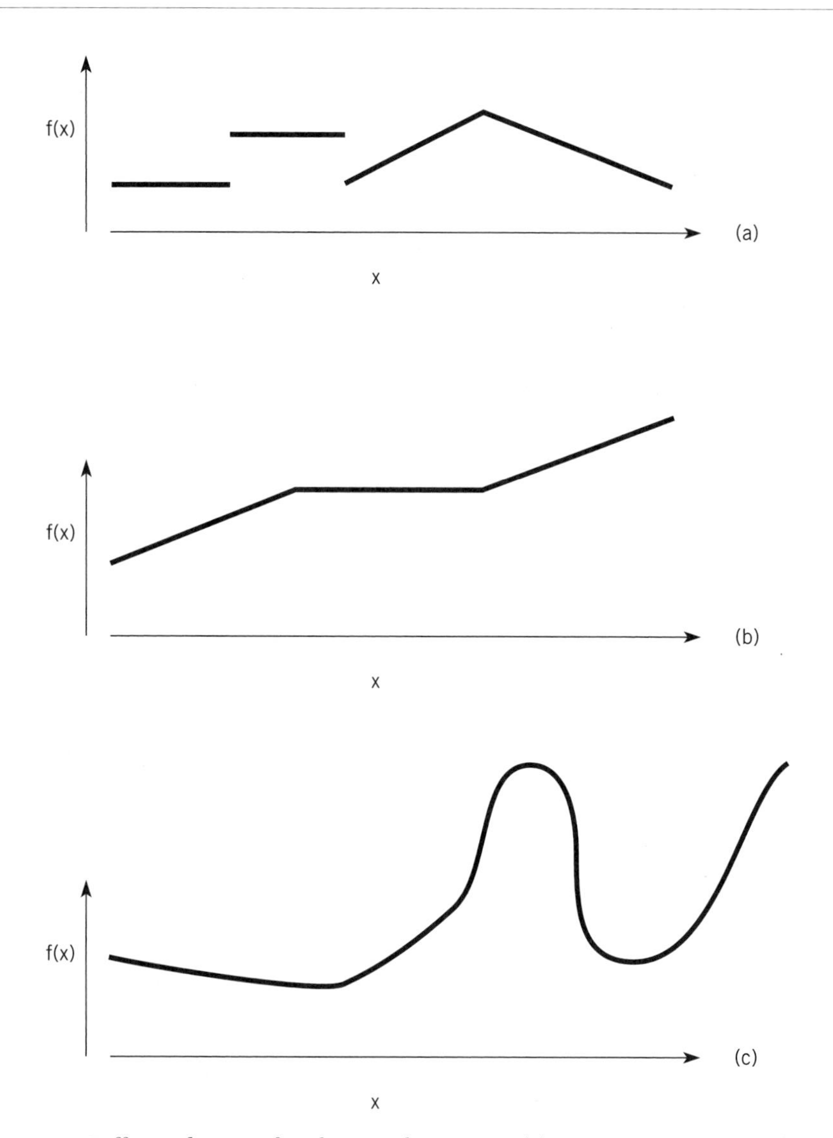

Figure 7-2: Different degrees of mathematical continuity: (a) A piecewise continuous function; (b) a continuous function with an abrupt change in slope; (c) a continuous function whose rate of change is also continuous. For simplicity these properties are shown for a function in a one-dimensional domain, not two dimensions as required for a geographical surface.

geographic example. Cayley (1859) provided early mathematical insights into the structure of surfaces. Warntz (1966) applied Cayley's structure to geographic analysis using the terminology presented here. The local behavior of slope provides a simple mathematical basis for many ancient terms for the landscape, though these relationships apply to any surface, not just physical relief. At most places on a surface, slopes

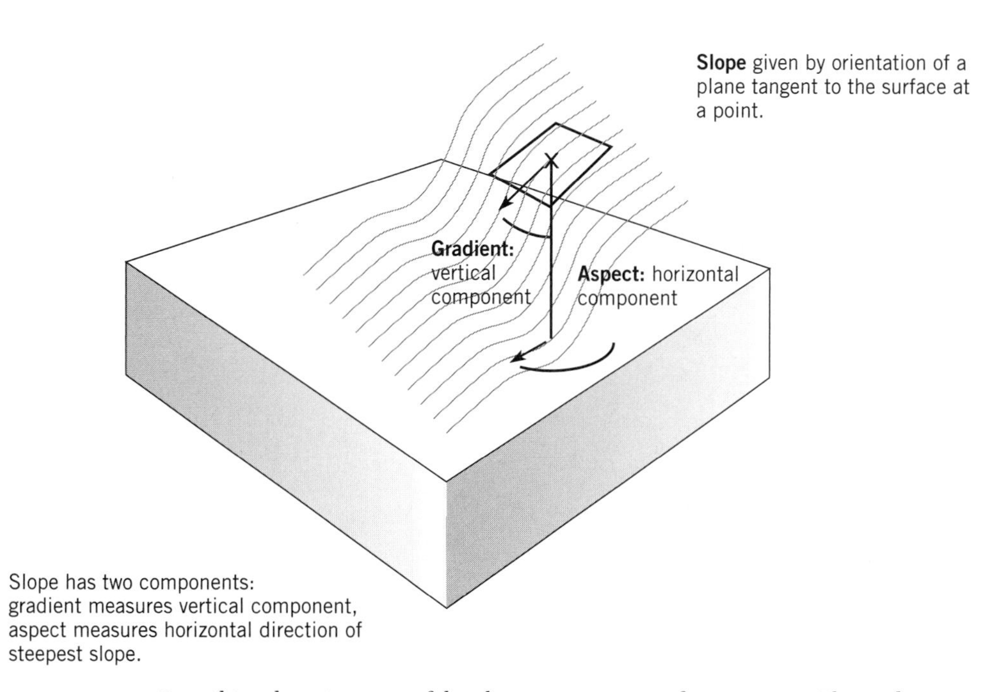

Figure 7-3: Describing the orientation of the plane tangent to a surface at a point. The gradient of the slope describes the rate of change as a function of the angle of the plane. The other component of slope is the direction of the steepest downhill slope, the "aspect," measured as an angle from some arbitrary bearing (like north or east).

are parallel, but at others the slopes converge and diverge (Figure 7-4). At the tops of hills (*peaks*), slopes diverge in every direction. As the slopes continue downhill, they eventually meet the slopes from another hill, causing a convergence. Warntz termed this a *course*, because in a water-eroded landscape, the streams (watercourses) would end up following this line. Similarly, from peak to peak, there is a line (a *ridge*) where the slopes diverge. The network of ridges divides the region into a set of areas, the dales or watersheds. The two networks—ridges around watersheds and courses around hills—constitute the **topology of the topography**. A student of geographic information must be able to use these two words correctly, particularly in this context.

The topology of a surface is defined by the local behavior of the surface—patterns of convergence and divergence. Gravity-powered flow of water over the surface will be strongly controlled by this structure. Water will remain contained by the watershed in which it falls. Ridges create drainage *divides* that separate river systems from each other. As water flows over the surface, it will converge into the course lines and eventually form streams. At the lower ends of the flow, the water will fill up any *pits*, form-

Topology of topography: Qualitative (ordinal) relationships in the structure of a surface; pattern of convergence and divergence of relief. Topography refers to the surface of the earth.

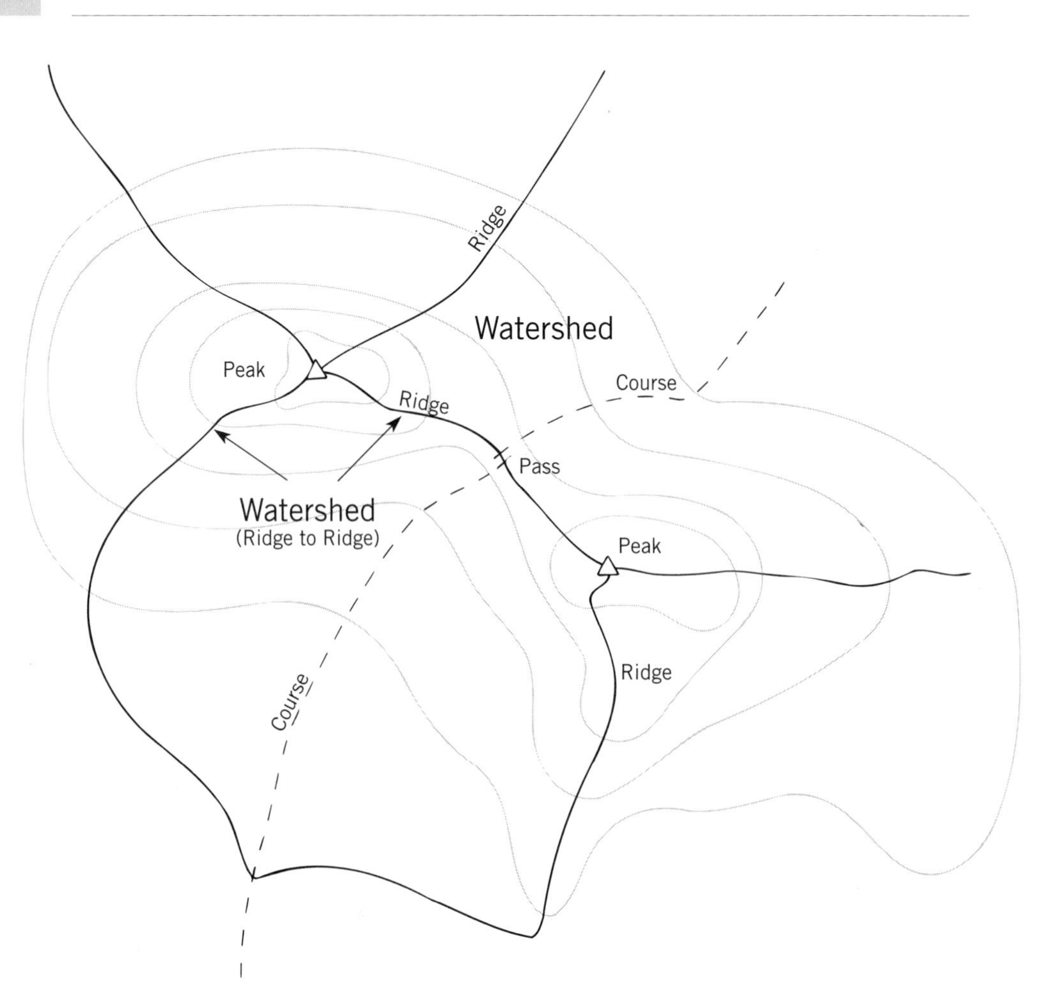

Figure 7-4: Topology of a surface, labeled with Warntz's (1966) terminology.

ing lakes. There is a worldwide covering of the lowest pits in the form of the world ocean. A similar interpretation can attach to surfaces other than topography. The analysis of surface structure introduces the logic created by relationships between neighboring values.

Computing Relationships on a Surface

A mathematically pure function can have a uniquely defined continuous derivative. If so, its slope would be defined analytically at all points. This relationship is of little practical consequence for geographic information, since so few landscapes can be described by smooth functions. For practical purposes, slope gradient and aspect are approximated by relationships between neighboring values.

Methods of computing slope depend on the representation system chosen. If the elevation is represented by a set of points as isolated objects, the slope cannot be computed directly because isolation precludes relationships. The information must first be transformed into a representation that provides some neighborhood information to compute the slope. If it seems that slope is being calculated directly from a scattered collection of points, then a transformation hides within the process. These transformations either construct a regular space-controlled structure by interpolation or they specify the neighborhood by some structure such as triangles. These methods will be covered in Chapter 9.

Once neighborhood relationships are established, the slope gradient for the line between two points can be described by a triangle that separates the horizontal orientation from the vertical difference (9). There are two conventional possibilities for a measure of gradient. One measures an angle of this triangle, typically the angle between the horizontal and the hypotenuse. The angle may be reported in degrees or grads or radians; each is simply a ratio measure with a particular scaling constant. The angle measure varies from 0° to 90°, 0 to 100 grad or 0 to $\pi/2$ radians. In all these cases, a 45° angle will have half the measure of a vertical angle. Angular measure of slope gradient is the convention in some countries (the British Commonwealth, for example) and in some geotechnical disciplines. The other common measure of slope is the ratio of the rise (vertical displacement) over the run (horizontal displacement), usually given as a percentage. This uses the tangent of the angle, which rises more and more rapidly compared with the angular measure. At 45° it attains 100%, and goes literally

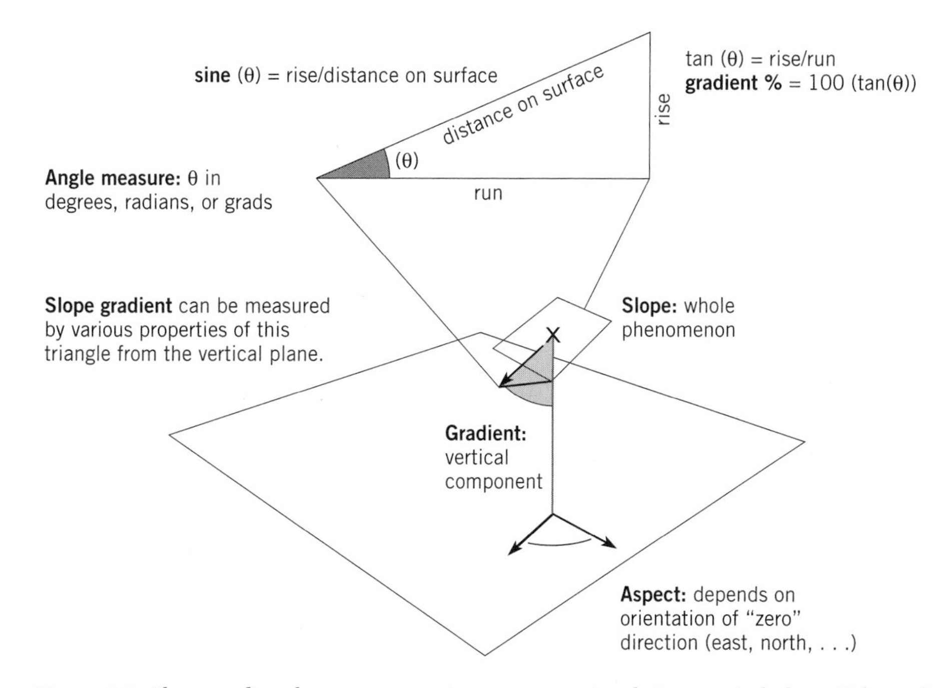

Figure 7-5: Slope gradient between two points creates a triangle in a vertical plane. Values of gradient may be given by some measure of the angle.

to infinity for 90°, far from a linear rescaling of angle. Twice the slope has a totally different meaning when dealing in these percentages. The halfway point between zero and the vertical is almost at the vertical—fairly meaningless due to the infinite value. The tangent percentage is firmly entrenched as the preferred measure of angle in the US and for highway engineering.

It is somewhat rare to measure slope as the sine of the angle (the rise over the distance traveled—the hypotenuse). The mechanical advantage of an inclined plane scales with the sine; hence it has a direct application in erosion models. Some software measures slope gradient on this scale (EPA Environmental Research Laboratory 1994). The sine measure sets the halfway point between 0° and 90° at 30°. As with tangent, sine is not a ratio rescaling of angle. Each of these measures is related by the relationships of trigonometry, of course. Each relationship has useful physical interpretations. The real solution is to provide the conversions between angles and the other measures to provide the appropriate value for each use. This requires distinct recognition of angles, cosines, sines, and tangents, just for a start.

Slope from Triangles The Triangulated Irregular Network (TIN) structure (introduced in Chapter 2) organizes the neighborhood relationships between a set of points as an exhaustive set of triangular facets. Each triangular facet has a constant slope, since it defines a plane in three-dimensional space. In some TIN implementations, the slope gradient and aspect are stored as attributes of the triangle. The triangle is essentially a device to establish the relationship between a specific set of points—a neighborhood. The slope properly belongs to the neighborhood (the facet between the points), not any of the points.

The TIN technique offers the capability to model the topological structure of the surface along with its measurements. The vertices of triangles should include all the critical points, like peaks and passes. With the flexibility of triangles, these points do not have to be approximated by regular spatial sampling. The network of ridges and courses should also provide the basic backbone for the edges of triangles, with more vertices inserted to locate them adequately. Finally, the triangles should represent areas of uniform slope (gradient and aspect) as closely as possible. For this reason, the slope can be treated as a uniform property of the triangular facet.

Slope from Matrices The more common representation for surfaces adopts a space-controlled method, usually a lattice of regular point samples, a DEM (as introduced in Chapter 2). Given the distance between sampling points, the slope can be estimated from the neighboring values in the matrix. There are two fundamental approaches to grid measurement frameworks: attributes stored in a grid can pertain to a point or to an area (see Figures 2-10 and 2-14). In the case of a DEM, it is common to use a point framework, considering the matrix to represent a lattice of lines rather than a coverage of pixels. Slope can similarly be considered as a point attribute, but it can also be applied to the facets between the points. This second alternative shifts the grid over by half a notch, implicitly. Some software produces point estimates of slope, while others make facet-based estimates; in both cases, the user manual is unlikely to provide sufficient information to tell the difference. Three different neighborhoods can be used to calculate slopes in a matrix (Figure 7-6). The point-based estimate can be derived from four direct neighbors along the rows and columns or from eight neighbors in all nearby cells including diagonal neighbors. An area estimate sees the cell as being bounded with four points at the corners of the cell. With the area estimate, the cell assigned the slope is not the same as the cells of

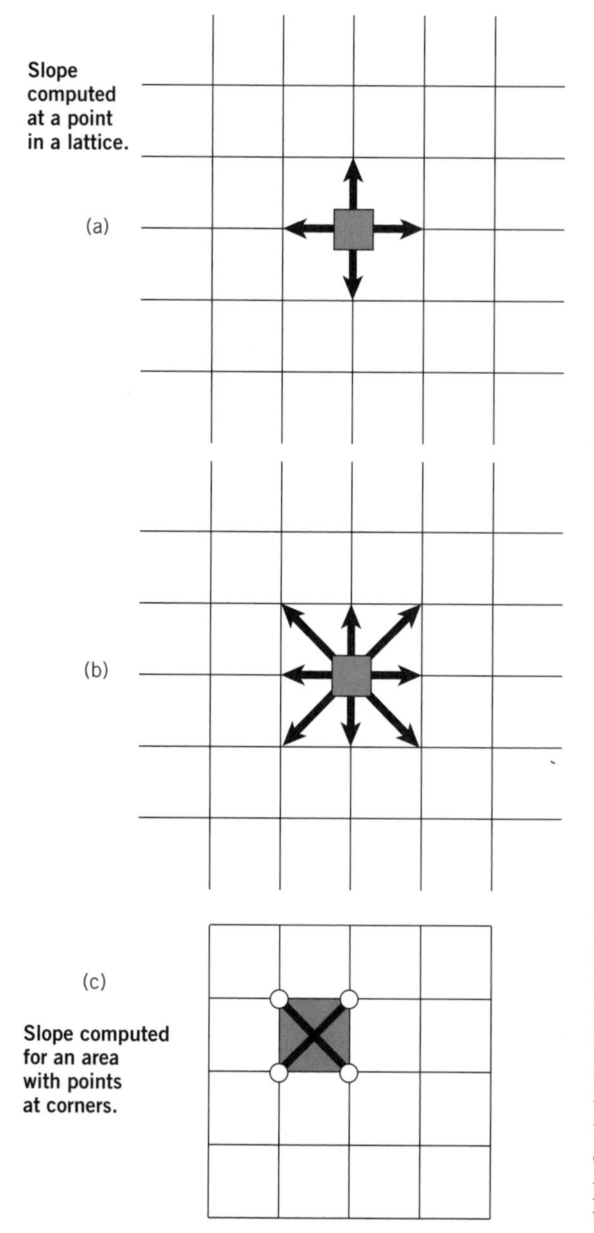

Figure 7-6: Neighborhoods in a grid for slope calculation: (a) and (b) concern point estimates; (c) makes an area estimate: (a) Four immediate neighbors along major axes of grid; (b) eight neighbors (a 3×3 kernel); (c) points with elevation values are considered corners of an area to triangulate. Four points create two possible pairs of triangles each with a distinct slope.

the original matrix. Thus, it is more common to treat DEM slope calculations as a point matter, assigning the slope to the same points.

With all arrangements of cellular neighborhoods, the estimates of slope are potentially ambiguous, since the four (or eight) points do not define a single plane. A slope calculation, then, requires choosing among possible relationships. The simplest point method computes all the eight adjacent slopes as distinct triangles, being careful to give the diagonal neighbors the longer distance: $\sqrt{2}$ times grid spacing. Then a dominance rule can choose the maximum slope as one of these eight. Another estimate of slope and aspect can be obtained by fitting the closest plane to the neighborhood. Equation 7-1 is one of many that fit a plane through the eight neighboring points (Horn 1981; Burrough 1986, p. 50). The terms *b* and *c* are essentially average gradients in the row and column directions. These two terms are combined trigonometrically to extract the vertical gradient component and the horizontal aspect component. The weight of two given to the direct (nondiagonal) neighbors in terms *b* and *c* counteracts the duplicate references to the diagonal elements between *b* and *c*. This plane minimizes squared deviations at the eight points, a common criterion for mathematical approximations but not the only possibility.

Equation 7-1: **Estimating slope by a best–fit plane**

Given a matrix Z with grid spacing S, the least square fit plane at z_{ij} can be written as:

$$z = a + bx + cy$$

where

$$a = (z_{i-1,j-1} + z_{i-1,j} + z_{i-1,j+1} + z_{i,j-1} + z_{i,j} + z_{i,j+1} + z_{i+1,j-1} + z_{i+1,j} + z_{i+1,j+1})/9$$

$$b = (z_{i-1,j+1} + 2\, z_{i,j+1} + z_{i+1,j+1}) - (z_{i-1,j-1} + 2\, z_{i,j-1} + z_{i+1,j-1}) \,/\, 8S$$

$$c = (z_{i+1,j-1} + 2\, z_{i+1,j} + z_{i+1,j+1}) - (z_{i-1,j-1} + 2\, z_{i-1,j} + z_{i-1,j+1}) \,/\, 8S$$

$$\text{Slope gradient tangent} = \sqrt{b^2 + c^2}$$

$$\text{Aspect angle} = \arctan\,(c \,/\, b)$$

$z_{i-1,j+1}$	$z_{i,j+1}$	$z_{i+1,j+1}$
$z_{i-1,j}$	$z_{i,j}$	$z_{i+1,j}$
$z_{i-1,j-1}$	$z_{i,j-1}$	$z_{i+1,j-1}$

Cells are indexed either in Cartesian form as shown here, or in row and column down and across. The direction of these coordinates influences the interpretation of the aspect angle.

The specific equations used to calculate the slope are not as important as understanding the use of neighboring values to obtain a result. Neighborhood operations seem particularly natural applied to surfaces, but they may be applied to all forms of attributes in many different measurement frameworks.

NEIGHBORHOOD OPERATIONS: THE SPATIAL COMPONENT

Neighborhood operations are more complicated than the overlay-based map combination operations described in Chapter 5, but there are many similarities. Overlay provides two values from different sources, and then some operation combines them to produce a result. In the neighborhood case, operations assemble the values within a specific neighborhood on a single source layer. The simplest ones define this neighborhood on the basis of distance, usually limited to a set of near neighbors. Slope, as defined above, is an example of such a neighborhood operation.

Tomlin (1990) divides neighborhood into immediate and extended. From a practical programmer's approach, such a distinction is quite useful. However, the purpose of some of the extended operations does not differ from the purpose of immediate ones, so I will group them with the neighborhood operations treated in this chapter. Other extended neighborhood operations serve quite different purposes. They are distinguished from the near neighborhood operations because they must operate iteratively, using relationships between neighboring values heavily in each iteration. Despite the use of neighborhood information in each iteration, there is a more global purpose. Iterative operations will be considered in Chapter 8 with other more complex operations.

Neighborhood Construction

There are two ways to define nearness. Nearness can be considered solely on the basis of Euclidean distance, a reasonable approach for a number of physical properties and models. Using physical distance adopts the Newtonian simplicity of space as a void, a container of material objects. Nearness can also be defined in terms of the connection between objects, not their size. Each link in a network can count as one, as in giving directions on the subway. Philosophically, this adopts a view of space as full of connected objects whose relationships create spatial properties. The tools for neighborhood construction depend upon a choice between these assumptions about nearness.

To assemble a collection of neighboring attribute values, the representation scheme must also support some kind of nearness relationship. For some common representations, this can be achieved only by a transformation into another framework. Thus, some forms of representation, particularly the isolated frameworks—objects and isolines—must be transformed to prepare them for the neighborhood procedure (Chapter 9 will develop these transformations).

Raster Neighbors A raster representation offers by far the most flexible implementation of near neighbor operators. Neighborhoods are implicit in the grid, and the amount of information is strictly uniform for a given distance. The topological connectivity of objects can be treated, but on a cell-by-cell basis. Within this approach, the major variants involve the definition of adjacent cells (the four neighbors of Figure 7-6a or the eight diagonal neighbors of Figure 7-6b). The cellular system unifies the concept of neighborhood. Because the objects are the same size, topological neighbors are distance neighbors of the same degree.

As the example of slopes demonstrates, the implementation of neighborhood varies for the measurement frameworks that share the matrix representation. Still, the neighborhoods of point and area frameworks are reasonably similar, as long as they are not mixed. When working in the discrete geometry of a lattice, the basic unit distance limits the choices available for distances. If the pixels are 10 meters, there is no difference between a 10 meter neighborhood and a 13 meter neighborhood. Most matrix representations are square (though some rectangular systems persist and some hexagonal and triangular meshes have been discussed). No matter what the arrangement, there will be certain directions that will be favored. Given a square 10 meter matrix, diagonal neighbors are 14.14 meters apart. Early software either considered this to be one cell (=10 meters) or two cells (20 meters), both of which create trouble. The distance to neighboring cells comes in rather discrete and uneven jumps (Figure 7-7).

5.7	5	4.5	4.1	4	4.1	4.5	5	5.7
5	4.2	3.6	3.1	3	3.1	3.6	4.2	5
4.5	3.6	2.8	2.2	2	2.2	2.8	3.6	4.5
4.1	3.1	2.2	1.4	1	1.4	2.2	3.1	4.1
4	3	2	1	0	1	2	3	4
4.1	3.1	2.2	1.4	1	1.4	2.2	3.1	4.1
4.5	3.6	2.8	2.2	2	2.2	2.8	3.6	4.5
5	4.2	3.6	3.1	3	3.1	3.6	4.2	5
5.7	5	4.5	4.1	4	4.1	4.5	5	5.7

Range of distances			Number of cells (cumulative)
0	–	0.999	1
1.0	–	1.414	5
1.414	–	1.999	9
2.0	–	2.236	13
2.236	–	2.828	21
2.828	–	2.999	25
3.0	–	3.162	29
3.162	–	3.606	37
3.606	–	3.999	45
4.0	–	4.123	49
4.123	–	4.243	57
4.243	–	4.472	61
4.472	–	4.999	69
5.0	–	5.7	81

Zones of distance shown by gray levels

Figure 7-7: Distance from a single cell in a square matrix as a factor of the cell width. Distance relationships in a square matrix are shown by gray tones for ranges of distance. Advances are uneven, eventually approximating a circle, but slowly.

The departures from circularity seem to be worse near the exact multiples of the grid spacing. The search radius eventually becomes effectively circular, but it may have rectangular edges in certain ranges of distances. When a raster representation is used for a continuous surface, some operations, like slope, are computed on the most local neighborhood. The immediate neighbors approximate the infinitesimal neighborhood of differential calculus as closely as possible.

Implementation is fairly direct for raster searching. Each cell sends out a search radius and assembles the relevant values, and then the result is assigned to the central cell. Some packages provide special treatment for the three-by-three *kernel* or *moving window*. In some software, the searching process is actually implemented using the same module that performs the operations of overlay. The data layer is simply offset by a row and column so that the adjacent cells temporarily overlay on the central cell. The overlay procedure actually performs the required combination of the values.

However implemented, the raster neighborhoods yield a discrete list of attribute values from the relevant cells. This list can become aspatial, or it can retain the relative distance and bearing to the cell from the central cell. Unlike map overlay, which can be considered pairwise, these lists of attributes can be relatively large even for short search radii. In a square matrix, the first increment outward captures a total neighborhood of five, then nine, then thirteen, and so on (as shown in Figure 7-7).

Vector Neighbors Neighborhood for vector representations comes in the two forms described above: geometric and topological. With a topological representation, the more direct neighborhood is the set of adjacent objects. A neighborhood based on distance is harder to extract. Despite the direct access to the topological neighbors, comparatively few GIS packages provide a set of operations to calculate a new attribute based on the attribute of one's topological neighbors. One major exception involves networks. The topology of the network can be used to route flows and other similar operations. Most network applications do not use the simple neighborhood operations discussed in this chapter, they require incremental techniques discussed in the next chapter.

Vector representations allow a distance search radius of any size—not limited by the grid spacing. Buffers (introduced in Chapter 6) are a special case of near neighborhood construction. To reproduce the same results as a raster algorithm in a vector-based system would require many steps. One could isolate each coverage into separate layers for each category, construct the buffers around each, and then overlay them all. The resulting overlay would record all the raw material (each polygon would record a unique combination of all the attributes found within the search radius) without the discrete limitations of the raster representation. Instead of a single command, these operations could require elaborate bookkeeping to ensure that all the categories were handled.

Vector representations also pose difficulties for data management and combination rules. In the raster case, neighborhoods capture a specific number of cells for a given distance. In the vector case, there are no limits to the potential number of distinct objects found in a search radius. Perhaps for this reason, distance-based operations for vector representations are less developed than they are for raster representations.

Edge Effects In either the vector or the raster case, neighborhood operators are designed to treat a location surrounded with information. Unless the subject matter covers the whole surface of the earth, there will be edges to the distribution. The neighborhood search will detect that the locations desired extend outside the region covered. There are a number of ways to treat this eventuality. Some operations can still give a result using the neighbors detected, but others will not give a comparable result. In some cases, the result covers less area; a raster result would be reduced by the number of rows and columns on each side where the full search is not possible. In a vector case, the external region (and other polygons without attributes) may also influence the results. It may be necessary to extend the database beyond the edge to ensure that operations can be performed on the whole study area.

Refinements of Neighborhoods A simple neighborhood can be modified with spatial criteria, attribute-based criteria, or both. Spatially, the search procedure can be applied at two different distances, forming a ring neighborhood. Alternatively, the search could be restricted to some arc, such as the southerly exposure that might shade a solar collector. If overlay logic combines with the search procedure, many attribute-based permutations become possible. For example, a mask from some other source may be used to exclude cells that fit some criterion. The search can also be restricted to an uphill or downhill direction on a surface. Each of these variations may be used relatively rarely, but each has its place. The neighborhood construction should model the interactions of the phenomenon.

COMBINING NEIGHBORHOOD ATTRIBUTES: A TAXONOMY

Operations applied to values assembled in a near neighborhood fall into two dimensions, forming the cross-tabulation of Table 7-1. The column dimension contains the same general groups of rules that applied to overlay operations (Chapter 5). A single value can be selected to dominate the others, or each value can be used in some formula to which all the values contribute. There are also some examples of operations sensitive to interactions between neighboring values. Near neighbor operations are different from overlay operations largely because the measurements collected within a neighborhood come from a single source, hence from one measurement scale. The rows of the Table 7-1 are organized from lower to higher levels of measurement.

Nominal Attributes

Unlike overlay, which assembles its values from different layers, the nominal categories in a neighborhood all derive from a common system of classification. The same

TABLE 7-1: Operations on near neighbors

	Dominance	*Contributory*	*Interaction*
Nominal	Buffer Drop-line aggregation	Voting tabulation majority filter diversity, etc.	Edge detectors, Explicit combination
Ordinal (at least)	Max/min neighbor	Percentile	Profile, drainage
Continuous (aspatial)	Max/min neighbor	Sum/average	Edge detectors
Continuous attribute with horizontal measures			
Slope	Maximum slope	Best fit plane	
Distance weights		Smoothing, filters	Autocorrelation

Gray cells seem not to occur in practice.

three groups of rules defined for overlay operations can be applied to produce a result.

Dominance Rules Applied in a geometric neighborhood, a dominance rule for a nominal attribute simply defines a buffer operation. The dominant category expands to the edge of its search radius. The discussion of buffers in the previous chapter need not be repeated here, except to note that the procedures to produce buffers varied between raster and vector representations. In both cases, the categories were converted to isolated object form in order to construct the buffers. This simplifies the distance search procedure because only one dominant value is required.

When handling vector representations, one form of topological neighborhood operation uses a very simple rule. If the area attributes on either side are the same, then the boundary is not required. This operation, called drop-line aggregation or dissolve, was mentioned above in Chapter 4, though it requires a form of neighborhood to operate. While it is not exactly a dominance rule, its simplicity places it in the lowest group. There is no direct analogue in the raster representations, since the raster does not manage connected entities in the same manner. To a degree, a run length encoding (introduced in Chapter 3) compresses a raster representation in a manner comparable to dissolve, but only along the axes of rows and columns, not as a generalized neighborhood operation. Similarly, the recognition of "zones" in certain raster packages leaves the cellular representation intact.

Contributory Rules The simplest contributory rule for nominal data assigns the most common category in the neighborhood to the center of the search. This form of voting performs the nominal smoothing often applied in removing isolated pixels from remote sensing classifications. Figure 7-8 shows two results of a *majority filter* for a simple land cover classification. Each cell simply records the majority vote over its

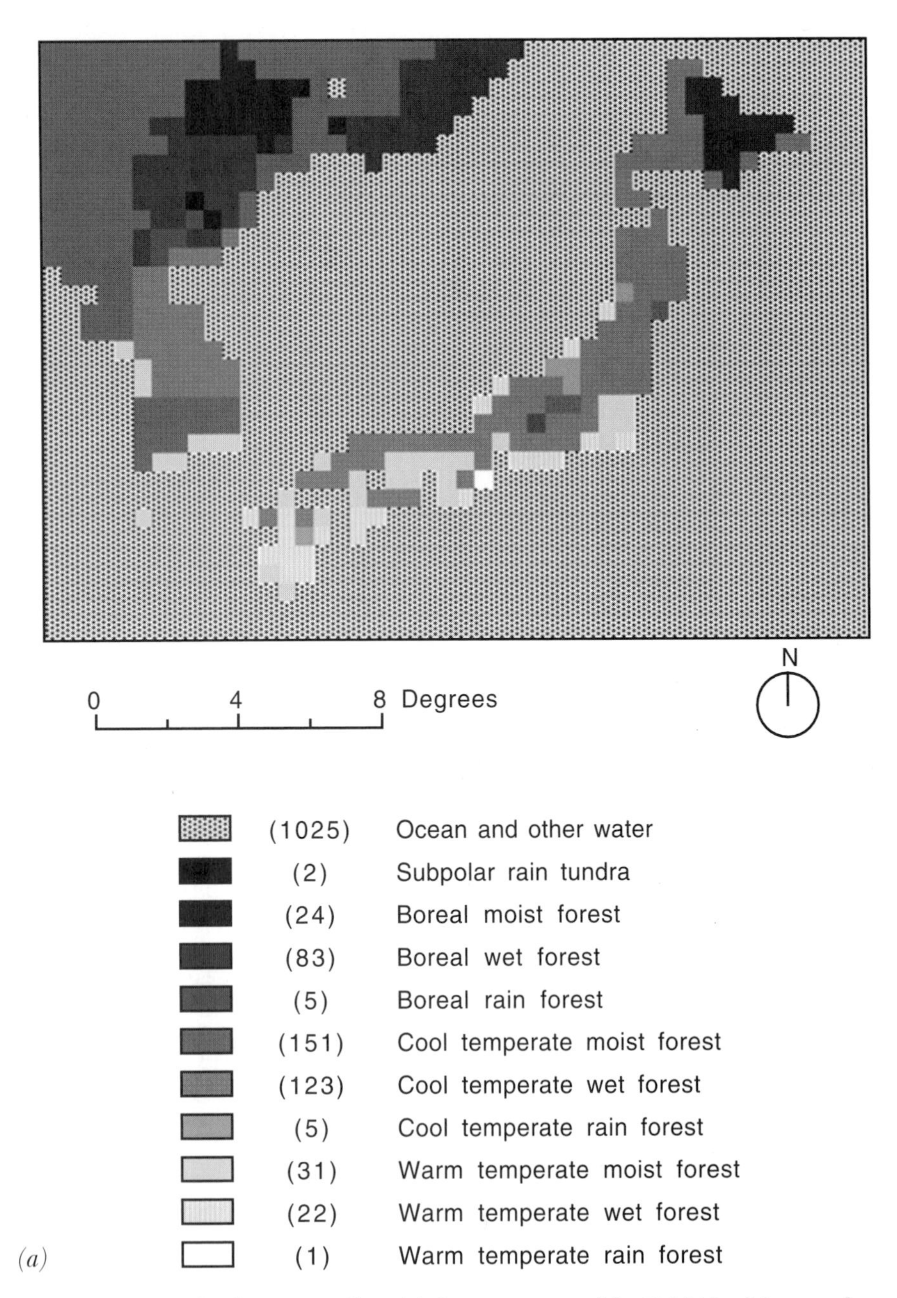

Figure 7-8: Example of a majority filter: (a) shows a portion of the Holdridge life zones dataset with half-degree cells; (b) shows the result after assigning each cell to the majority category within 1 degree; (c) shows the result of the same filter on a 3 degree neighborhood. Number of cells in each category shown in parentheses.

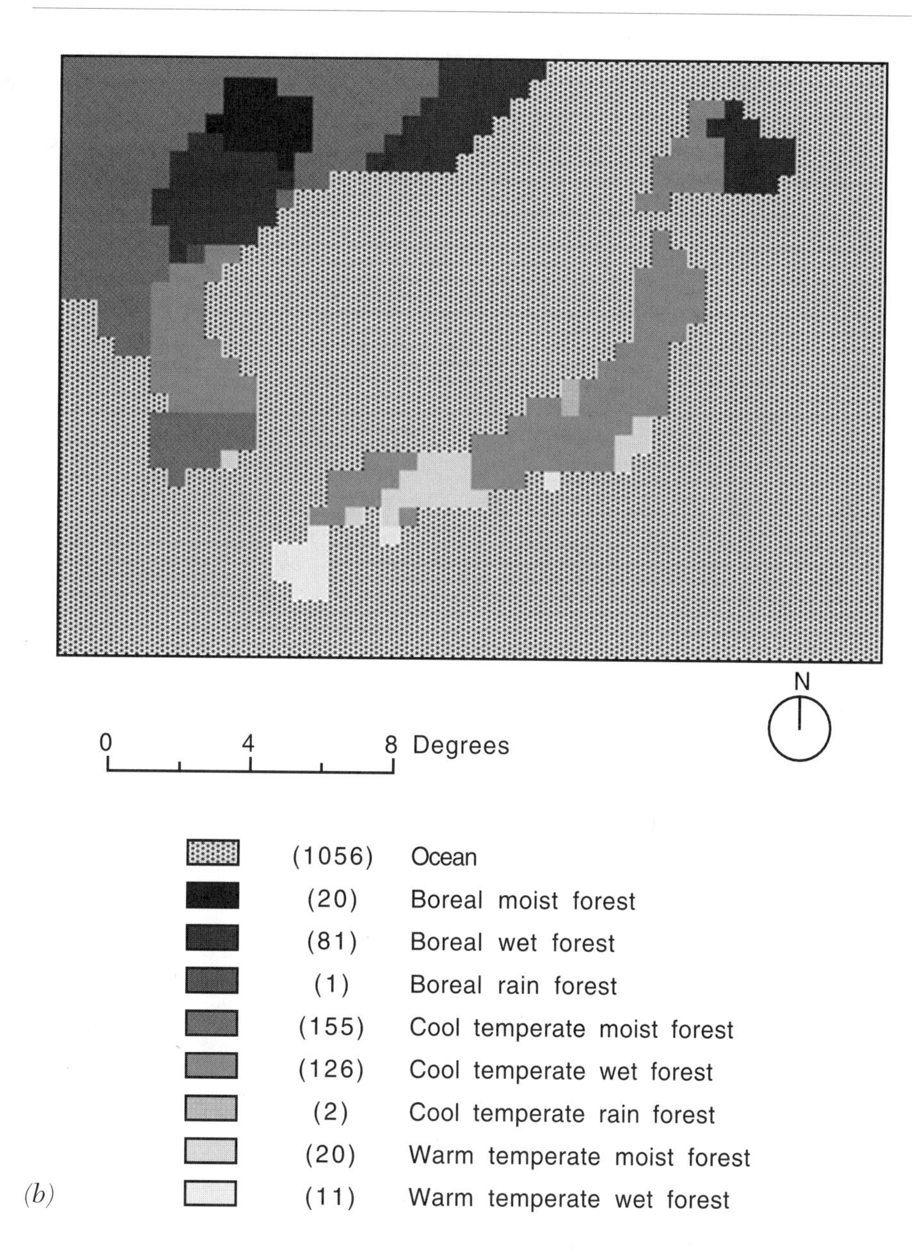

(b)

neighborhood. Notice that thin features disappear and the number of categories is usually reduced by this kind of operation. This kind of filter alters the scale of the classification system, and changes the measurement associated with the specific cell.

Neighborhood voting tabulation operations follow the logic of the voting tabula-

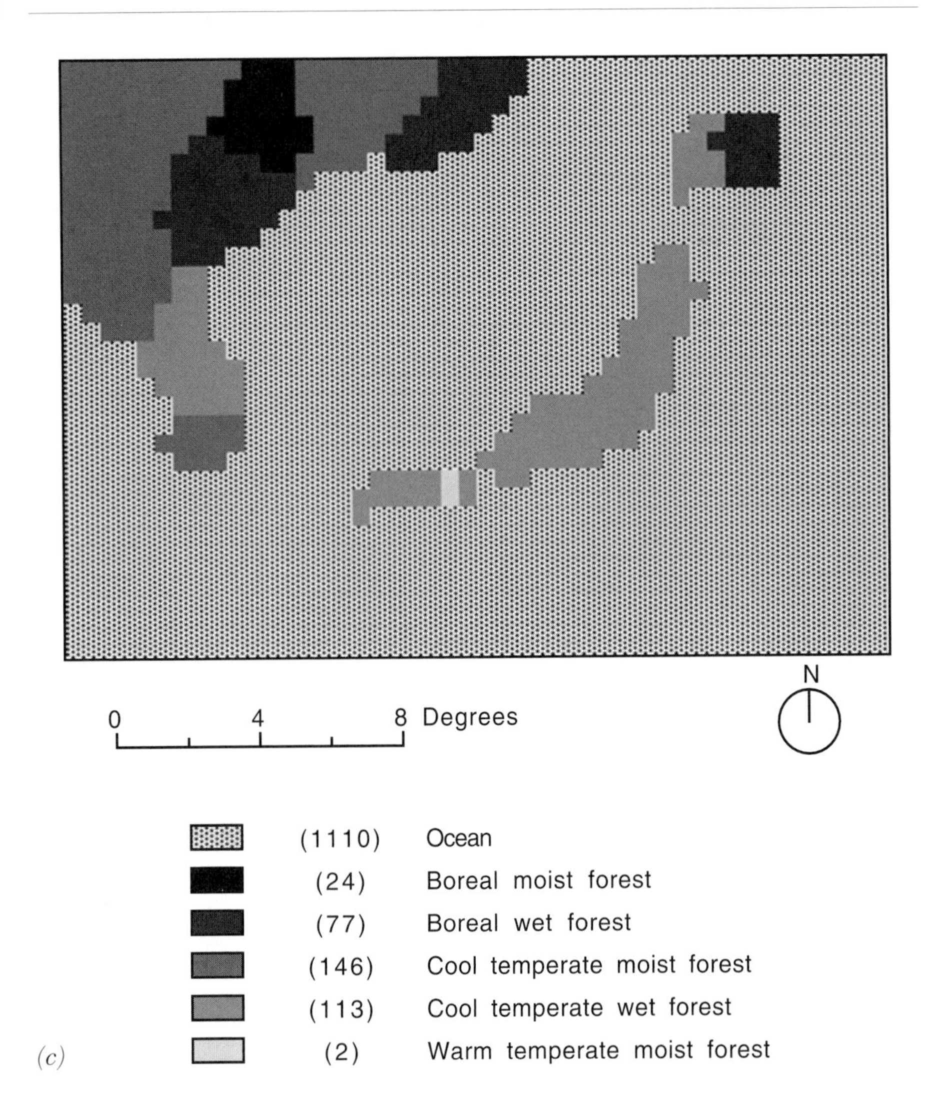

(c)

tions discussed in the overlay section, but with a significant new opportunity. In a neighborhood case, the central cell provides a reference. Counting the number of cells that share the central value gives a measure of homogeneity in the surroundings. These counts are usually converted to a proportion of the cells in the search radius (Figure 7-9). High values indicate the interiors of large blocks of one category; low values indicate isolated cells. Note that the discrete nature of a raster representation means that the proportion is not really continuous, particularly for limited neighborhoods. A somewhat similar measure counts the number of different values in the search radius (Figure 7-10). This provides an index of diversity that detects the edges of blocks that share the same value, much as the proportion filter. Unlike the propor-

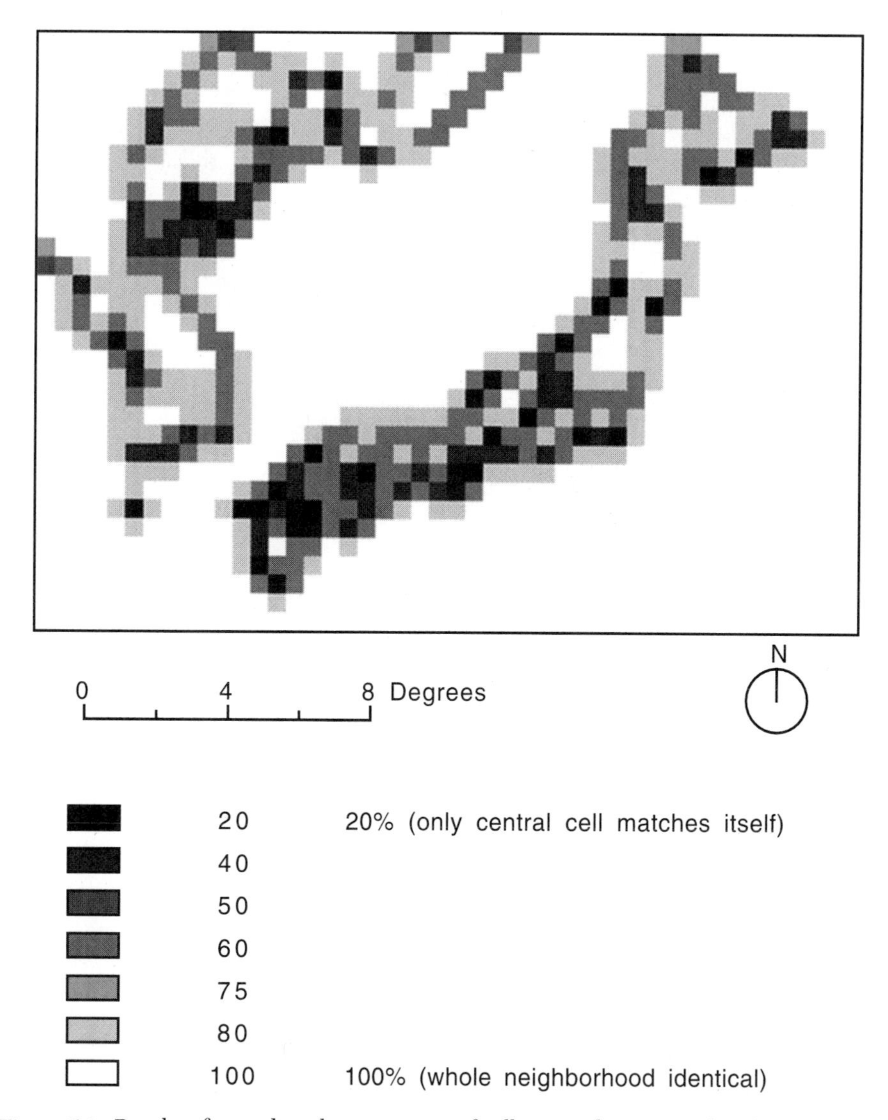

Figure 7-9: Results of recording the proportion of cells in a 1 degree search radius that share the central value (proportion filter) applied to the life zone data shown in Figure 7-8a.

tion filter, the diversity index records all the variety. Its values rise near boundaries of thin features. The diversity index can be applied in landscape ecology to model habitat. Run at different distances, these measures can reflect the different ranges of various organisms (owls may move farther than lizards). These various neighborhood operations begin to address the need to handle the *situation* (surroundings) instead of simply the immediate characteristics of a *site*.

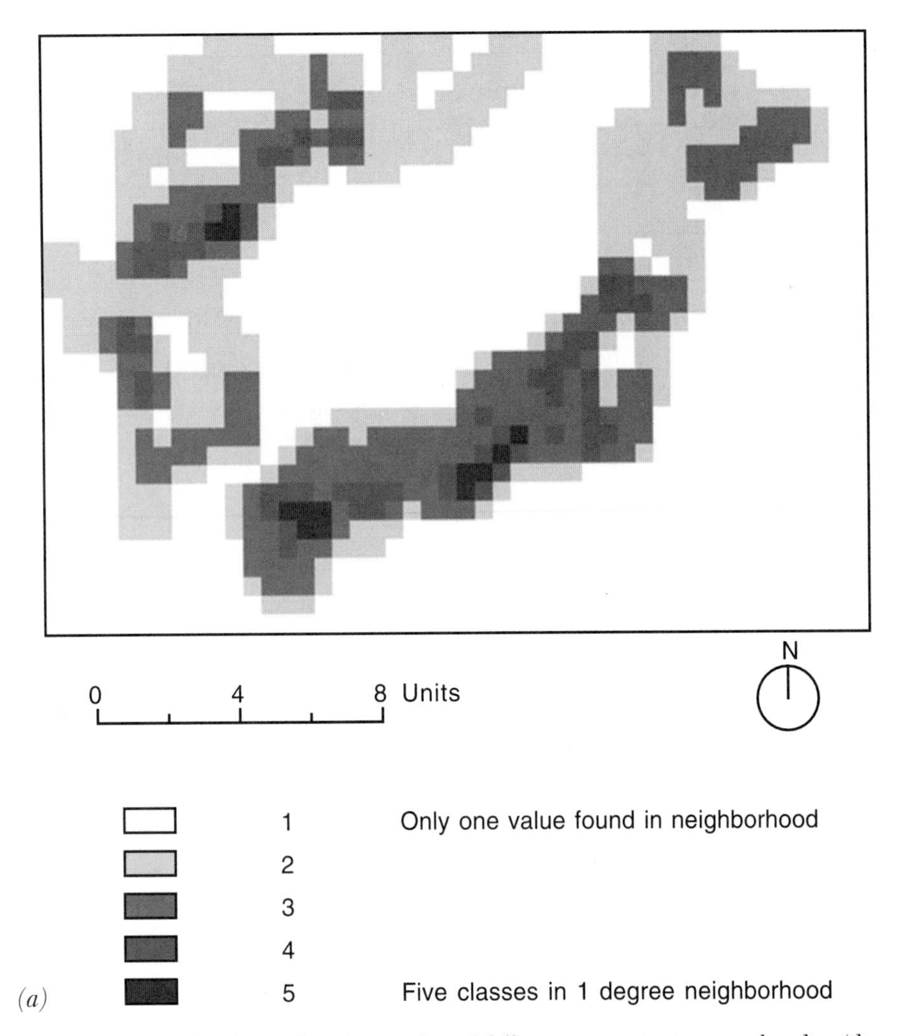

Figure 7-10: Results of recording the number of different categories in a search radius (diversity index): (a) 1 degree search radius; (b) 5 degree search radius.

Diversity scores are most easily implemented in a raster representation or in a choropleth framework. In the choropleth situation, neighboring polygons may share identical categories. The *join-count statistic*, a basic measure of spatial autocorrelation, summarizes the diversity of neighboring polygons. The similarity of raster and choropleth comes from the space-controlled nature of the attribute assignments. By contrast, the polygon objects created in the categorical coverage framework are, by definition, the largest contiguous unit of their category. Hence all their topological neighbors will be different. A count of different neighbors will not show the geometric neighbor structure detected by the raster diversity count. The area of the polygon gives an accurate measure of the size of patches, which the raster representation does not handle directly.

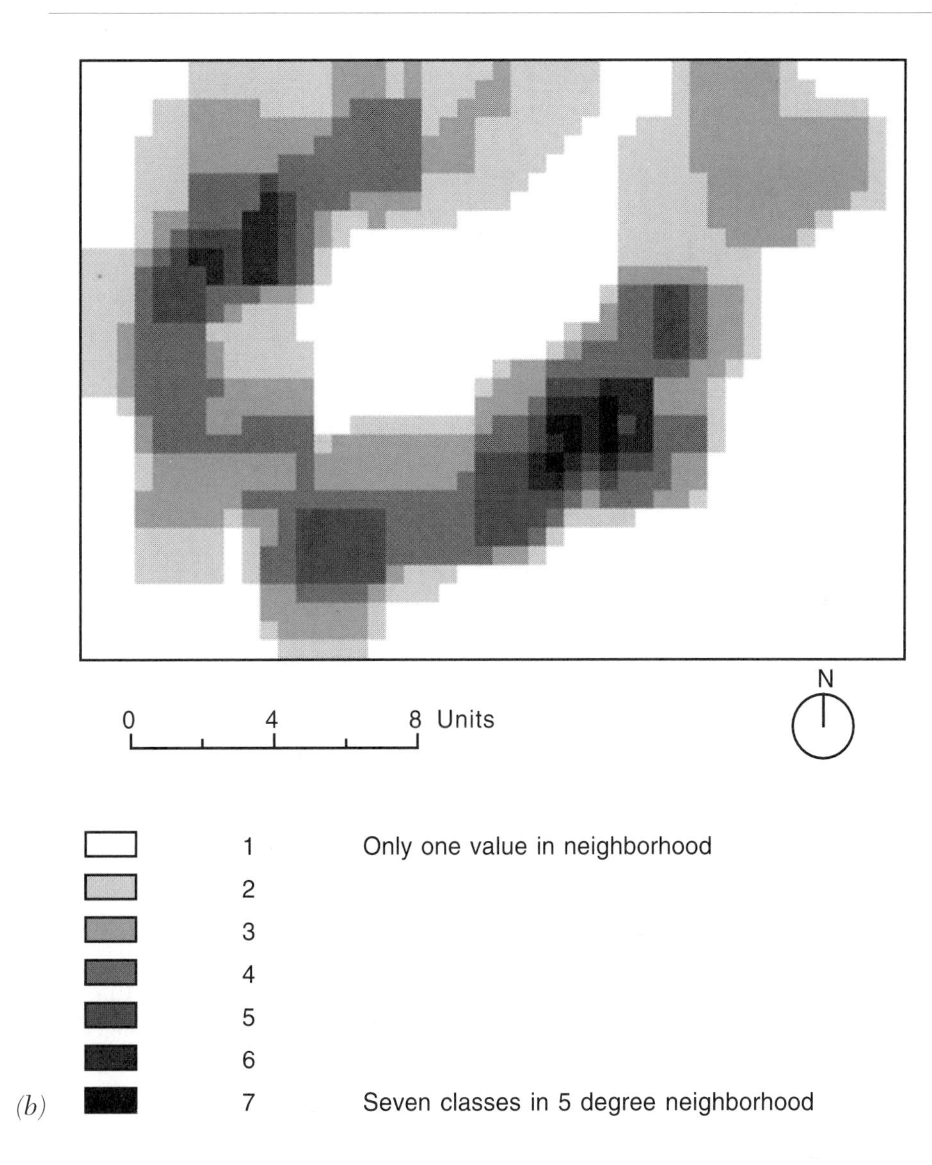

(*b*)

Interaction Rules The final kind of rule addresses interactions between the categories found in the neighborhood. Certain combinations of categories may be more important than a simple count of different categories. The most direct method gives a value for specific combinations of the original categories. For example, the combination of water near a forest gets a one, water near a wetland gets a two, and so on. These interactions remain manageable if limited to specific pairs. A general solution would seem to imply carrying out the factor combination method in the neighborhood context. The number of category combinations increases even faster in a neighborhood context than it does in overlay. At modest search distances, a large number of category combinations can occur. Both vector and raster software have the same difficulty in presenting such a messy situation to a user for a solution.

The contributory tools described above as voting tabulations can recognize edges in a mosaic of categories. A full–strength edge detector must move beyond counting up neighbors to include spatial arrangements. Spatial interaction rules are more common than attribute interaction rules. The specific location of the various values around the central cell become critical in tracing a boundary through an image. This information forms a major part of image processing techniques (Rosenfeld and Kak 1976; Lillesand and Kiefer 1994). For example, Rosenfeld's thinning algorithm works on binary images of linework. The center cell of a three–by–three neighborhood is changed to nonline if changing its value will not disconnect the other line cells from each other. These techniques depend upon the regular spatial structure of the raster to manage the operation. Of course, a vector representation does not need an edge detector that operates from the attribute values. The edges are at the core of the representation, and they can be treated geometrically, independently from the attributes.

Operations Based on Ranking Attributes

The next set of operations treats ordinal data or continuous data for its ordinal characteristics. The simplest rules, as in the overlay case, give dominance to some single value. With ordinal data, a dominance rule can select the highest or lowest value within the neighborhood. This value is then assigned to the origin of the search. By performing both maximum and minimum, a range of local variation can be obtained. This kind of dominance rule can easily apply to networks or polygon neighborhoods. For example, a link in a network that restricts some class of traffic (such as dangerous cargo) will constrain any route that passes through it.

The basic contributory rule follows the majority filter in its motivation. When processing nominal attributes, the majority filter counts the number of neighboring cells that share the same value. The basic topology of surfaces introduced at the start of this chapter works at the ordinal level through convergence and divergence. With an ordinal scale, one can ask the proportion that is below the origin of the search. On a topographic surface, the lines of ridges and courses appear with high and low values respectively after this operation.

More sophisticated detection of surface structure involves analysis of the interaction of neighboring values (Mark 1984; Marks and others 1984; Band 1986). One method to categorize topography is to distinguish shapes of profiles based on ordinal relationships. A set of possible profiles can be drawn, ranging from peaks (all neighbors are lower), ridges (sequence of lower, higher, lower, higher), regular midslopes (lower, higher), and so on. A cell would be coded according to its profile type, an interaction rule.

Continuous Attributes

Near neighbor operations, like overlay operations, assemble a collection of attribute values and produce a result based on rules. In the case of map overlay, the attributes come from different sources whose attributes could only be associated by collocation. Hence, much of the trouble with continuous attributes arose from the lack of a common scale for measures between layers. In the case of neighborhood operations, the attributes all come from one source, so that this objection disappears.

The ordinal operations discussed in the previous section pertain to all forms of continuous data, because all continuous scales can be treated as ordinal. The discussion of continuous attributes will be divided into two groups of operations that go beyond those discussed above. The first will cover the simpler case where the neighborhood simply collects the values and the set of scalar values is treated aspatially. The second group will deal with the inclusion of horizontal measures attached to each of the attribute values, considering them as multicomponent attributes.

Because the vector data model implies attribute control at some stage of measurement, continuous measures attached as attributes are controlled in their variation. This control limits the ability to model continuous changes in neighborhoods. For example, a choropleth map, as an indirect measurement, has been aggregated to some degree. In a choropleth framework, it may be difficult to separate the breaks induced by the arbitrary boundaries of collection zones from variations in the attribute distribution. Hence the raster implementation for neighborhood operations provides a clearer explanation of the procedures.

Aspatial Treatment of Continuous Attributes in a Neighborhood Dominance operations that treat continuous attributes must be based on ordinal properties discussed above. The new contributory rules for continuous measures mobilize the arithmetic not available for ordinal scales. While a simple sum or average may raise questions in an overlay, these operations make sense for measurements from a common source. Here the distinctions between different measures at Stevens' ratio level become important. The sum of a derived ratio (like density) is not meaningful. While a sum of an extensive measure is permissible, interpretation requires some caution. Even with distance, the archetype extensive measure, the sum of the elevations around a point does not produce another elevation. By contrast, a sum of counts within a distance is more useful. Such a sum could record the total number of schoolchildren who could walk to a given site. Such a result should be carefully labeled, because any further transformations cannot treat the sum of all these measures as the total population. The population of schoolchildren will be duplicated, and the total will far exceed the original quantity. An average is gentler in its effect. Counts and extensive or intensive ratios can all be averaged. Near the edges of data coverage, the average can simply be computed on fewer cases, so that the coverage need not shrink inward. An average is just one kind of smoothing that gives equal weight to all surrounding values. If closer points are more important, some form of distance weighting (see below) may be appropriate. Another contributory rule can move a cell away from its surroundings, hence increasing local contrasts. For instance, in a gray-scale image, a neighborhood operation can calculate the mean brightness of the neighbors. The central value may be above or below that mean. If the value is moved away from the mean (either darker or lighter) by enhancing the local difference, then the largest changes will occur at the sharpest edges. This procedure is often a part of the hidden steps inside the operations of line recognition algorithms.

A few methods use interactions between the neighboring values. For example, raster display devices include an anti-aliasing feature, a form of edge-sensitive neighborhood operation linked to brightness of the pixels. Similar edge enhancement methods can be applied to any form of surface data. More advanced image processing techniques use edge detectors sensitive to local interactions, not just contributory numerical properties.

Continuous Attributes with Horizontal Measures Though mathematically suited to the numbers, the aspatial neighborhood operators for continuous attributes are of limited usefulness. Since the information is all from the same source, combining local values will usually reduce the resolution. By contrast, the neighborhood operations that incorporate spatial information can, under some circumstances, discover new relationships. The simplest case of an operation that involves both the attribute and its horizontal measurement is slope calculation. The discussion earlier in this chapter presented various methods for slope calculation. One rule computes eight values and selects the steepest gradient. In this case, the attribute difference between the target cell and each of the eight neighbors is not enough. The distance between each had to be known in the same metric as the elevations. This measurement comes from a scale factor for the whole matrix, plus the relationships between neighbors in a matrix. In a TIN triangle, the relationships are also specified, using a less discrete approach.

The procedure of fitting a plane to the surface involves some form of a linear equation. Each neighboring point contributes independently to the equation. In some software, only two or four neighboring elevations are used, but others use eight (as shown in Equation 7-1). The fitted plane will produce *lower* gradients than the dominance rule at peaks and along ridges. The plane fit at a peak may be close to flat, despite substantial slope in all directions. Mathematically, the differential at a peak is zero, but that has little practical value. One millimeter off the peak, there is a well-defined slope. Away from peaks and ridges, slope gradient calculated by fitting a plane will often be *greater* than that calculated by the dominance rule, when the aspect is not exactly on one of the eight directions of the neighbors used in the dominance method. Some software packages use the term `maximum` for this dominance rule, but do not be fooled. It is simply the use of a maximum rule among discrete choices, not the maximum under another set of assumptions.

The topographic slope gradient can be reported as an angle, as a tangent (rise/run), or as a sine (rise / distance traveled) as discussed above. However, this depends upon measuring the rise and the run in the same units. The tangent is dimensionless (distance/distance) when the scale is correct. Conversions to angles will be invalid if the scale is misspecified. It is quite possible to represent surfaces of phenomena that are not measured vertically in the same units as horizontal distance units. Weather maps show barometric pressure, temperature, and other properties. We can also construct surfaces that describe land rent in dollars per month or biophysical oxygen demand in molecular units per second, and many more. Each of these surfaces has a slope that can be estimated using these procedures, but the result must remain in original units per distance, using the tangent form. Conversion to angles is meaningless. The concept of "distance traveled on the surface" cannot be expressed in distance units, so the sine form is not appropriate either.

The calculation of slope is just one example of a rich set of operations that involve some form of horizontal measurement. Few of these use a dominance rule. It is much more common to introduce distances and bearings into the contributory rules of smoothing and filtering. Instead of simply adding up all the neighboring values, it often makes more sense to compute a distance-weighted average or some development based on the same logic. Physical forces such as gravity decrease with the inverse of distance squared, so this often serves as a starting point, though the gravity model is not the only basis for spatial interactions. Some filters can accentuate certain directional trends in the data or enhance edges using the orientation of the neighbors as well as the value (developing the simple edge enhancement discussed above). These filters may cross over from strictly contributory calculations into the use of interactions between adjacent values.

Another form of interaction involves the relationships between values of the surface at different distances. A general rule of geographic surfaces paraphrases *Tobler's First Law*: "Everything is related to everything else, but near things are more related than others." (Tobler 1979a). A number of statistical procedures, generally called **geostatistics**, have developed measures to assess the degree to which Tobler's Law seems to apply. For example, local differences in value, plotted against a distance axis, trace a curve called a semi-variogram (Burrough 1986). The summary measures, such as **spatial autocorrelation**, provide a single value for the whole region (the subject of Chapter 8), but these techniques imply neighborhood operations at the base. Few operational GIS packages include even the summary measures, let alone the possible neighborhood interaction rules.

DATA QUALITY APPLICATIONS OF NEIGHBORHOOD OPERATIONS

The operations presented in this chapter provide tools to study some important aspects of data quality, particularly those based on internal evidence such as logical consistency. A surface representation implies some degree of continuous behavior in the attribute, thus an upper limit to the local differences between adjacent measurements. For instance, a slope map was computed for a matrix of elevations entered by a student project (Figure 7-11). Some huge slopes appear in a suspicious pattern around a single cell. This is near a gravel pit outside Issaquah, Washington, so some

Geostatistics: Branch of statistical estimation concentrating on the incorporation of spatial measures, particularly distance and neighborhood, into models.

Spatial autocorrelation: Degree of correlation between a surface value and the values of its neighbors; propensity of spatial data to vary smoothly with distance.

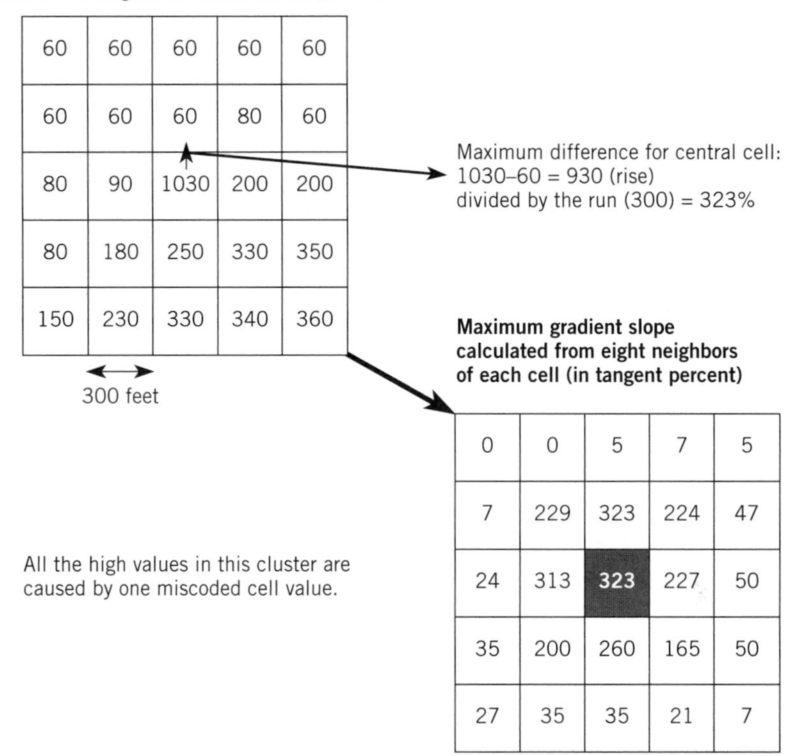

Figure 7-11: Use of near neighborhood calculations for local consistency checking: Maximum slope gradient picks out cells whose values differ from their neighbors. Data source: elevation matrix near Issaquah, Washington, entered by a class project; central cell should have the value of 130, not 1030. Slopes over 100% are quite unlikely in this glaciated terrain.

steepness might be reasonable. However, on inspection, the elevation of the cell was miscoded by clerical error and displaced upward by coding 1030 instead of 130. Perhaps this is a gross and simple error, but this elevation value occurred correctly a few kilometers away. This error had to be detected by a neighborhood operation, not by a simple check of valid values.

For categories, a neighborhood operation can check for prohibited or unreasonable combinations. For instance, glaciers and permanent snowfields are less likely to be adjacent to residential or agricultural land uses than tundra and barren lands. Beaches should be adjacent to some form of water, and so on. Vector software based on the topological model should provide easy access to the raw contiguity information because each chain has left and right identifiers. However, most major GIS packages provide access to this information through a relational database, and relational databases find it difficult to deal with ordered pairs. Consequently, it takes significant labor to construct this relationship.

SUMMARY

Neighborhood relationships play an important role in understanding surfaces. The calculation of slope demonstrates distinct approaches to surface representation. Neighborhood operations require two components: a neighborhood and a combination rule. Physical distance usually provides the basis for neighborhoods, but topological connections can provide similar relationships. Once the attribute values in a neighborhood are assembled, the general classes of rules discussed for overlay (dominance, contributory, interaction) apply. Since the neighborhood assembles values from a single source, the concerns differ. Level of measurement plays an important role in choosing a rule. In addition, some computations use the spatial measurements not available through overlay.

CHAPTER 8

COMPREHENSIVE OPERATIONS

CHAPTER OVERVIEW

- Develop iterative operations that use neighborhoods to reach more distant goals.
- Discuss linkage between location-allocation methods and GIS.
- Consider different perspectives of statistical analysis and GIS.

The last set of operations to consider are those that are not confined to limited neighborhoods. These operations are the most diverse group presented so far. Some are simple summaries that reduce the spatial information to a single report, while others create detailed spatial descriptions of their own. In many applications, a goal can be phrased in deceptively simple language. For example, it sounds simple to ask for the minimum distance path to visit all nodes in a network. There may be strong pressure to produce such an answer. After all, it represents the least-cost solution for routine maintenance and other needs. Yet, mathematically, this *traveling salesman problem* and other related problems are extremely costly to compute. Some seemingly similar problems have quite attainable answers, if attacked in the right manner.

As a common thread, this chapter deals with comprehensive measures. In all previous chapters, the result of an operation could be determined in some kind of restricted locality; the rest of the database did not matter. In Chapter 4, the operations were the most local, restricted to one map, entity by entity. Chapter 5 introduced multiple maps, integrated through overlay, but the spatial scope remained restricted. Chapters 6 and 7 expanded the spatial scope to neighborhoods, but the operations produced an answer at a location without needing to know the result elsewhere. The operations discussed in these previous chapters include most of the functions provided by current GIS software systems. Comprehensive operations now complete the progression from the simple to the complex.

The diverse operations discussed in this chapter are also unified by another thread. Various critics assert that these operations should be better handled by GIS software. This chapter will try to show how a GIS can address some of these issues and how the nature of spatial data impedes solution of others.

There are three groups of operations covered in this chapter, and because they are so different, there will be no summary. The first two groups follow directly from the neighborhood operations of the previous chapter. Iterative operations are extensions of neighborhood operations in which a result at one place can propagate to influence the results elsewhere. However, the decision at each spot can be made on the local information. These iterative solutions may take substantial computing, but they are feasible for reasonable applications. The second, tougher group includes many seemingly different problems whose solution can be restructured into a general set of location-allocation models. The strategy for these problems is often to devise an iterative procedure that leads to a practical solution, though perhaps not the optimal one. Following this section, there is a warning about types of graph problems whose solution may not be attainable. The chapter ends with a short reference to operations that summarize the whole distribution, usually in the form of statistical models. These operations may be completely aspatial or include some spatial measures in developing the model.

ITERATIVE OPERATIONS

Some of the most interesting operations on geographic information do not fit into any of the groups presented in previous chapters. As mentioned in the previous chapter, Tomlin's extended neighborhood category provides the next logical step. But some operations change more than the scope of the neighborhood: they also change the internal logic. The first set of operations covered in this chapter derives from an iterative application of the near neighborhood operations presented in Chapter 7. Although many are possible, only a few are in common use. The following sections describe viewshed operations that compute visibility on a surface, cost accumulation that propagates cost to neighbors, drainage that accumulates flow downhill, and network operations such as shortest path.

Viewshed

Intervisibility—predicting what sites can see each other—is important to a number of different applications. Some communications systems, particularly those that use FM radio transmissions, are limited to line of sight. The providers of cellular telephone services or the proposed personal communications networks need to determine the areas served by transmission facilities and then need to locate new sites to serve the blind spots. A similar analysis applies to military or border control authorities. While each of these uses contributed, the calculation of intervisibility was developed early in the GIS period to model the impact of forest harvest on scenic beauty (Travis and others 1975). Each of these widely different purposes can be served by much the same tool.

The area visible from a point can be modeled by radiating a set of rays outward. If the situation involves complex three-dimensional forms with reflection, refraction, and other optical effects, then the process of **ray-casting** can become one of the most complex computational procedures known. However, geographic intervisibility problems occur on a surface, which is much less complicated. The surface provides an ordering to what is seen and not seen; near things can block far things. The basic geometry of a line of sight can be shown by a diagram in two dimensions; distance along the line of sight plotted against elevation (Figure 8-1). Each object seen sets a new threshold of the vertical angle along that line of sight. Any object farther away

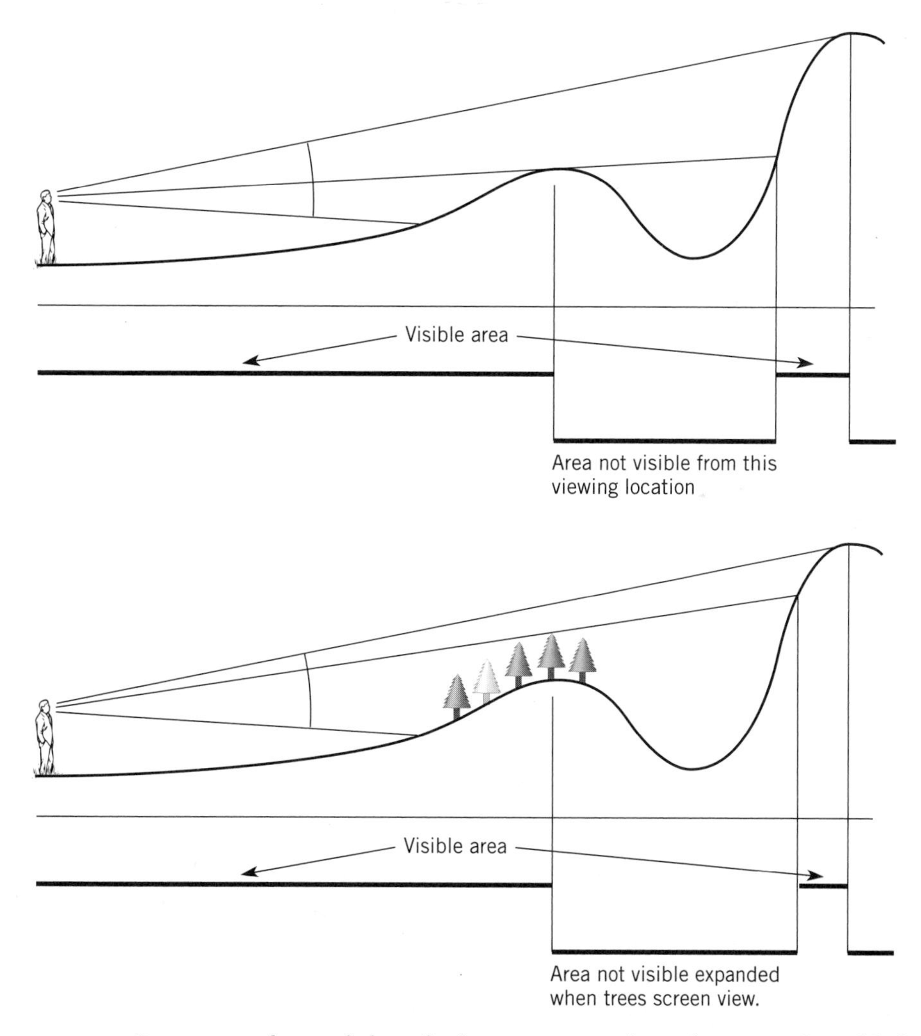

Figure 8-1: Cross-sectional view of a line of sight as it intersects the surface. Near objects block objects that fall below the previous angle. Obstructions on the surface can be modeled by increasing the height of the surface.

Ray-casting: Computational technique used to simulate a visual scene with optical effects, variations in light sources and other effects; usually developed for arbitrary 3-D objects, not just a single surface.

must be at a higher vertical angle. Thus, an iterative procedure works outward from a vantage point, carrying along the solution from previously seen locations.

The geometry of visibility along a line of sight is simple. The locations seen can be arranged to form a logical panorama around the vantage point. New locations along somewhat different lines of sight may need to be interpolated along the horizon. This *horizon line* can be updated as the processing moves farther away. It records the highest angle of view, a form of a dominance rule. A similar procedure applies to triangles, but it is most commonly programmed on a grid (Travis and others 1975). The surface is then marked as "seen" or "not seen," though it may also be possible to characterize the angle between the surface and the line of sight (the *exposure angle*). A glancing view may not reveal much scenic detail, though it may provide adequate cellular telephone service. With triangles, a part of the triangle may be seen, which gets tricky to represent within the triangle structure. Cells are simply marked "seen" or "not seen" based on the central point.

The normal binary result is often called a *viewshed*, meaning the area seen from a given vantage point (Figure 8-2). This concept is then extended to handle the area seen from multiple vantage points, simulating the view from a given highway or river. The individual viewsheds can be combined using a dominance rule (one sighting is adequate), or the individual viewsheds can be treated as votes to tabulate for the most visible elements of the landscape. A viewshed result (Figure 8-3) was used to construct a case for acquisition of development rights for a given parcel because it was seen so many times from a recreation area. The tabulation of number of times seen could be done using overlay logic, but it is often built into the visibility procedure.

For some purposes, the viewshed satisfies the analytical goal, but some other applications might seek to design the network of the fewest radio transmitters to cover a given area or something more complex. These operations require the logic developed in the next section for location-allocation in combination with the viewshed algorithm.

Viewshed calculations are strongly dependent on the choice of the height of the observer. Moving from 2 meters to 20 meters extends the horizon dramatically (Figure 8-4). Furthermore, errors in the surface can produce very severe consequences on the viewshed. Recent studies based on error simulation indicate that errors tend to reduce the viewshed (Fisher 1993).

The ordinary DEM may not capture the information required for a realistic model for all visibility applications. Human vision cannot see through vegetation such as trees; yet, these trees may be effectively transparent to FM radio transmission. Most viewshed algorithms provide the ability to add obstructions on top of the topographic surface (Figure 8-5), making these procedures some of the most feature-laden in the GIS tool kit. Often the obstructions and other parameters come from the most flimsy of measurements. For instance, all trees are deemed to be so many meters tall. Certainly such an assumption can help illustrate the impact of a radical change such as a clear cut in a forest, but the accuracy of the measurement of obstructions must be similar to the accuracy of the elevation values. Traditionally, photogrammetry has tended to take the highway builder's view of topography. Contour maps traditionally map the

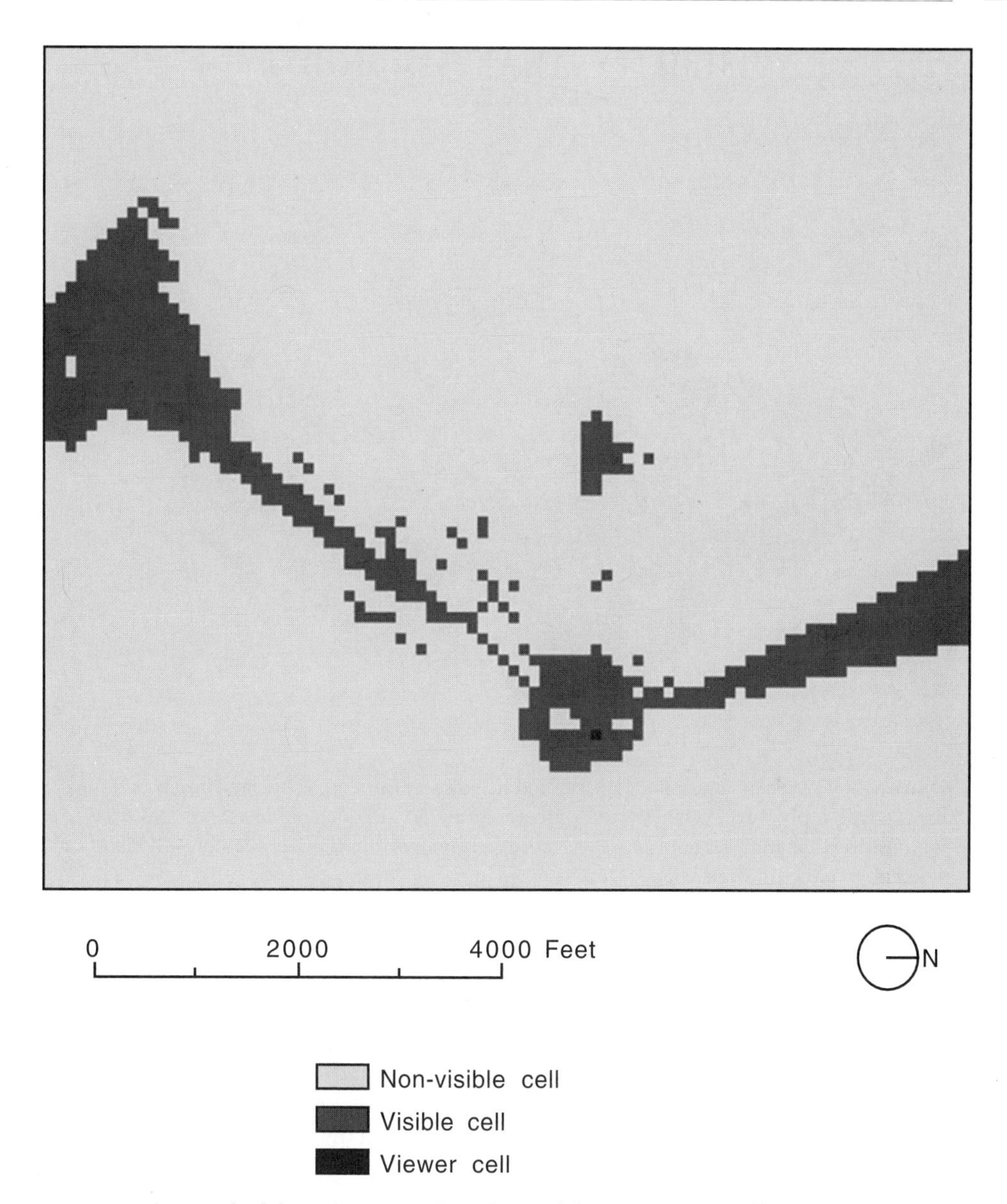

Figure 8-2: Viewshed from a position along the road from Preston to Falls City, Washington. Calculated without vegetation at a height of 2 meters.

forest floor even when it is not visible at all. Some recent databases, such as the French BD-Topo, will provide accurate measurement of the elevations at treetops and rooftops, the obstructions for a viewshed analysis.

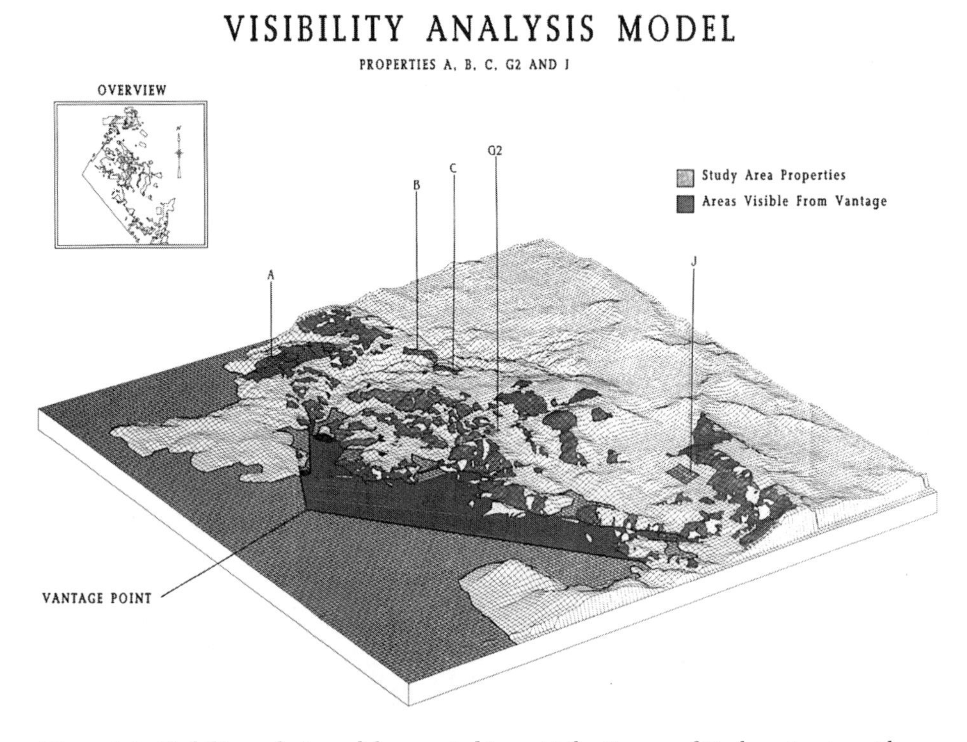

Figure 8-3: Visibility analysis model generated to assist the Bureau of Reclamation in settling a dispute with a property owner (of the parcel marked A). A series of viewsheds was used to demonstrate that parcel A could be seen more frequently from the New Melones Recreation Area. (Source: ESRI Maps book 1991).

Cost Accumulation

Viewsheds, as well as the neighborhood operations described in the previous chapter, rely upon physical distance as the measurement of separation. Distance is often not the real subject of analytical interest; it acts as a surrogate for some other measure of separation. An ambulance dispatcher cares about elapsed time to a particular location, not the physical distance traveled. A highway engineer aims to minimize construction cost in choosing different corridors. An environmental critic may counter with a highway corridor to minimize environmental impact—the original problem addressed by McHarg (1969). Overlay, by itself, does not solve the problem of locating a highway with minimum environmental impact. The highway cannot simply occupy the least impact locations, because it must also link origin to destination inside a host of geometric constraints.

The common thread in cost accumulation problems involves differential rates of movement. People have been solving such problems approximately for centuries,

Figure 8-4: Viewshed from same position as Figure 8-2 at a viewing height of 20 meters. to stimulate a tall sign.

using experience and knowledge of local conditions. The wind and current diagrams of the Solomon Islanders (Blakemore and Harley 1981) are a Stone Age example of a minimum cost solution. These problems can be solved formally using an *accumulated*

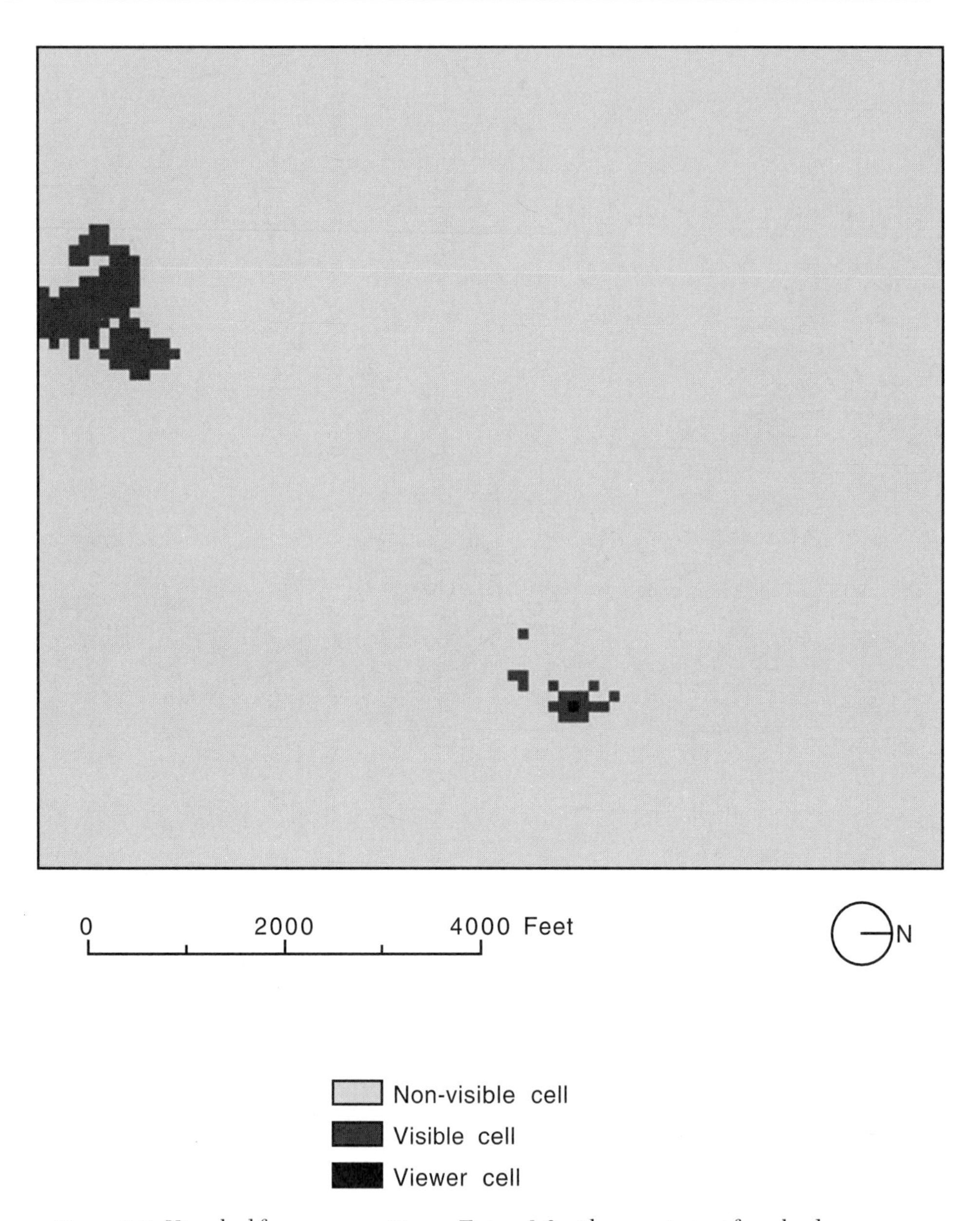

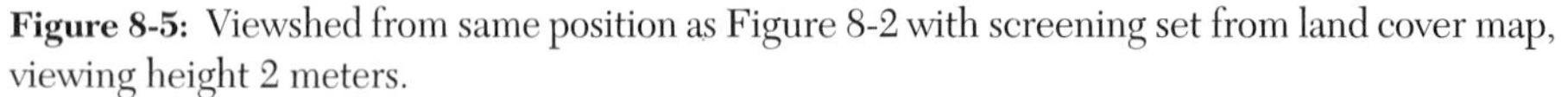

Figure 8-5: Viewshed from same position as Figure 8-2 with screening set from land cover map, viewing height 2 meters.

cost surface. Although Warntz (1965) sketched the basic concept, this capability remained in a theoretical realm as the hardware and software of the era were not fully prepared to tackle such sophisticated problems.

On an accumulated cost surface, the costs increase outward from starting points, and the rates of increase vary depending on site conditions. The first requirement is to generate an attribute that measures the difficulty (or cost) to travel over each link. Many variations on this cost factor can be constructed. In a highway network, travel time can be measured or inferred from a presumed speed times the length of the segment (Figure 8-6a). In the hexagonal grid and controlled reality of many strategy games, the game

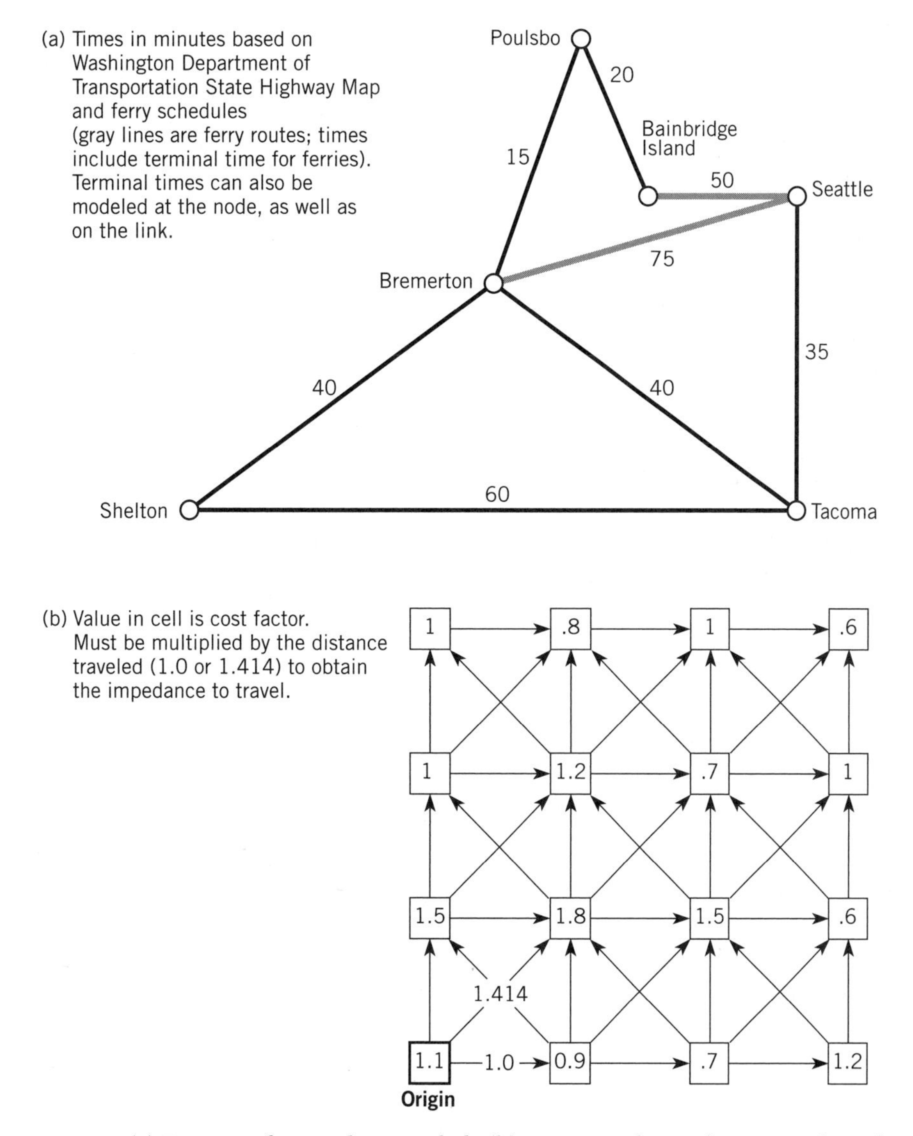

Figure 8-6: (a) Time cost for traveling on a link; (b) Incremental costs for passing through a cell.

rules specify a cost factor for each category of cell on the game board. In a raster representation, the cost information has to be a difficulty factor (or a coefficient of friction) to multiply by the physical distance to account for diagonal or right-angle paths across the same cell (Figure 8-6b). This method applies to connected networks and rasters because both define topological structure for a graph of possible connections.

Using cost information, an incremental operation can accumulate the cost surface by moving iteratively from cell to cell in a matrix, or moving from node to node along the links of a network. At each step, the incremental cost for a cell or a link can be added onto the cost previously incurred to give a value to a neighboring node or cell. This is a form of a weighted sum for the neighbors. If the cost surface is sufficiently variable, the first solution discovered at a spot may not be as low as an alternative that finds a lower cost path (Figure 8-7). Searching moves outward based on connectivity, but later arrivals may offer a lower cost. The cost to each location takes the value of the lowest path discovered using a form of dominance rule.

A cost accumulation algorithm starts by setting all locations as some huge value to flag them as unvisited. The starting points are given their initial values and marked as changed. Then each step takes the values changed in the last step and propagates their cost plus the incremental cost to all their neighbors. These neighbors accept the new value if it is lower than the previous value. Changed cells are marked for the next round. These operations are executed iteratively until the whole search space has been exhausted or some upper threshold of cost has been reached. Cost accumulation fits inside the larger procedure diagrammed in Figure 8-8. This procedure includes isolation, simple arithmetic operations, and contributory overlay combination as prepara-

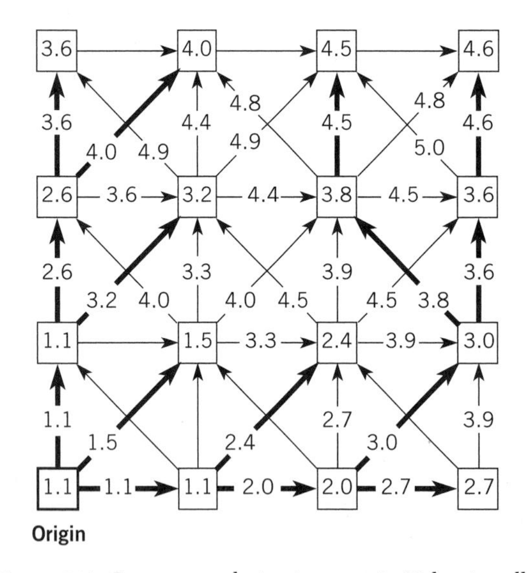

Value in cell is accumulated cost by lowest cost route through network given in Figure 8-6b. Lowest cost paths shown with dark arrows.

Figure 8-7: Cost accumulation in a matrix. Value in cell is accumulated cost through network shown in Figure 8-6a. Lowest cost shown with thick arrows. Longer paths may cost less; lower value replaces higher costs discovered earlier.

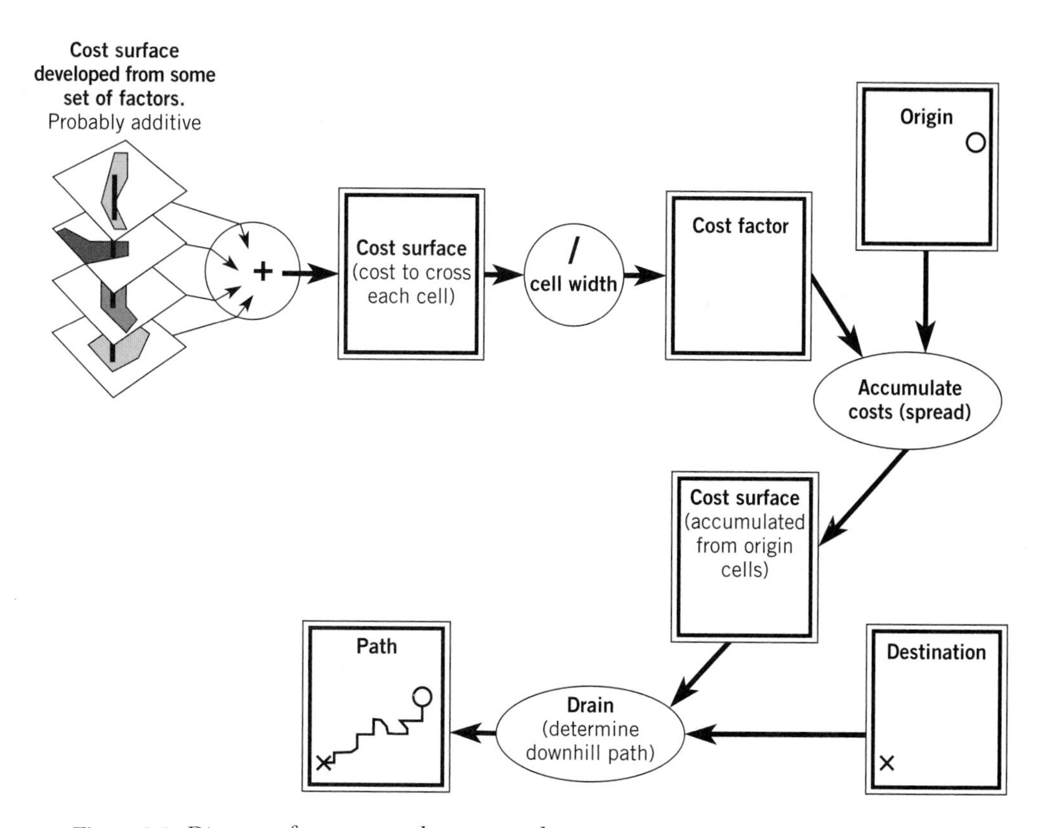

Figure 8-8: Diagram of cost accumulation procedure.

tion for the two comprehensive operations. An example of a pipeline will be used to illustrate the whole procedure.

An Example of Cost Surface Construction: A Water Pipeline News accounts described a perched aquifer in the Snoqualmie Valley east of Seattle. A student team developed a hypothetical plan to link a possible wellhead to the water distribution network along the I-90 corridor. The cost estimate for each cell (Figure 8-9a) was developed from a number of factors. Some of these were relatively easy to measure. The cost of laying pipe in flat terrain can be modified to accommodate slope and other negative factors. The cost of land acquisition was based on a layer of public and private landholdings; it could have been based on tax assessments, market appraisals, or some other source. A very high cost was added to any cell over a certain elevation, to account for the cost of building extra pumping stations. The pipeline would have to cross a few rivers, and a relatively high increment was added for crossing a river cell. A detailed engineering study could have given more reasonable costs to these factors, but the principle is clear. Each of these costs can be measured on a common extensive scale—money. In order to have the route avoid wetlands and other critical

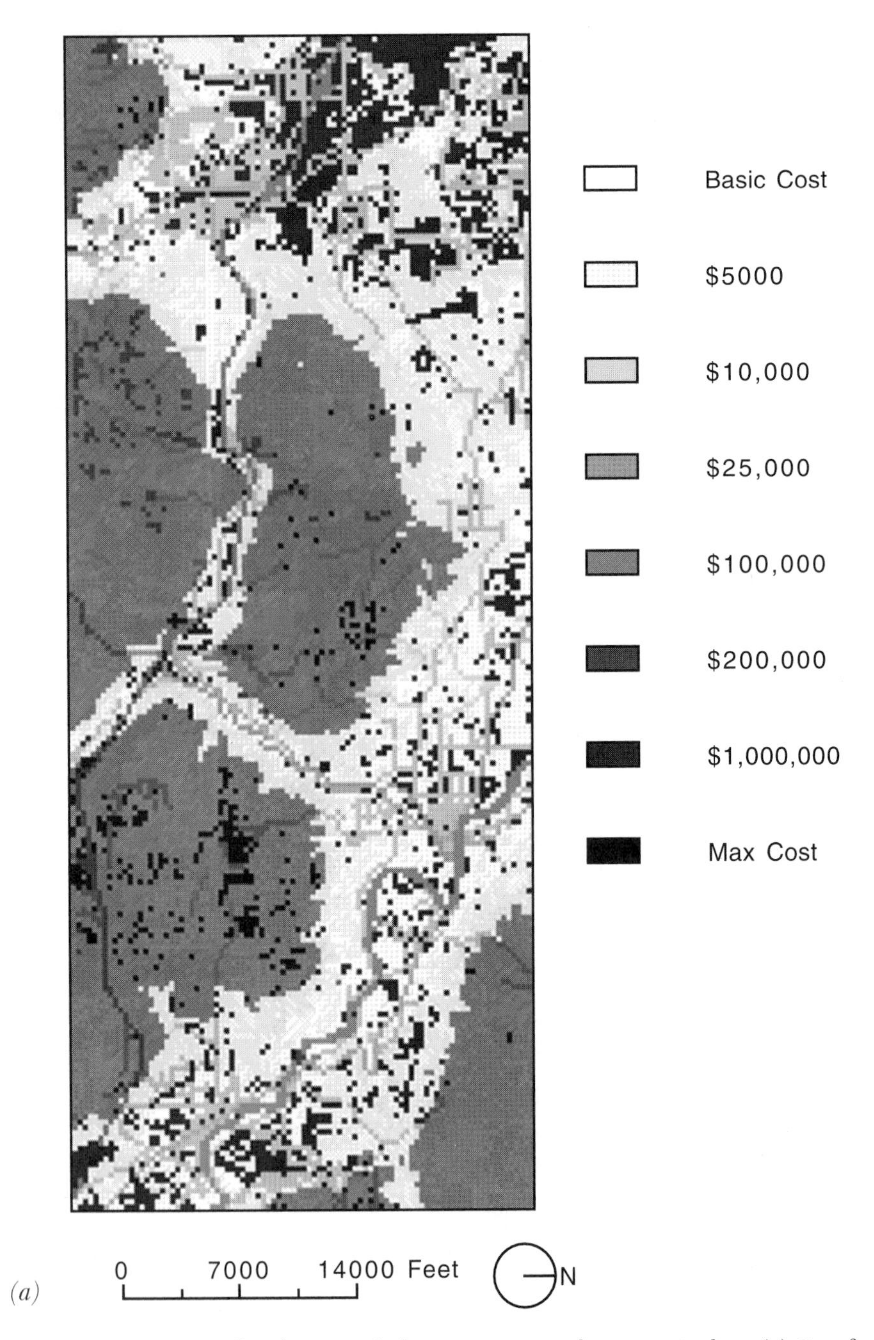

Figure 8-9: An accumulated cost study for construction of a water pipeline: (a) Cost factor assigned to cells based on hypothetical land acquisition costs, construction costs, operation costs, and environmental protections; (b) cost accumulated from possible wellhead, using procedure in Figure 8-8. Project performed by Mark Wilbert and Robert Gray during Summer Quarter 1992.

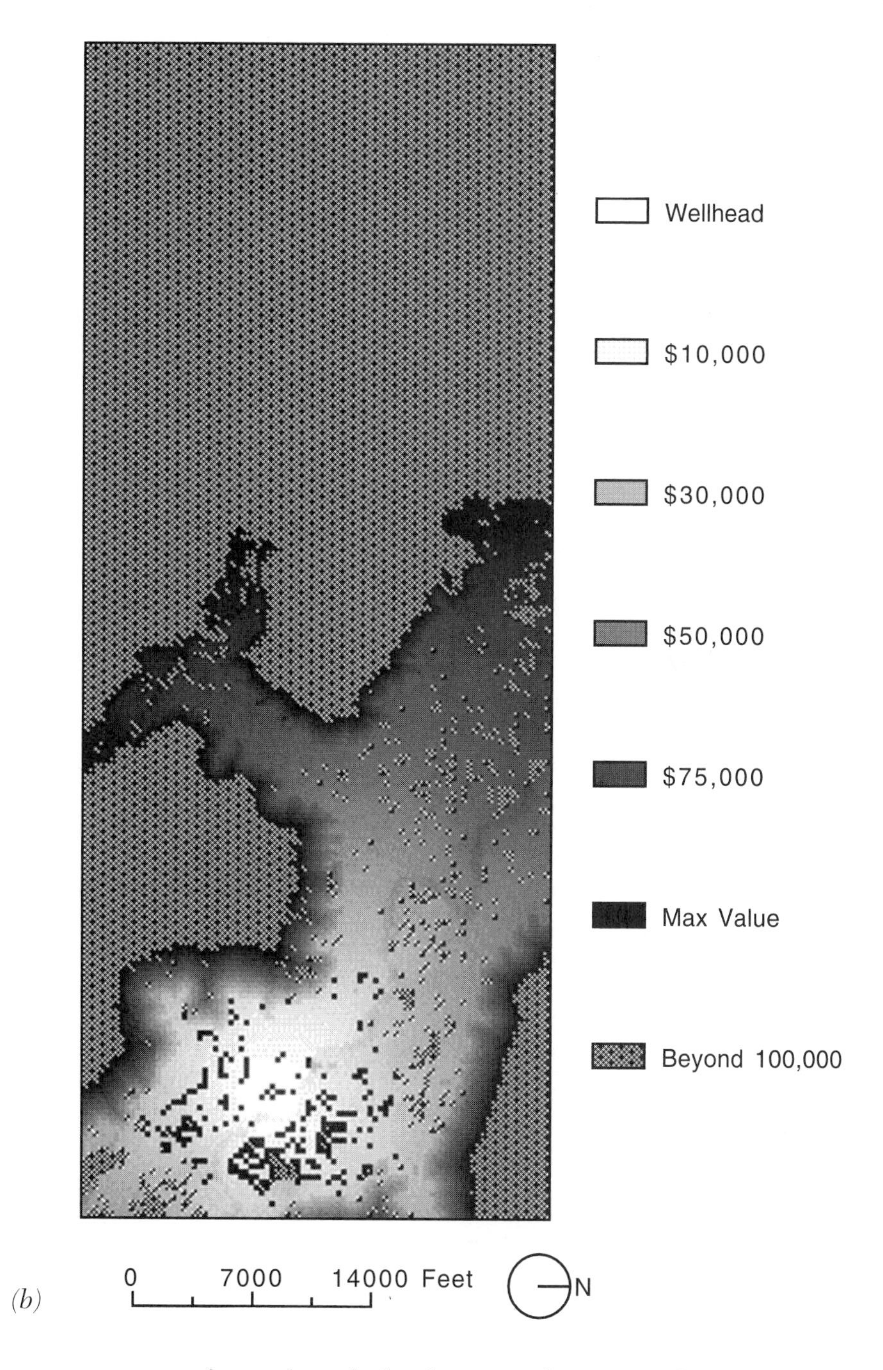

(*b*)

areas, some extra financial cost had to be assigned. Here, as always, it may be difficult to provide a financial value on habitat and esthetic values. The choice would have to be justified within the purposes of this project. It is easy to prohibit certain paths by simply setting a huge price to cross certain kinds of features.

Once all the costs were assigned for each relevant overlay, the composite difficulty factor was obtained by adding all the component costs. For this particular raster package, the cost had to take the form of a factor to multiply by the cell width (or diagonal). Thus the cost was converted to cost per distance so that the cost surface would end up in cost units. Sometimes software packages require such indirect maneuvers.

The accumulated surface was obtained after very long computer runs—all weekend on the vintage computers in the instructional lab (Figure 8-9b). Douglas (1993) has recently unveiled a new approach to accumulating cost surfaces, which may reduce these time requirements substantially. Still, old code has a way of surviving far past its obsolescence in both public domain and commercial packages. Each cell in the resulting surface gives the total cost to construct a water pipeline from the wellhead to that cell. If the surface had been constructed from a set of alternative starting sites, the accumulated surface would give the minimum cost to one of the alternatives. The cost surface is an interim product for the next step.

Drainage Operations

Another iterative problem simulates the drainage of water on a surface. Some area receives a quantity that we will call rainfall, to make the analogue concrete, but it could be any material. In concept, the quantity delivered to each spot is sent downhill, and the quantity drained through each spot is accumulated. The drainage operation works by summation in an orderly manner from headwaters downstream for a treelike stream network. First, of course, the water must flow across the surface before it reaches the stream channels.

A raster representation can model flow reasonably efficiently. If cells are processed from highest elevation downward, the drainage accumulation can occur without needing to model it as a topological network. Each cell simply transmits its accumulated quantity downward, sometimes splitting it if the steepest downhill direction is ambiguous. There is no particular distinction between surface and channel flow because the cellular approximation works from point to point.

A TIN can handle both overland and channel flow. The triangles cover the whole area and can route flow from areas downhill unambiguously. When triangles converge on a courseline, the flow moves from the areas onto the network of triangle edges. This division is also useful hydrologically because the regimes of overland flow are quite different from stream channels (Silfer and others 1987).

All drainage methods are quite sensitive to errors in the measurement of the surface. In particular, places that seem to be local pits will stop the downhill propagation. Badly structured triangles can create these problems, but not if the courselines provide the basic skeleton for the triangulation. The raster representation is more prone to unwanted pits. Depending on the orientation of the landscape relative to the grid points, it is possible to create a pit even without any errors in the measurement. Special de-pitting operations modify the surface to drain downhill at the expense of the original measurement. As usual, accuracy is sacrificed for logical consistency, because the software demands the relationship.

Drainage procedures can be used for nontopographic surfaces as well. Figure

8-10 shows the result of depositing a value of one on the cost surface for water pipelines (Figure 8-9b). The drainage procedure routes this material downhill on the cost surface, and thus the path follows the minimum cost from a potential terminus of

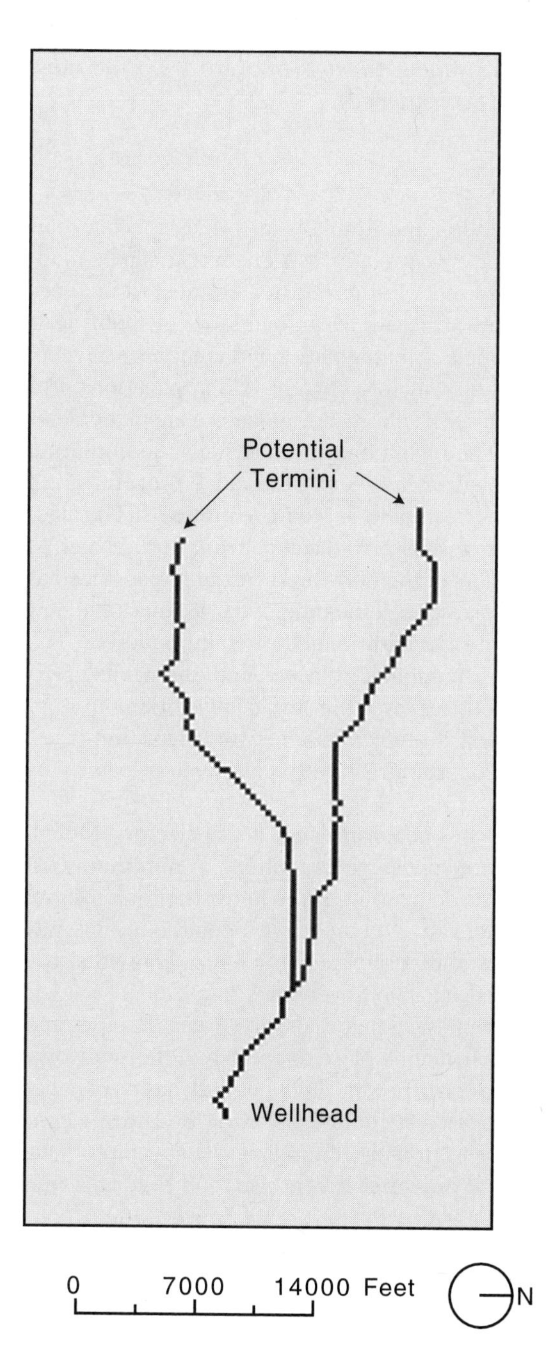

Figure 8-10: Minimum cost path across the cost surface shown in Figure 8-9 to two potential termini.

the water pipeline back to the wellhead. The sum of the cost values for the cells on the path should predict the cost of construction.

The cost surface method requires these two procedures: first, cost accumulation works outward; then drainage works backward. The water pipeline could have been formulated in reverse, from possible connections to the water distribution system instead of the wellhead. Typically, it is prudent to work outward from the most restricted of the set (or the most physically constrained).

Network Operations

The iterative operations of cost accumulation and drainage introduce special, relatively limited cases of a general class of network operations. All networks can be modeled as connected graphs, essentially the model inherent in a connected network model represented using a topological data structure. A large number of sophisticated network algorithms have been developed in mathematics and computer science, and relatively few of them have found their way into current GIS applications. For instance, the method used to accumulate cost in a raster above is a special case of Dijkstra's (1959) pivot point algorithm. The pivot point method finds the minimum cost path through an arbitrary network by using a two-stage iterative procedure. All nodes in the graph have temporary distances, initially set to huge numbers. The algorithm begins with the starting point set to a final zero distance from itself. The first phase consists of relabeling all the neighbors of the node most recently recognized as final with new temporary distances. The lowest of all the temporary distances can now be made final, since no other distance plus an additional cost could be lower. The algorithm then pivots to this newly labeled point and proceeds until the whole graph is explored. Dijkstra's pivot point essentially explores the surface by working upward from the lowest levels using a **breadth-first** strategy. Other network algorithms use a more depth-first approach, but these may require backtracking when lower cost routes are discovered later in the search.

With some careful programming, the pivot algorithm can be satisfactory, though faster methods can be developed for more specialized situations. A minimum cost path from any node is the starting point for many solutions to network problems, but this assumes that the network's costs do not vary with the traffic upon them. This may apply for some purposes, but most automobile drivers in large cities know that the choices of other network users can seriously change travel times.

Many networks are used to transport physical materials, like water (in pipes and in rivers), natural gas, sewage, or vehicles from one place to another. Other networks serve as pathways for communication or electric power. Each network has its peculiar set of rules. For example, a domestic water distribution network tends to be constructed as a tree, with a few closed loops to provide redundancy. Water pipes have limited capacity, controlled by supply and pressure. Rivers also tend to form a tree

Breadth first: Algorithms to traverse a tree that explore the structure by enumerating each level totally before moving to a finer level, as opposed to enumerating each path from the root to the leaves (depth first).

structure, controlled by the terrain and gravity. Channel storage of the river is much more variable, because it interconnects to a floodplain. The electricity distribution system can have many more interconnections, but at any one time, each link must flow in one direction. A short circuit can trigger circuit breakers in microseconds. Highways act in a very different way. The network contains many duplicate paths from one place to another. The units of highway traffic (vehicles) are discrete and interfere with each other in moving from place to place. Capacity of these networks is measured in radically different terms. With all these different axioms, the introduction of capacity constraints must be modeled differently.

Some generic capabilities can be provided to solve network problems. At the base, each application-specific model tends to be built by performing many local decisions, and then propagating the results outward to their neighbors. The approach used above to accumulate cost can be applied to find a least-cost path through a network, but it cannot account for the changes in cost due to the simultaneous interactions with other influences on the network. Adding capacity issues converts network problems to the next level of difficulty.

LOCATION-ALLOCATION PROBLEMS: A FAMILY OF PROBLEMS WITH A COMMON APPROACH

Some problems involve a global measure or constraint. An **objective function** might seek the minimum overall transport cost and a constraint might insist on equality of service. Without some careful construction of the problem, there is no way to process a particular pair of objects on strictly local criteria. Fortunately, many seemingly different problems can be modeled by a few **heuristics**.

The heuristics to confront these problems have been established for a number of years, dating back to their origins in operations research during World War II. Applications in geography were first developed under unrealistic conditions to permit easy solution. Modern computing permitted increasingly realistic networks to be entered and solved (Rushton 1979). While the development of location-allocation methods occurred inside geography and regional science, GIS tools developed somewhat separately. Often a series of inelegant reformatting interfaces create what is euphemistically called loose coupling between the two packages. Eventually, these two groups of applications must grow together. The allocation software will have to develop more flexibility for huge networks, probably by using generalization methods to simplify the representation to the essentials required. The GIS software will also have to adjust to operations that are not so easily localized in the database.

These location-allocation problems go far beyond basic GIS tools like map overlay, but they often include some of the primary needs in practical application. In

Objective function: A formula giving the goal in an optimization problem.

Heuristic: A procedure that attacks a problem in a way directed toward the goal, but not guaranteed to attain it exactly.

extolling the virtues of GIS for the City of Milwaukee, Huxold (1991) lists a series of applications; most of them require solution of location-allocation problems. One prominent application in Milwaukee involved reapportionment of aldermanic districts. Each district should have equal population and satisfy various political expectations. Milwaukee also needed to balance the workload assigned to building inspectors. This process used to be performed in a laborious session, trading file cards back and forth until each inspector had an even stack. More formally, the objective was to equalize assignments. The transportation cost of visiting all the building sites would make a better objective function (minimize aggregate travel time). Huxold also mentions a design of solid waste collection routes. There are only seven days in the garbage cycle of most American cities. The number of trucks must be able to complete the whole city in the period to be able to repeat it again at the end of the cycle. The constraints on solid waste involve the capacity of trucks and the distance to and from the receiving stations. A model based on a fixed number of days and a variable number of trucks to minimize the distance traveled subject to capacity limits would probably work. Huxold reports that every one of these applications was done by human operators working on cartographic products. The allocation procedures were never entrusted to the computer, so the solutions were probably far from optimal.

Some of Huxold's applications could be solved by adding constraints to the transportation problem. The simple statement of the transportation involves a collection of fixed demand, such as consumers who might buy a product or need a certain administrative service. A certain number of suppliers must be located to serve the demand. The problem can be constrained in a number of ways:

- Equalize the number assigned to each facility
- Ensure that the number assigned is greater than a threshold
- Ensure that the number assigned does not exceed capacity
- Route all traffic from origin to destination along the least-cost path subject to the capacity constraints of the network and the interaction between capacity and cost

In this formulation, the terms *demand* or *supply* are attached by convention and historical reasons. Sometimes, the purpose is not really to locate any facility, but simply to group the objects into a district or cluster. The last one, the capacitated network problem, does not seek to locate either origin or destination but to allocate travelers among a set of possible routes. All of these constraints can be built onto the iterative procedures introduced in the previous section. The heuristics to solve these problems are beyond the scope of this book.

Some other location-allocation problems involve objective functions that cannot be attacked with the transportation problem. For example:

- Minimize the total distance between demand and supply
- Minimize the maximum distance to the closest facility

These basic possibilities can be combined or inverted to create an even larger set of possibilities:

- Minimize the total distance subject to an upper limit on distance
- Set the minimum number of facilities such that the maximum distance is less than some upper limit

The "p-median" heuristic can be shown to approximate the optimal solution rather closely (Rosing and others 1979). Most of the different formulations, including the constraints applied to the transportation problem, can be rewritten in the form of a unified linear model for solution by the p-median approach.

When attacked analytically, all these problems can be solved using a common set of numerical tools. The statistical problem of **cluster analysis** raises similar issues, often in much more complex spaces than simple planar geographic space. The procedure assigns objects to clusters to minimize some measure of distance. This process can work either by dividing the whole space or by clumping the individuals. The most efficient solution depends on the relative numbers of supply and demand points. Like many of the location-allocation procedures, cluster analysis works iteratively.

TOUGHER PROBLEMS

The problems discussed in this chapter so far involve some global measure, but they rely on the special character of geographic networks to provide a solution. Incremental techniques can decide locally, then allow those decisions to propagate. Location-allocation problems pose more global questions but can be attacked using local heuristics. The median method applied to a graph can approximate the exact solution in most realistic situations. There are some questions, however, that will not resolve themselves into a solvable form without exploding to examine all the possible combinations of all the objects. One such set of graph problems explored by mathematicians and computer scientists whose solutions all connect is called **NP-Complete**. If an exact solution exists for any one, it will solve all the others. No such solution has been discovered, and algorithmic theorists suspect that a solution is unlikely.

A few members of NP-Complete relate to geographic problems. The most directly geographic is the **traveling salesman** problem. This is phrased as the minimum-cost circuit of a graph that visits each node once. There is no local rule that tells if a given link will form a part of this circuit. Another interesting reformulation is the knapsack problem: chose a set of integers (blocks or objects) such that they sum exactly to a given integer value (the capacity of the knapsack). This seems like an easy mathematical problem, but there is no way to know if a given block goes in the knapsack until

Cluster analysis: A procedure that groups points in a multidimensional space (attribute measurements) into clusters that minimize the distance from cluster centers (or some other objective function).

NP-Complete: Class of algorithms conjectured to exhibit worst-case performance that cannot be written as a polynomial equation of the number of objects processed.

Traveling salesman problem: Given a graph connecting a set of nodes, devise a route that visits each node in the graph exactly once and minimizes the total cost accumulated.

the last one is in place. The knapsack problem lurks whenever there is a constraint on capacity. Since this presents substantial mathematical trouble, any approach to these problems must be approximate. The spatial structure of most geographic problems can help avoid the worst impasses of NP-Complete, but sometimes it is necessary to deflect from a goal because it comes too close to this difficult class of problems. The heuristics developed for the location-allocation problems can often find a solution which is quite close to the theoretical minimum without all the exhaustive searching.

STATISTICAL MODELING OF SPATIAL DATA

It is just a bit presumptuous to insert statistical methods here as a small part of the processing of spatial data. Statistical models were often equated with the scientific method and seen as the core of spatial data analysis. Many geographers of a quantitative stripe berate GIS software for paying too little attention to statistical methods, while other geographers mistakenly consider GIS as an outgrowth of a quantitative movement they judge as failed. Both criticisms are partially true but also incomplete. This book has no space to review all the analytical operations in the statistical tool kit and no space to review all the arguments about quantitative methods. Still, it is important to explore the connection between the measurement of geographic information and the application of statistical models.

Statistical methods can be characterized inside the scheme adopted by this book: objects, relationships, and axioms. A statistical model creates the framework for relationships between measurements with some assumptions about an error term. The parameters of the model are estimated from the information available to minimize the error. The axioms of statistics introduce some relationships not required so far in the discussion of geographic information. Statistics rest on concepts of populations of individuals whose attributes exhibit certain regularities and samples drawn from those populations that exhibit certain distributions of their attributes. Usually, to make the mathematics of estimation tractable, observations are taken as independent samples from the larger population. This assumption may be difficult to accommodate with the interrelationships so common in spatial data.

Unlike most of the GIS operations described so far, a statistical procedure also evaluates the fit to the data. Statistical operations depend on assumptions about the nature of the objects, the measurements, and the relationships. To construct an example, a study of household spending could be interested in the relationship between income and shopping behavior. The proportion of income spent on food may decrease with increasing income, once the size of household and other factors are taken into account. This substantive relationship is represented by an equation, and then the strength of the relationship is estimated from the data. This exercise depends on a number of assumptions about the objects of analysis.

The data model for most statistical methods in the social sciences is still the flat file of the geographical matrix introduced in Chapter 2. Measurements attach to indivisible objects, and the relationships are incorporated in the statistical models. These models specify relationships between attributes, taken at an aggregate level. Thus the goal of a regression analysis is to estimate the parameters of a chosen equation between two variables, using the cases in the data matrix. A linear regression equation is just a form of weighted linear combination, with weights derived from a least squares fit of all the cases. In some odd inversion, the classical statistical procedures, though they seek a comprehensive solution, can be estimated with the simplest attribute-based procedures.

The model of error for standard regression depends on a particular measurement framework. Cases must be independent realizations from the same stochastic process which is the same as saying independent samples from a single population. The process (or the population) must exhibit a Gaussian normal distribution and must have no correlation between the error and the value of the attributes. These axioms are hidden in the techniques used to derive regression estimates and in the statements of probability assigned to the measures of goodness of fit.

Geographic information can occasionally fit these conditions. Some isolated objects may be considered a part of some larger population whose distribution conforms to the normal model and thus is subject to ordinary statistical assumption. For instance, households could be considered as separate units, without regard to their membership in a community or their contact with certain environmental factors. Yet, it is very hard to conceive of a geographic analysis that does not run up against Tobler's First Law (stated in Chapter 7). The variation between nearby objects will be different from that of distant objects. More distressingly, the size of the objects that we use for measurement will influence the patterns we can detect. Thus, the regular axioms of statistical techniques do not apply very well to spatial analysis. Much of the data in a GIS is not sampled in the statistical sense, and many geographic frameworks emphasize exhaustive coverage. Geographers have worked with statisticians to develop some techniques for estimation that can be applied to a few measurement frameworks. In particular, spatial autocorrelation can estimate the relationship between distance and values of the attribute (Cliff and Ord 1981). This method applies most directly to the field view of continuous surfaces (Oliver and Webster 1990), but it can also be applied to some of the effects of choropleth data collection (Arbia 1989). Usually these methods describe the behavior of the whole distribution in a comprehensive summary, recent research tools estimate a spatially variable result. Tools developed in the geostatistical research community are rarely incorporated into operational GIS packages, though some progress is beginning to occur. Current geostatistical methods do not address the full variety of measurement frameworks possible in a GIS. There is a need to develop a stronger linkage between the various subdisciplines that treat spatial data so that existing tools connect better.

CHAPTER 9

TRANSFORMATIONS

CHAPTER OVERVIEW

- Review prior approaches to transformations in analytical cartography.
- Describe operations to transform surface information between measurement frameworks.
- Develop taxonomy for transformations: neighborhood and attribute assumptions.

Each of the preceding chapters has emphasized the central role of the measurement framework in understanding the operations on geographic information. Some measurement frameworks are carefully tuned to permit certain operations but make others more difficult. Because the differences between measurement frameworks arise from diametrically different views of the information content, there is no single choice that can satisfy all needs. The diversity of frameworks arises for good technical reasons and will connect to equally good institutional, social, and cultural reasons (to be discussed in Part 3).

As a result of this diversity, any reasonably sized project will require transformation of some information from one form to another. This chapter will consider transformations between geographic measurement frameworks. First, it will review the approaches to transformations in the fields of cartography and GIS. This will lead to a short consideration of the transformations between different measurement frameworks applied to surfaces. From this consideration will develop a scheme to organize transformations.

PRIOR APPROACHES TO TRANSFORMATIONS

While the dominant school of cartography views cartography as a communication process, there has always been another group focused on transformations. The most classic transformation involves the mathematical conundrum of transferring the nearly spherical earth onto a flat piece of paper, the process of map projection. No cartographic education is complete without a thorough understanding of projections (Maling 1973; Snyder 1987). For centuries, a cartographer could ensure a place in posterity by inventing another solution to the map projection problem. Even Arthur Robinson, whose career was dedicated to thematic cartography and map communication, may have greater recognition for his compromise world projection through its adoption by the National Geographic Society.

Understanding map projections remains a critical element in managing geographic information. Most of the source material for modern information systems exists on flat maps in some projection. Any project with multiple sources is likely to have to convert between map projections (at least in the United States). This topic was raised earlier while introducing the problem of registration as a preparation for digitizing (Chapter 3) and map overlay (Chapter 5).

The key importance of a map projection is not in the mathematical details. Projections demonstrate how measurements taken on one kind of geometric model can be transferred to another model, subject to certain constraints. For instance, in moving from the earth to a plane, it is possible to preserve either the geometric relationships of angles (conformality) or of area (equivalence), but not both. While projections are purely geometric and leave attribute information intact, other transformations convert between systems of measurement.

The standard view, dating back to Robinson's (1953; Robinson and others 1995) textbook and the classics of analytical geography (for example, Bunge 1962; Haggett 1965) begin by dividing geographic entities into points, lines, areas, and surfaces. Attributes are divided using Stevens' framework. Taken together, the dimensional primitives and the levels of measurement are used to describe maps and the operations on them (see Unwin 1981 as a well-constructed example). These concepts are so firmly entrenched that it is hard to recognize where the system fails. As demonstrated in the previous chapters, measurement frameworks provide a more useful model than point–line–area.

Waldo Tobler has made a career of developing an analytical cartography built upon transformations. While much of his work dealt with projections, he did advance a "transformational view of cartography" (Tobler 1979b) that considered all operations as transformations of information content. This book continues Tobler's direction and that of other analytical cartographers (Nyerges 1991; Clarke 1995), with the altered emphasis on measurement as a core. Indeed, all the operations presented so far could be considered some kind of transformation, but this chapter will focus on those that convert between measurement frameworks.

TRANSFORMATIONS FOR SURFACES

The core concept of a transformation involves changing from some form to another. If information can be represented in a certain number of ways (N), then logically there will be a transformation between each possible pair. The total number of transformations will be N^2. Some software systems have embarked on this ruinous approach, creating a seemingly separate command for the conversion between each pair of possibilities. For simple taxonomies, like point–line–area, a matrix of size nine presents no trouble, but this book has already introduced 12 measurement frameworks, implying 144 required transformations. A brute-force pairwise matrix also creates difficulties in organizing this presentation. With so many cells, the book could not keep a clear focus, even if it could afford the space for individual discussion. In order to demonstrate that a gigantic matrix is not needed, I will present a tractable matrix (Table 9-1) by limiting the focus to surfaces.

The rows and columns of Table 9-1 list some of the major alternatives for the representation of surfaces. The first "Points with Z" refers to an object view where a continuous surface value is measured at an isolated point feature. This is the dominant form of surface measurement in a number of fields such as geochemistry, geodetic surveying, and archaeology. Photogrammetric instruments can measure spot heights for any identified feature on stereo photography, a major source of mapping at many scales. The second representation is isolines, closed contours that measure the location of a given surface value. Digital Elevation Model/Matrix (DEM) refers to a raster structure with a regular, spatially controlled measurement of elevations. The fourth is the Triangulated Irregular Network (TIN) whose triangles establish relationships of slope between spot heights (described in Chapters 2 and 7). These four are the primary alternatives for representation of surfaces.

The cells in the four-by-four matrix (Table 9-1) give a label for the procedure that converts information in the row dimension to the column dimension. The following sections will explain each transformation. The three-by-three matrix in the upper left (light gray) is filled with one form or another of interpolation. This operation provides

TABLE 9-1: Surface-oriented transformations

Input \ Output	*Points (w. Z)*	*Isoline*	*DEM*	*TIN*
Points (w. Z)	Interpolation	Interp. & trace	Interpolation	Triangulation
Isoline	Interpolation	Interp. & trace	Interpolation	Triangulation*
DEM	Interpolation	Interp. & trace	Resampling	Triangulation*
TIN	Extraction	Tracing	Extraction	Simplify/Refine

* Denotes a triangulation operation that may produce overly dense triangles without some filtering.

a good example of how a transformation combines relationships and assumptions (axioms) to produce new information.

Interpolation

The simplest definition of interpolation involves a process to determine the value of a continuous attribute at some location intermediate between known points. Part of this process requires relationships—knowing which points are the appropriate neighbors. The other part involves axioms—assumptions about the behavior of the surface between measured locations. The balance between these two can vary. Some methods impose a global model, such as fitting a *trend surface* to all the points. Most methods work more locally. The top left cell in the matrix poses the classical problem: given a set of point measurements, assign values to another set of points. This requires two steps (Figure 9-1). First one must discover the set of neighboring points for each desired location. Then one must apply some rule to determine the result.

Interpolation from Scattered Points If the input points are unorganized, finding the neighbors can involve fairly wasteful searching. If the process will be repeated, it is common to introduce some spatial indexing system. One common indexing system

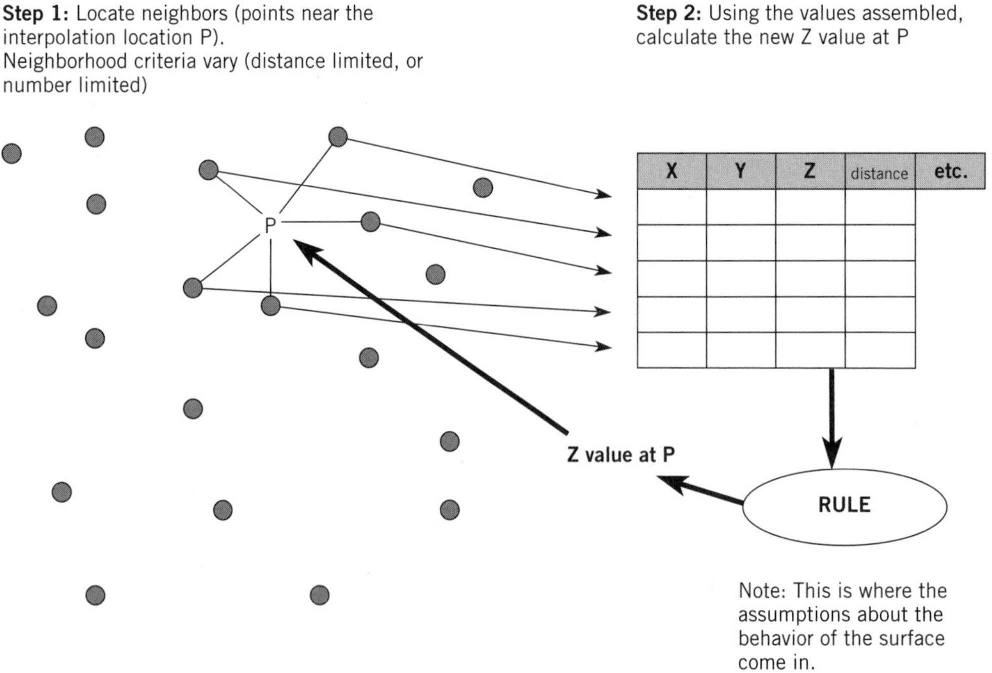

Figure 9-1: Interpolation from a set of points to another point. Step 1: locate neighbors. Step 2: assign values based on some rule.

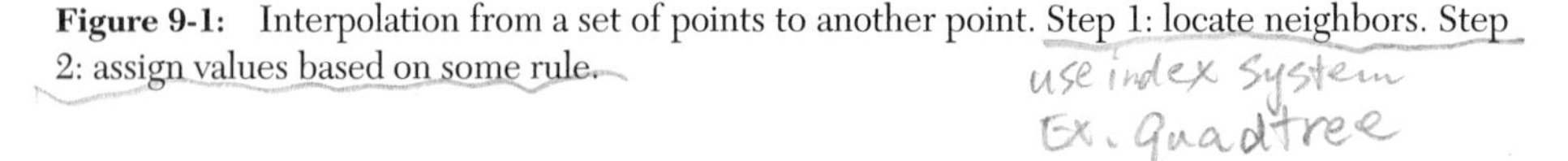

organizes the points into a hierarchical system of nested cells, organized as a quadtree (Figure 9-2; defined in Chapter 3). As long ago as **SYMAP**, software packages have used some kind of searching mechanism to construct a neighborhood for interpolation. The cells of an index, such as a quadtree, provide an approximate neighborhood, not a complete conversion to matrix representation. An index usually delivers more points than required; additional processing is required to determine which ones are the actual neighbors. Various rules can be applied in this selection. Some take the nearest points, while others try to ensure a sufficiently even distribution around the target location.

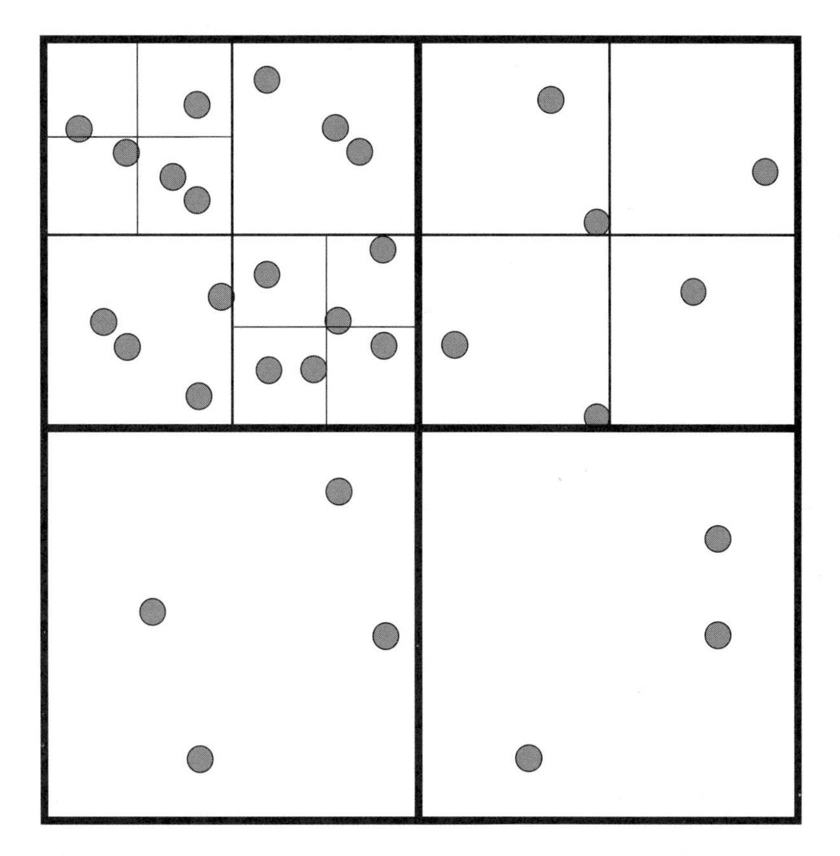

Figure 9-2: Quadtree indexing system for points. The region is covered by one large square. If there are too many in the square, it is divided into four, and the contents partitioned. The process is applied recursively until no cell has too many points (could be one or more). A search process can locate the desired point in the quadtree, then move upward in the tree until enough neighbors are found. This naive algorithm would not guarantee all the nearest neighbors when near the divides between subtrees.

SYMAP: SYnagraphic MAPping package, developed under the direction of Howard Fisher at Harvard Laboratory for Computer Graphics beginning in 1966.

A more precise method of interpolation involves creating a triangulation. Ironically, this approach involves a transformation from points to a TIN in order to produce point output. With a triangulation, any point will fall into one specific triangle. One of the vertices of the triangle will be the closest point, if the triangulation used the Delaunay network discussed in Chapter 6. A neighborhood of closest points can be assembled from neighboring triangles, or some circular search can guarantee a distribution around the point.

Once the neighbors are known, the problem of assigning a value resolves itself into the same choices described for neighborhood operations in Chapter 7. A dominance rule will not yield a smooth surface because it will assign the same value to a neighborhood (usually the Voronoi polygon). A contributory rule usually involves a distance-weighted average of the neighbors (Davis 1973). Various forms of interaction rules are in use as well. SYMAP had a much-copied interpolation system that weighted points so that distance and orientation to other points were considered. A spline adopts a model of a thin spring with a particular modulus of elasticity. **Kriging** produces an "optimal" interpolation by using a statistical model of spatial interdependence (Burrough 1986; Oliver and Webster 1990). Each method operates by using certain relationships, plus some assumptions about the distribution of values between points. For example, kriging is optimal if its assumptions of a "stationary" (spatially uniform) error structure apply. Some methods (such as weighted averages) make the surface pass through all the original data points, while others (like splines) put more emphasis on the neighborhood fit. The differences between various forms of interpolation reflect various assumptions about the nature of the attributes.

The process of producing a DEM with uniformly spaced points is just a special case of interpolation for scattered points. Sometimes the searching structure can be optimized to accommodate the coherence of the sequence of requests. Otherwise, the procedure is the same.

To produce isolines, interpolation is somewhat different. Instead of requesting a value at some arbitrary point, the contour specifies the height, and the interpolation discovers the location. Functionally, this is not very different, because the procedure for a weighted average can be algebraically restructured to give a coordinate where the surface has a given value. To make this work directly from scattered points, the geometry must be kept fairly simple. The manual procedures for contour drawing involved linear interpolation on what amounts to a triangulation (Raisz 1948). More than the other interpolation problems, the output of isolines may involve a hidden conversion to triangles first. In addition to the interpolation, the construction of isolines requires *tracing*, the process of following the contour from neighborhood to neighborhood. Usually, this procedure involves some assumptions about the smoothness of the surface, since the shape of the contour cannot be really estimated from the original point measurements. Tracing also involves relationships between adjacent contours, even those not created with the same neighborhood of points. Parallel con-

Kriging: A geostatistical technique for interpolation that uses information about the spatial autocorrelation in the vicinity of each point to provide "optimal" interpolation (in the sense of greater use of the information provided by the spatial arrangement).

tours imply slope gradient and aspect properties, along with other interactions caused by ridges and courselines. Thus, tracing contours involves many more relationships than a simple decision about the value at a point.

Interpolation from Isolines If the input consists of a set of contour lines, the procedure for scattered points still applies. Interpolation will need to establish neighbors, but neighbors between adjacent contours as well as along the lines. Finding the nearest point on the two adjacent contours does not ensure a correct reading of features such as ridges or courselines. However, this geometric criterion does provide a place to start.

In manual map reading, the normal procedure is to locate the two contours on either side of the point in question. A straight line is drawn that tries to cross the contours at right angles. This straight line is a simplification for the line of steepest descent. Linear interpolation then proportions the value between the two contour values (Figure 9-3). This procedure can be automated, but the human map-reader can construct these form lines in a more nuanced manner than most algorithms.

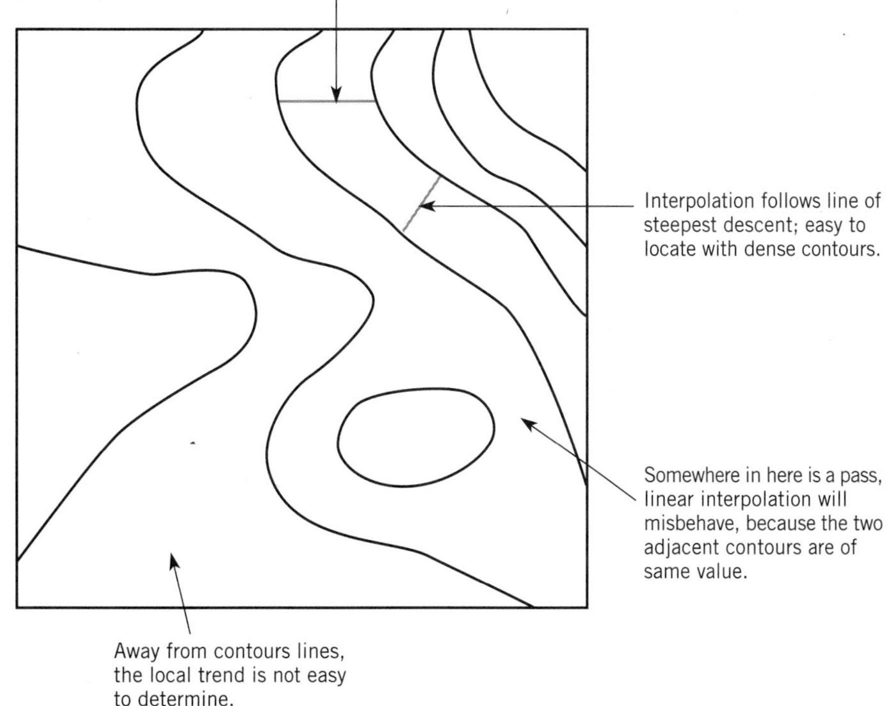

Figure 9-3: Linear interpolation between adjacent contour lines.

Interpolation from a DEM When the input values are organized in a grid structure, the matrix provides the means to access neighbors directly. The interpolation process is very similar to the neighborhood operations described in Chapter 7, except that the output value does not apply to a grid cell. To produce output for scattered points, the rules can be applied on the immediate neighbors in the grid. To trace contours, the grid values are used to estimate values in the area between them. This situation was introduced in Figure 7-6c as the transformation from a point view to an area view of the raster representation.

Producing a matrix output from a matrix input is a common requirement. A matrix of double or half the rows and columns can be produced using neighborhood rules, but other changes are more complex. Unlike the vector method, where the coordinates can be transformed fairly directly, a matrix is delineated orthogonal to a given spatial reference system with a given spacing. If a different cell size or orientation is needed, the values will have to be converted in a process called "resampling." Furthermore, if a grid in Universal Transverse Mercator must be used with other information stored in State Plane Coordinates (often a Lambert Conformal Conic or a different transverse Mercator), the grids will not transfer directly. Resampling is required for any of these changes. For continuous variables, there is no real difference between resampling and interpolation. Sometimes, a simple dominance rule is used: each new grid cell gets the value of the nearest input grid cell. As long as the spacing is not wildly mismatched, this may produce a reasonable representation. For remotely sensed sources, the "nearest neighbor" interpolation retains a combination of spectral values actually measured by the sensor. It does mean that each value has been shifted from the position at which it was measured by as much as 0.707 times the original pixel distance. Alternatively, it is common to use a contributory method to weight the change over distance using a various formulae, such as **bilinear**, **cubic convolution**, or higher order polynomials. Each function imposes different assumptions about the continuity of the surface.

To and From Triangles

Transformations involving TIN representations are significantly different from those discussed previously. The TIN is organized in triangles that specify the neighborhood relationship unambiguously. While the points of a grid neighborhood can lie on a number of different planes, a triangle specifies exactly one. Thus, for the simple job of producing a value at a specific point, there is no ambiguity in the TIN representation. There is no need for alternative rules; it is a matter of extraction of a value implicitly stored in the equation for the plane. The TIN implies linear interpolation between the vertices of the triangles. Producing a DEM is essentially identical to extracting scat-

Bilinear interpolation: Interpolation method where the value is obtained by linear interpolation on the two axes (row and column). Uses four neighbors from a raster representation of the surface, and averages out the inconsistency between the four possible triangles.

Cubic convolution: Interpolation method where the value is interpolated by fitting a third-order equation to the 16 grid points surrounding the desired location.

tered points. To produce contours, there is still a tracing stage, and usually a step to smooth the linear sections into a more conventional curved form.

Producing triangles from the other three systems involves discovering relationships latent in the source. If starting with scattered points, then a triangulation must be found to connect neighbors. The constraints on TINs are rather special, and the ideal TIN must minimize the volume between the true surface and the *finite element* approximation. Nearest neighbors in two dimensions may not create the best fit to the topology of the surface (Peucker and Chrisman 1975; Males and Gates 1978), but without other information, the assumption is most likely.

Using simple procedures, it is possible to construct a TIN from contours (Bello-García and others 1992) fairly directly, but these TINs are likely to be needlessly bulky. First, it is important to remember that the points along a contour line are not the best material to feed into the general purpose method for scattered points described above. Adjacent points along a contour line are most likely to be nearest neighbors. In flat terrain, the next contour can be far away, so that triangles may connect points along a single contour. The resulting triangulation produces large areas with absolutely flat terrain. A method to convert contours must ensure points of a triangles come from two different contour levels, never entirely from the same contour line. This can be very tricky in relatively flat terrain where contours snake around without any correlation to the shapes on the adjacent contours (Figure 9-4, case a).

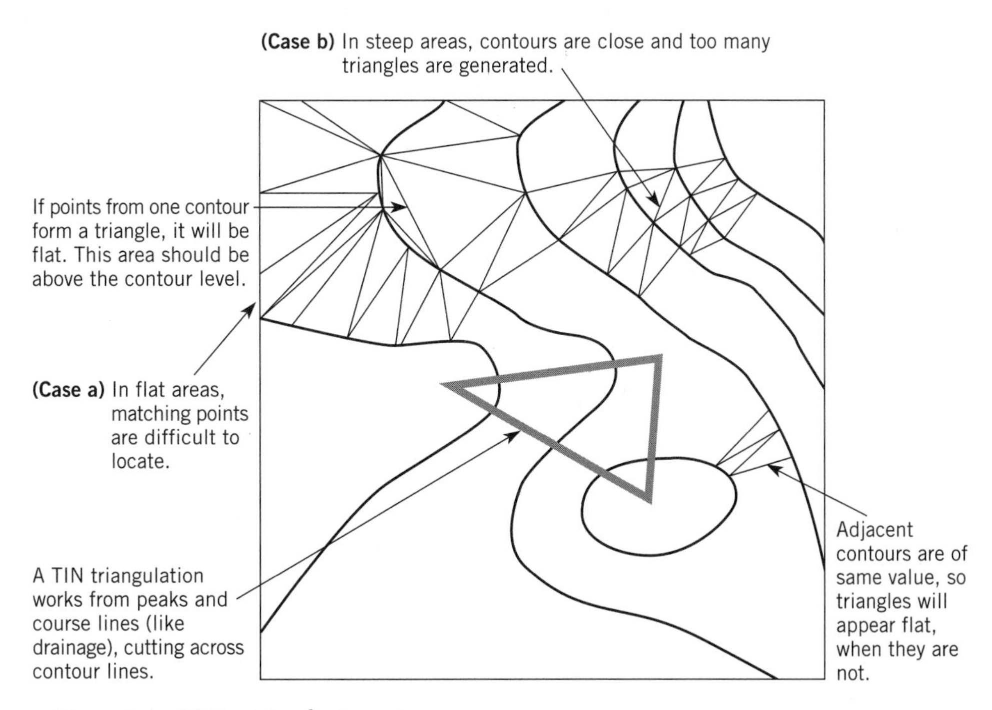

Figure 9-4: Fitting triangles to contours.

Additionally, where contours are closely spaced, the power of a TIN comes from building a triangle that crosses many contours lines (Figure 9-4, case b). This analysis is much more complex than pairing adjacent contour levels. To produce a triangulation from contours, the change in structure is quite radical. The density of points needs to change dramatically, removing many points from steep areas while adding other information that can only be inferred indirectly from the contour shapes. The automated procedures for this transformation may not be as effective as visual interpretation.

Producing a TIN from a DEM has some of the same problems, though not as severe as it is with contours. The density of the DEM is constant, while the TIN should be variable. Some method is needed to detect areas of essentially constant gradient and aspect, so that they can become a common triangular facet. Often, however, this process works through another path. The DEM is culled to produce a set of salient points, and then these are triangulated as scattered points, using the Delaunay neighborhood. This loses some surface information, but it ensures a sparse enough set of vertices to make the TIN competitive with the DEM.

Producing a TIN from another TIN is difficult to describe. The entry in Table 9-1 mentions both simplification and refinement, but the process would involve many issues. If the change is simply a change in coordinate systems, the TIN can be transformed by a projection without any restructuring (unlike the DEM's resampling). Technically, the straight lines on one projection are not the same as straight lines on another projection, but TINs tend to involve lines so short that these differences are not substantial. The main distinction between alternative TINs would involve the density of points and the attendant changes in accuracy in fitting the surface. A TIN is a tightly designed network of points, with many constraints on the contiguity of the triangles. The conversion from one TIN to another will involve coarsening a dense TIN or inserting new points to refine a coarse one. In any case, these are local operations that are rather specialized for the purposes of this introduction. In either case, the TIN provides the local topological structure to guide the process.

Network Information and Surfaces

Beyond the matrix of transformations between the four surface frameworks, the network of surface topology can also participate in the transformation process. In Chapters 7 and 8, various operations extracted the topological structure of a surface. Some near neighbor operations tag pits, peaks, and ridges. Iterative drainage operations can produce a network of courselines. Similar operations can define the ridges, simply by temporarily turning the landscape upside down and using the same tool. All these operations require access to neighborhoods. In a DEM, flow is modeled cell by cell, and certain cells are found to represent the convergence of drainage. As a result, cells in the raster represent the network. On a TIN, the flow is modeled across the triangles until it reaches the edges where two triangles converge. Then, the flow follows the lines. The network of courses or ridges will be a line network. Neither isolines nor isolated points provide enough neighborhood information to perform the drainage analysis; it must be done indirectly. These representations would have to be converted to raster or TIN to do the work. Some representations permit the operation directly while others require a prior transformation.

Network information can also be used to build certain relationships in a surface representation. For instance, contour lines should cross courselines with a concavity uphill – "notching." Usually, a contour tracing operation will not produce these relationships, because the courselines were not available during the interpolation process. A DEM may have false pits along stream courses. One way to ensure better results is to impose the drainage relationship externally. The stream network could be used to enforce a downhill drainage along the network. Similarly, in constructing a TIN, certain relationships should receive higher priority. If the lines of ridges and courses are known, then no triangle should cross them. While these relationships were considered important in the original definition of TINs (Peucker and Chrisman 1975; Males and Gates 1978), they will not be respected if the TIN is constructed strictly on planar criteria. If a TIN module has no way to input a ridge network or a stream network, something critical is missing.

Summary of Surface Transformations

In review, the transformations applied to surfaces can be simplified into decisions about geometric neighborhood construction and about attribute relationships (Table 9-2). For instance, the TIN, within its data structure, provides the required information to extract any point. The only operation required is to determine the triangle in which a point lies—a geometric relationship between the two sets of objects. Then,

TABLE 9-2: Surface-oriented transformations reclassified

Input \ Output	*Points (w. Z)*	*Isoline*	*DEM*	*TIN*
Points (w. Z)	Found / Ext.	Found / Ext.	Found / Ext.	Found / Ext.
Isoline	Found / Ext.	Found / Ext.	Found / Ext.	Found / Ext.
DEM	Implied / Ext.	Implied / Ext.	Implied / Ext.	Found / Ext.
TIN	Pt. in Δ / Inh.	Implied / Smo.	Pt. in Δ / Inh.	Implied / Inh.

Each cell contains two elements: neighborhood relationships / attribute assumptions.

Key: Neighborhood relationships

Implied	Implied by matrix or Δ neighbors
Found	Must be found by geometric process
Pt. in Δ	Point in triangle (simplified overlay)

Key: Attribute Assumptions

Inh.	Inherent in structure
Ext.	External, must be imported
Smo.	Interpolation inherent, smoothing external

using its established relationships and its built-in assumptions, the triangle provides a means to answer the question. Of course, this does not mean that the answer obtained would be exactly the answer obtained by measuring the surface at that point independently. Triangles create flat planes where the surface may curve in more complex forms. Yet, within its system of assumptions, the TIN needs no further information to produce an answer.

The other transformations, despite much diversity, all require more information. In some cases, the structure does not provide the relationships, so the first step is to construct the neighborhood required. The geometric processing to relate adjacent contours is more complicated than a simple point-in-triangle required for the TIN. In some cases, mostly involving the DEM, the neighborhood can be constructed using implicit relationships in the matrix representation. In either case, once the neighborhood is assembled, some further assumptions must be provided for the final step.

These transformations for surfaces are usually described using terms specific to the application. Under this distinct terminology, the interpolation operations are simply special adaptations of the neighborhood operations described in Chapter 7. Most surface transformations can be performed using a general purpose tool kit. Thus, it is not necessary to understand each possible transformation but to understand how the process of transformation works.

A TAXONOMY FOR TRANSFORMATIONS

Most of the research on transformations focuses on the geometric components. Interpolation is described in terms of distance. Projections are described in terms of angles and areas. Yet, the metric geometry is not the only ingredient. The key difference between interpolation from a cloud of points and interpolation in a DEM involves assembling the neighborhood. With a cloud of points, it takes some computation, and usually some indexing structure, to discover the neighbors. A DEM has implicit adjacency relationships. Thus, the common divisions by point, line, and area fail to recognize important commonalities between transformations.

To generalize from the examination of surface representations, a transformation involves two basic components: a neighborhood and a rule to apply to the attributes. Each of these have two fundamental possibilities: they may be implicit in the source material or they must be constructed. In the case of a neighborhood, the construction process involves geometric processing. In the case of attribute rules, they cannot be discovered by processing; they must come from the external world as axioms. Table 9-3 summarizes this set of distinctions in a two-by-two matrix. The code is not meant to be cryptic; it identifies how many relationships must be constructed to perform the transformations. Zero means that the information needed is implicit in the source structure, and the transformation is just an extraction. Two means that both geometric and attribute relationships need to be constructed; 1A means that the attributes require some assumptions; and 1N means that the neighborhood must be created. These four categories generalize from the specific situations raised by surface transformations.

TABLE 9-3: Taxonomy of transformations

		Attribute assumptions	
		Implicit	*External*
Neighborhood construction	*Implicit*	0	1A
	Discovered	1N	2

The next sections will consider each of the four cases in Table 9-3 briefly, moving from simple to complex. To a large extent, this is an artificial order, since the same issues discussed in the simpler cases apply to the more complex. Following this introduction, a few examples of transformations will be presented.

Transformation by Extraction (Case 0)

The transformations that require no new information are relatively rare. Most of these transformations involve pairs of representations where the source is superior to the result. For example, a topological structure contains the information encoded in simpler object-based vector structures. Thus, the isolation operations discussed in Chapter 4 are effectively one kind of transformation by extraction. These transformations are not as easy in the inverse direction; an assemblage of isolated objects does not necessarily exhaust the whole space to create a consistent comprehensive coverage.

In general, the transformations from one framework to itself (the diagonal of matrices such as Tables 9-1 and 9-2) do not fall into the extraction category. A transformation implies a change in the objects and/or the attributes. Much of the research in analytical cartography has concerned *cartographic generalization*. There are various schemes to categorize generalization operations (Brassel and Weibel 1988; McMaster and Shea 1992), but at the core they would all agree that these operations require a spatial criterion, an attribute criterion or both.

Transformations Based on Attribute Assumptions (Case 1A)

The upper right cell of Table 9-3 (labeled 1A) refers to situations where the transformation does not require geometric processing, only alterations in the attributes. The operations in Chapter 4 are a simple example of these transformations; the geometry remains identical and the attributes are changed. Not all these transformations are local, as demonstrated in the discussion of neighborhood operations for surfaces in Chapter 7, reviewed at the start of this chapter.

Classification of Remotely Sensed Imagery One example of an attribute transformation is very common in the application of remote sensing tools. Satellite imagery measures reflectance in a number of bands of the electromagnetic spectrum (commonly called "light" but sometimes outside human ranges of vision). Reflectance is encoded as integer values (usually from 0–255), discretized from a continuous axis. Reflectance is not an end in itself. It serves as a raw resource for extracting other information. Some measures, particularly ratios and differences of various bands like the **Normalized Difference Vegetation Index (NDVI)** convert multispectral reflectance into some related space. The most common process applied to satellite images is classification—the conversion from continuous reflectance values to categories. Classification moves downward on Stevens' hierarchy, but often it is useful to trade a reflectance in a specific band at a given instant (with all the conditional issues of atmosphere and climate) for a land cover class that should be less transitory.

It is far beyond the scope of this book to give a thorough treatment of remote sensing algorithms and procedures. However, it is important to show that, though specialized for their purpose, remote sensing tools fit into the range of functions performed by a GIS. There are two major approaches to classification in remote sensing. A *supervised* classification starts with a few spots (the "training set") whose classification is known. The algorithm tries to group the rest of the image into the class that fits best. An *unsupervised* classification first clusters those values that seem distinct based on their multispectral characteristics. Mathematically, this clustering process is performed using an iterative procedure that shares certain basic methods with the location-allocation algorithms discussed in Chapter 8, although remote sensing clusters usually work in spectral "space," not the coordinate space used by location-allocation. Traditional category labels may not fit these clusters, whose identity comes from numerical properties of the multispectral space. The association between clusters and categories is based on the concept of prototypes (see Chapter 2), not necessarily a formal taxonomy with clear-cut rules.

Transformations Based on Geometric Processing Only (Case 1N)

The only surface transformations that only require geometric processing (Table 9-2) originated with TIN representations. The TIN contains a complete model for the behavior of the surface attribute, so the process is nearly one of extraction. Given a point, the transformation must determine the relevant triangle. In Table 9-1 this was described as point-in-triangle, a special case of general-purpose geometric intersection processing.

Outside the realm of surfaces, there are analytical procedures that involve only geometric processing, such as the direct analysis of overlay described in Chapter 5. However, most transformations involve some additional assumptions about the

NDVI: Difference between two bands (near infrared minus visible red) divided by the sum of the two bands; high values indicate active vegetation growth; often applied to NOAA AVHRR data.

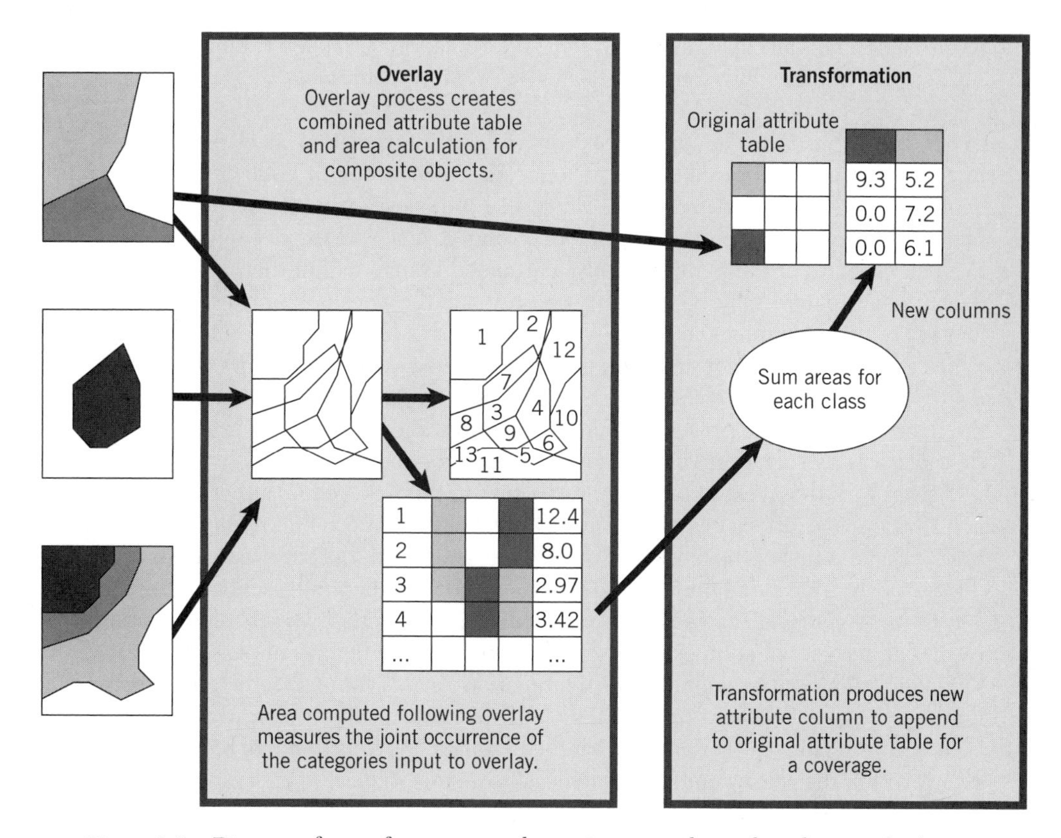

Figure 9-5: Diagram of area of one category becoming an attribute of another set of polygons.

attribute distributions. One attribute that requires no assumptions is the measure of area. Many applications have a simple goal such as characterizing the amount of habitat or the amount of potentially developable land. A transformation between two categorical coverages begins with the geometric process of finding the intersection of all the objects through overlay. Then one attribute would be the area of one category inside the second (Figure 9-5). This is a form of summary, a sum of the areas of the components inside one object that satisfy some criteria on the other factor. The second, output layer becomes a kind of choropleth because the area attribute has lost its original home. This form of areal cross-tabulation transforms from one set of objects to another using geometry only.

Complete Transformations (Case 2)

The most interesting examples of transformations involve the most radical changes in geometry and attribute structures. In considering surface processing, most types of interpolation fall into this class. The term "interpolation" has been extended beyond surfaces with a procedure sometimes called *areal interpolation*.

Areal Interpolation The core definition of areal interpolation involves the transformation of an attribute attached to one kind of choropleth collection zones to another set of choropleth zones. This process provides a very clear application of many of the measurement concepts. For instance, a school district might need to have an estimate of the population of a certain age range within their school attendance zones. There may be a recent census, but the census tabulates results for another set of geographic units. The school district is in luck if it can define the attendance zones as aggregations of basic units in the census. Then the process simplifies nearly to a process of extraction (Case 0). A nested administrative hierarchy removes the need for geometric processing, and the aggregation of population is also implicit (because it is already a sum of a more basic phenomenon).

Even with the finest of census units (blocks in the US), school attendance zones may divide space differently, thus requiring areal interpolation (Figure 9-6). On Bainbridge Island, some of the blocks are quite large, and the school zones cut across them due to bus routing along the roads. To estimate the population by school zone, first overlay must discover the geometric differences between the two sets of zones. The result will be a set of composite zones falling into a specific pair of census unit and school unit. The census will not have a tabulation for this subunit; some assumptions must fill the gap. Typically, the population of each census unit is allocated to its subunits according to area. This means that the population is considered to be distributed evenly within the census unit. In the general case, this seems the only possibility, but in local context the results may seem foolish. A park will be attached to some census tract, diluting its average density, or a group facility like a nursing home will make density seem higher. In the simple version of areal interpolation, the park will get a share of the population and the nursing home will be spread over the whole unit. More careful assignments can be made with improved spatial information. Once the population is assigned to the composite units, the last step is less troublesome. It makes sense to add up the populations of subunits to obtain the estimate for the whole. The addition operation is implied in the logic of population, because a population count is already a sum.

Other forms of measurements such as means, rates, or densities require more assumptions. If the census variable is median family income, then the procedure for a count does not apply. It may make sense to consider the distribution to be uniform within the census unit. Then there is no need to apportion the median income out to each subunit; each subunit gets the same median as the whole. To obtain the estimate for the school unit, there will be a collection of median family incomes for the constituent subunits. These values do not add together as populations did above. Ideally, a median should be obtained by pooling the whole distribution, but that has been lost in the census reporting process. An area- or population-weighted average might give a reasonable estimate, but each one involves some hidden assumptions. In most GIS packages, all forms of attributes are just treated as numbers, without maintaining the

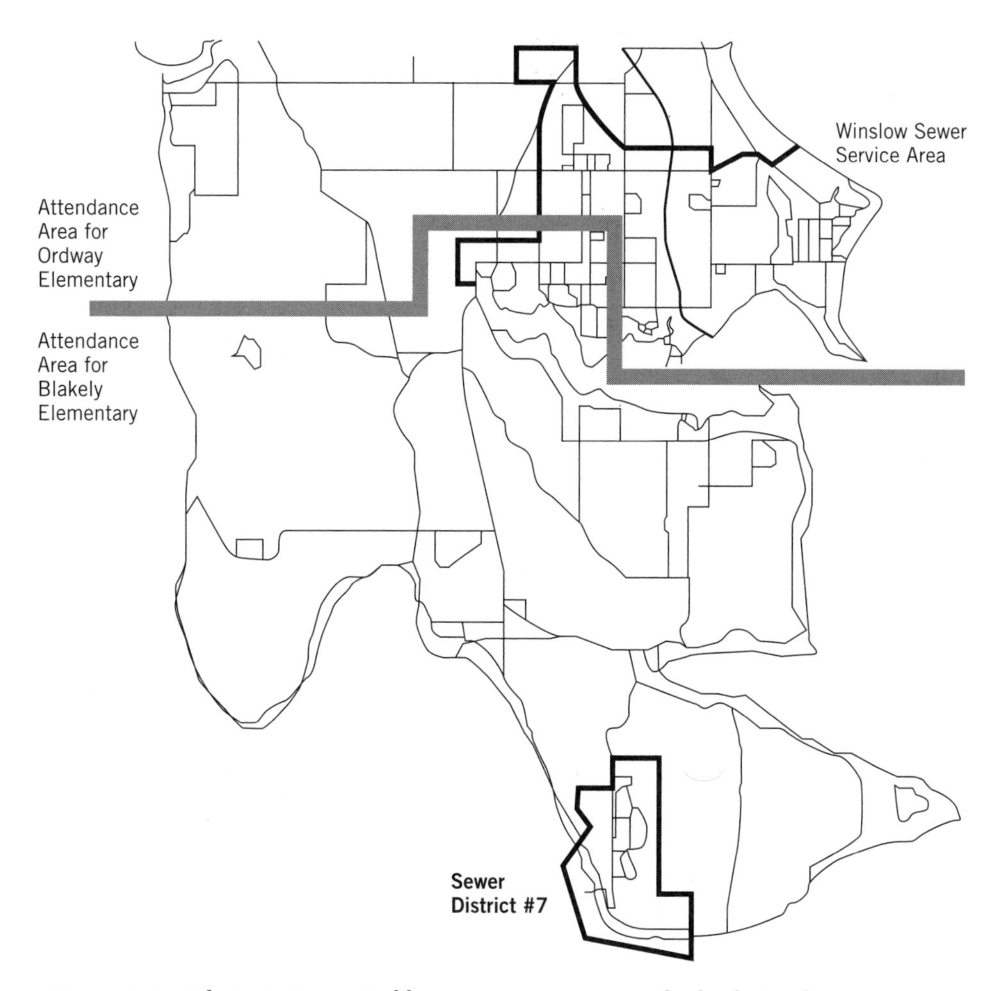

Figure 9-6: Administrative units like sewer service areas and school attendance areas cut across all other objects, particularly census blocks.

differences of various units of measure. The user must keep track of the operations that makes sense. The dasymetric method in the next example can introduce some realism into this process.

Areal interpolation works most directly in the relational tables associated with a choropleth representation, but there are some analogous operations in certain raster-based packages. These operations work on the basis of "zones," so that the cell simply becomes a representative of the larger areal unit to which it belongs. Zone operations take some careful thinking about the nature of the attributes. It is easier to think about the process for attributes of choropleth zones, and then implement them with the raster tools, if required.

EXAMPLES OF TRANSFORMATIONS

There are so many possible transformations that this book cannot give an example of each. However, a few brief examples may give a sense of the transformation process.

Dasymetric Mapping of Population Density

In 1936, John K. Wright, a geographer working at the American Geographical Society, produced a map of the population distribution on Cape Cod, Massachusetts. The source material was the census. Population counts were assigned to the cities and towns in the standard choropleth procedure (Figure 9-7a). Wright decided that the choropleth method masked some important elements of the distribution. The density of population in each unit depended more on the amount of marsh and wasteland assigned to the town than to the density in the habitable areas. First, he produced another density map, removing areas he deemed "uninhabited" and recalculated the density for the town on the reduced area (Figure 9-7b). Then, using what he called "controlled guesswork," he divided each town into regions based on land use and settlement. He was able to estimate some of the densities from the simpler towns that had just a few land use types. Once he had assigned densities to all but one, the density for the last land use could be calculated from the residual population in the town. Thus, he constructed a puzzle of densities from a whole series of assumptions and relationships. The boundaries on the map reflect breaks in the land use system, not just arbitrary political boundaries (Figure 9-7c). With a detailed land use map and disaggregated census data, the intentions of Wright's measurement could be carried out much more exactly than could be done in the 1930s.

Wright termed his method **dasymetric**. For many years, textbooks discussed dasymetric maps as some mystical process whereby the breaks in a distribution were related to some other source. The analytical clarity of Wright's method was obscured because the process was treated as map construction, not as a form of areal interpolation. Modern GIS tools provide the capacity to use multiple layers to produce a more reasonable interpretation of a measurement contained in a choropleth census map. Some current applications show the possibilities to reform the assumptions usually applied to areal interpolation (Flowerdew and Green 1992). Perhaps Wright's vision of integrating different sources could become a standard, rather than a footnote to cartographic history.

Wetland Regulation and Wasteland Assessment in Westport, Wisconsin (Area Cross-tabulation)

Sullivan, and others (1985) produced an analysis that hinged upon transforming a categorical coverage into an area attribute for another coverage. The problem originates

Dasymetric: A method proposed by John K. Wright to estimate densities using areal interpolation and quantitative puzzle solving; more generally, a type of thematic map whose boundaries are conditioned by some other distribution.

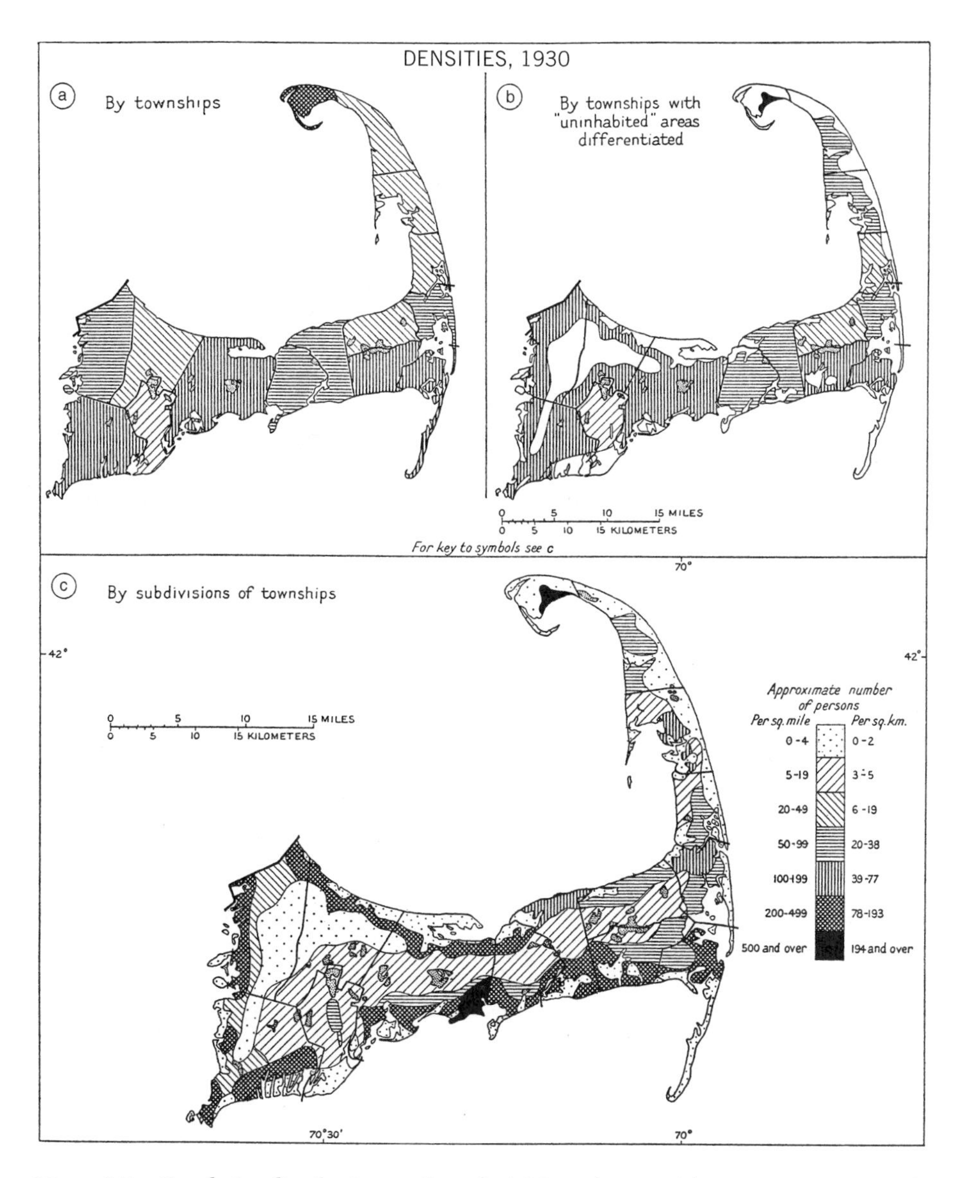

Figure 9-7: Population distribution on Cape Cod, Massachusetts: (a) persons per square mile by township, regular choropleth method; (b) choropleth method with "uninhabited" areas of towns removed; (c) dasymetric result. Source: Wright (1936).

in the social and administrative context. The State of Wisconsin has an ambiguous history in its treatment of wetlands. From the period of European settlement until recently, wetlands were not considered to be valuable. Eventually, the taxation of land developed so that "swamp and waste" lands were assessed at a much lower rate than normally developable land. This offered the opportunity for the state to recognize the diminished economic potential of wetlands. The responsibility for assessment is very local in Wisconsin, though operated under statewide guidelines issued by the Department of Revenue. Taxation is assessed by units of ownership—land **parcels**—because owners pay the taxes. Thus, an assessor must judge how much of the parcel is swamp and waste. Unlike other uses that are under human control, wetlands do not follow property boundaries. In Westport, Wisconsin, as in other jurisdictions, the local assessor had produced estimates of the area of swamp and waste for parcels. The figures tended to be in even units of acres—2.000, 4.000, and so on. In 1983, the typical value of a wetlands acre fell between $200 and $400 while agricultural land was valued at $1000 and more. The tax reduction of a wastelands assessment is directly proportional to the decrease in assessed value. Figure 9-8 shows the parcels whose assessments include swamp and waste for all or part of the parcel.

More recently, the environmental movement developed. Protection of wetlands was always a strong concern, ever since Aldo Leopold's *Sand County Almanac* (1949). Eventually, the state decided to protect wetlands from development. The Department of Natural Resources (WiDNR) was charged with producing a statewide inventory of wetlands, then working with each city and town to adopt zoning regulations to ensure that no further development occurred on wetlands. Wetlands are delineated from aerial photographs, using a categorical coverage approach. This project produced a digital coverage of the state, a pioneer in GIS implementation. Figure 9-9 shows the wetlands inventory for Westport, Wisconsin. The Wisconsin Wetland Inventory became a model for a nationwide effort called the National Wetland Inventory, though the national effort involves mapping but no zoning regulation.

Thus, Wisconsin has two programs based on wetlands. One gives a substantial reduction in property tax, and the other regulates activities. These two programs would serve as the carrot and stick, but only as long as they coincide geographically. In an era of manual record keeping, there was no way to ensure consistency between two programs based on such radically different measurement frameworks. Once the digital versions of parcels and wetlands existed, it took little effort to check the consistency.

To perform the analysis, the wetland inventory had to be converted into an attribute of the parcel. First, the two coverages had to be transformed into the same coordinate space. WiDNR operates on its own transverse Mercator projection, while the parcels were referenced to state plane coordinates (Wisconsin South Zone). The projection transformation is routine but a sign of the incompatibility of these two databases. The two layers were overlaid, producing the area of each composite polygon. The total area of wetlands could then be summed for each parcel. In principle, the assessor's estimate should match the area from the wetland inventory. In practice, the two areas differ significantly (Table 9-4). Interestingly, more parcels seem to be regulated without the tax break than the reverse. The differences are not just a matter of accuracy of measurement, because half the parcels were simply missing from one source or the other.

This very simple transformation shows some real environmental and political con-

Parcels: Contiguous unit of the earth's surface defined by a common collection of legal rights.

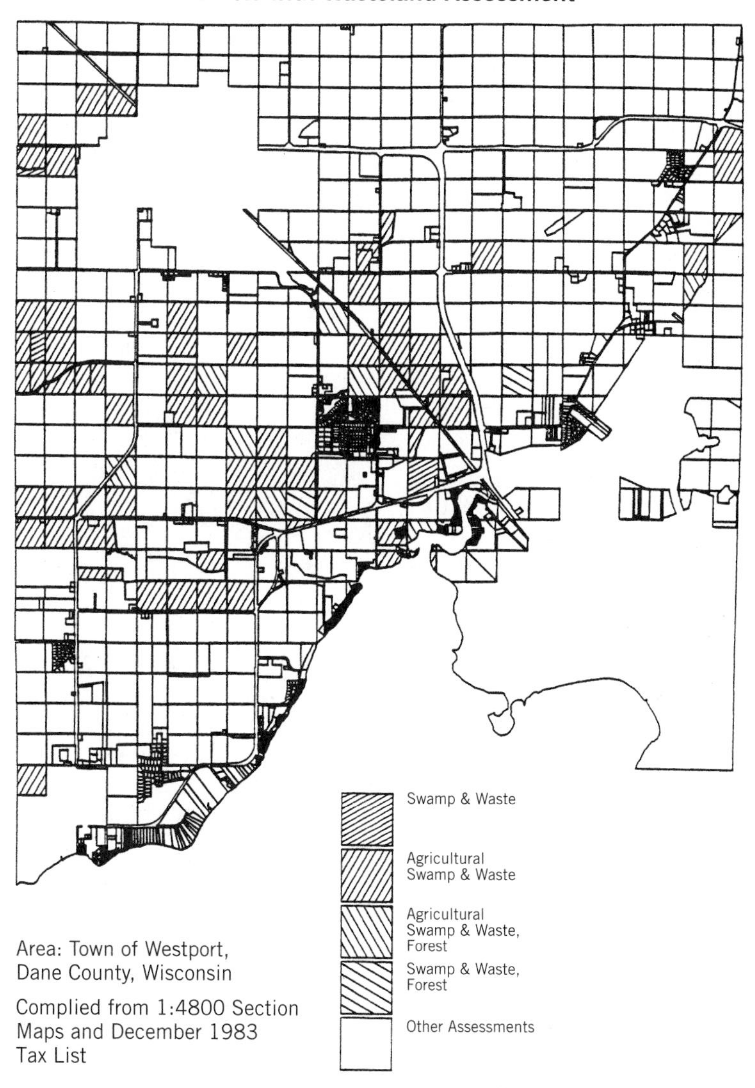

Figure 9-8: Swamp and waste classification for parcels in Westport, Wisconsin. The classification is often mixed with other uses, such as agriculture, to account for mixtures over the parcel. Parcel maps compiled from originals by Dane County Land Records Project. Attributes from 1983 tax list. Blank areas are incorporated municipalities with a different tax assessment. Source: Sullivan and others (1985).

sequences from the lack of integrated information. Two components of wetland policy did not affect the same landowners. Some became free-riders, while others were not getting tax relief they deserved.

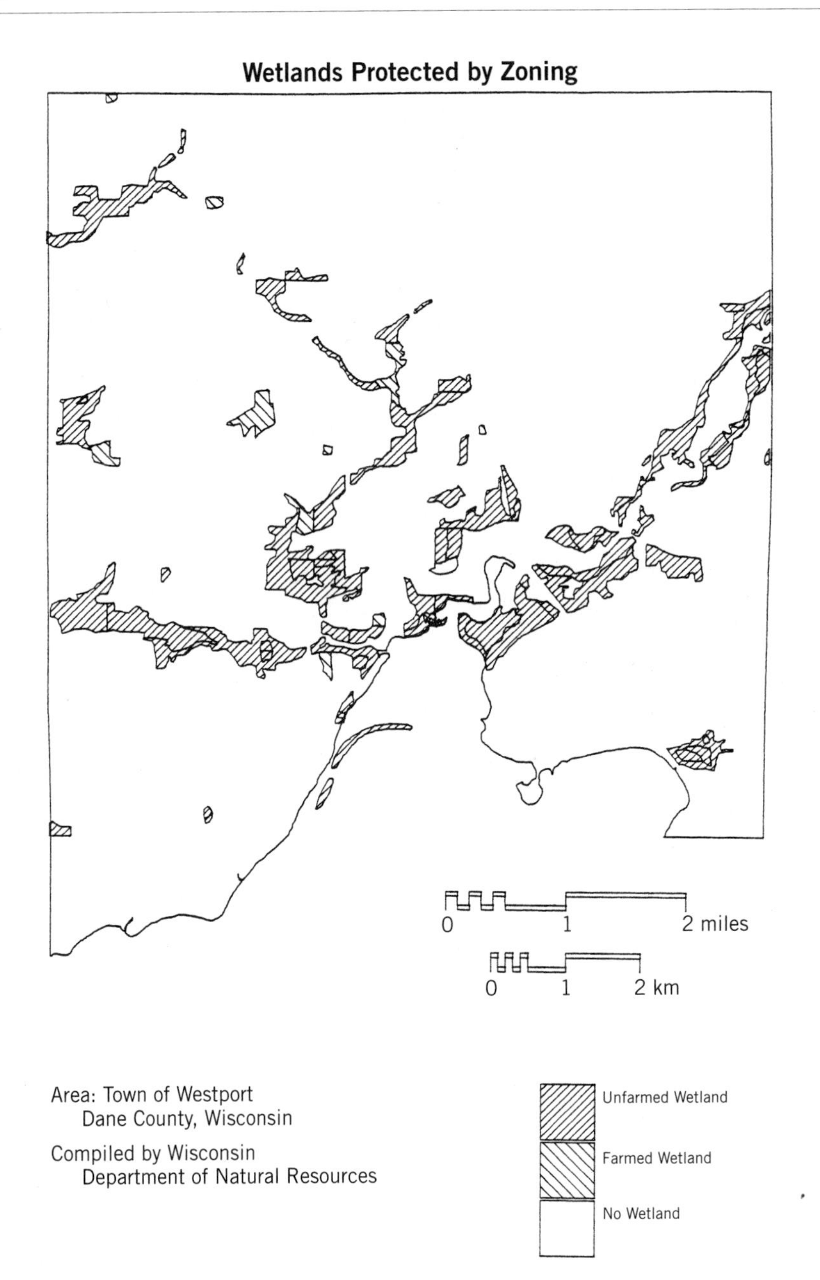

Figure 9-9: Wetlands Inventory for Westport, Wisconsin. Source: Digital data from Wisconsin DNR. The original inventory used a dozen classes, but the regulation aggregates them. Only unfarmed wetlands are regulated. Farmed wetlands are shown to determine whether they might be mistakenly given a tax reduction. Source: Sullivan and others (1985).

TABLE 9-4: Comparison of wetland inventory acreage to wasteland assessment, by parcel

Result of comparison	*Number of parcels*
Wastelands assessed, Wetlands not mapped	16
Wastelands assessed, only Farmed Wetlands mapped	3
Wastelands exceed Wetlands by more than 20%	36
Wastelands match Wetlands within 20%	19
Wetlands exceed Wastelands by more than 20%	7
Wetlands mapped, Wastelands not assessed	38
Total nonexempt parcels with Wetlands or Wastelands	119

Source: Sullivan and others (1985) Table 1

Forest Mapping for the United States (Resampling and Conversion of Imagery)

Covering large areas is critical to providing a consistent basis for administration and to providing information for understanding global environmental change. Recently, many projects have attempted to provide biophysical coverage at the continental scale, or for whole countries. The US Forest Service project to map the forest lands of the United States (Zhu and Evans 1994) provides a clear example of the role of transformations in such a project.

The US Forest Service produces a nationwide assessment of forest resources on a regular basis, but this process normally involves statistical tabulations, rather than map information. Comprehensive land cover coverage such as a nationwide mapping of the GIRAS program (Anderson and others 1976) or a satellite-based classification (Loveland and others, 1991) ends up with a few simple classes for forests and no information on the intensity of forest cover. These deficiencies are tied to the categorical coverage framework adopted for land cover mapping. The last nationwide forest type map (USGS 1970, pp. 154–155) dated from 1967, a part of the National Atlas.

To cover a region the size of the United States, the team chose the NOAA **AVHRR** satellite as the primary source. For a national map of forest cover, the resolution of 1 km seems appropriate. The project acquired nine composite image data sets from the USGS EROS Data Center. Each composite consists of pixels from a range of dates, not the instantaneous snapshot of an aerial photograph. Each pixel was

AVHRR: Advanced Very High Resolution Radiometer, a satellite sensing system operated by the US National Oceanic and Atmospheric Administration (NOAA) with a resolution of about 1.1 km, twice-daily repeat cycle, broad scenes (2400 km), and four bands of spectral data.

selected by having the highest NDVI from all the available measurements during that period. In part, this dominance rule tends to remove cloud-covered pixels and to select for the strongest vegetation response. To construct the composites, the raw measurements must also be registered and resampled into a uniform 1 km pixel system, based on a Lambert equal area projection. All these transformations are implicit inside the composite image database.

Forests, however, do not come in square kilometer units. To use the AVHRR spectral data, the project started with a nested design. The country was divided into 15 physiographic regions along the lines of Hammond's (1964) Land-Surface Form map, using a mixture of ecoregion concepts from various sources. In each region, a scene of Landsat **TM** was acquired. The Landsat TM has a resolution of 30 m and significant spectral resolution. Each TM scene was classified (using methods not described) into forest and nonforest. Then the TM classification was registered to the AVHRR database, giving at least 1089 TM pixels per AVHRR pixel. The binary forest/nonforest classification of the TM cells was tabulated into a percent forest cover for the cruder AVHRR cells. A regression analysis selected the best combination of AVHRR measurements to predict the variations in forest cover derived from the TM scene.

A common strategy in various sciences is to use an intense (and expensive) study nested inside a comprehensive one. Each regression model applies in its particular physiographic region. In some scenes, the models explained 85% of the variance in percent forest cover. Other regions went as low as 51%. The regression models were applied to the image database to produce the final percentage forest cover estimate (Figure 9-10). Some of these physiographic regions on Hammond's map have reasonably sharp boundaries, at least at the scale of kilometers, but others are much less distinct. More important, the ecological relationships implied by these regions may show gradual transitions over hundreds of kilometers. A single scene of TM data (covering an area 185 km on each side) was used to characterize the whole physiographic region. Thus what looks like a measurement of percent forest cover is a composite of a number of connected measurements, each with distinct assumptions. When dealing with continents or the whole globe, comprehensive coverage must be built from various nested sources and assumptions about relationships. This example uses one particular methodology. The field of global data is developing rapidly, mostly in finding ways to use the limited data sources and limited budgets to provide better value.

SUMMARY OF TRANSFORMATIONS

These few examples should show that practical projects with geographic information require many transformations between measurement frameworks. The assumptions required for these conversions are a critical part of the analysis and thus link to the scientific questions asked.

TM: Thematic Mapper, a satellite sensing system with resolution of 30 meters, 16 day repeat cycle, 185 km scene width, and seven bands of spectral data; launched by NASA.

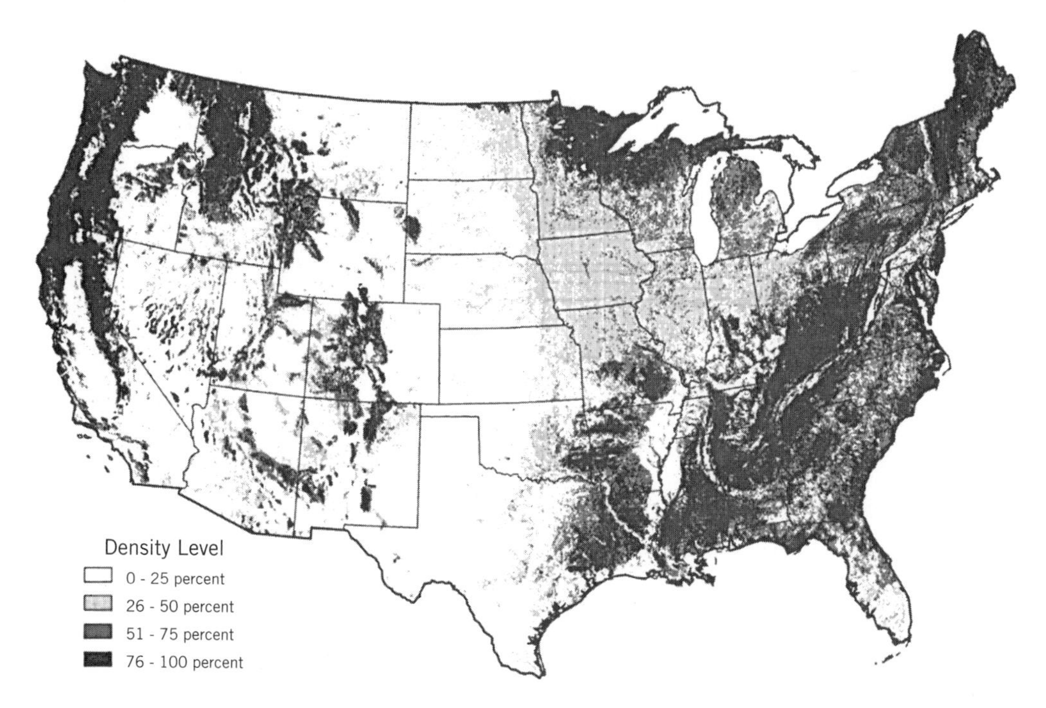

Figure 9-10: Predicted percent forest cover for the United States, produced by US Forest Service Southern Forest Experiment Station from 1991 AVHRR imagery and Landsat TM training sites. Source: Zhu (1994).

Overall, transformations require a geometric (neighborhood) component and an attribute component, the basic concerns in constructing measurement frameworks. Compared to the point–line–area taxonomy, measurement frameworks elucidate what is simple and what is more complicated. Looking back over Part 2, the operations presented in the earlier chapters can be seen as kinds of transformation. While Chapter 4 dealt exclusively with attribute operations, some of these altered the measurement frameworks. Overlay (Chapter 5) either works inside the attributes of a geometrically registered raster system, or it requires significant geometric processing to place vector coverages on a common geometric base. Buffers (Chapter 6) construct new spatial data from distance relationships. Surfaces and neighborhoods (Chapters 7 and 8) continue the connection between neighborhood and attribute handling toward more comprehensive objectives. Overall, the operations of a GIS should be seen as active transformations in manipulating the geographic information.

PART 3

THE BROADER CONTEXT

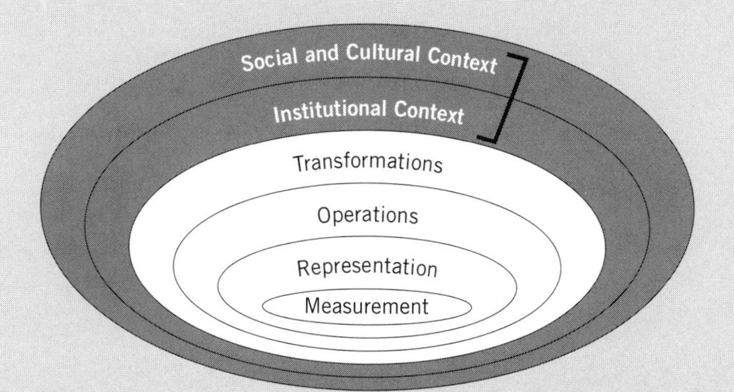

The measurement and representation of Part 1 and the operations and transformations of Part 2 make sense only when put to some purpose. Part 3 will address the issues of this broader context. Chapter 10 will consider the process that evaluates the operation of a system. Then it considers the process of implementation, how a system moves from concepts to practice. Chapter 11 will sketch out the social and institutional context that constrains the production and consumption of geographic information. These two chapters progress rapidly through the outer rings of the concentric diagram for the book presented in Figure P-1.

CHAPTER 10

EVALUATION AND IMPLEMENTATION

CHAPTER OVERVIEW

- Review measures used to evaluate geographic information systems.
- Describe database design procedures to implement a new GIS.

So far, this book has moved from the details of specific measurements through the tools that operate upon them. This tool kit will remain lifeless, however, until put to some use. In a realistic situation, this process involves considerations far beyond those discussed. In part, a GIS must operate within constraints of budgets and technical resources available. In addition, the technical process must be managed in a way that accomplishes the objectives of institutions and the people who work within them. All these issues influence the design of a geographic information system. The design process occurs throughout the life cycle of a system, but it is most apparent when implementing a new system. This chapter will thus concentrate on the implementation process, but first it must consider the various ways to evaluate a geographic information system.

TECHNICAL EVALUATION

Evaluation compares results to intended goals. There are many possible ways to evaluate geographic information, from technical details to broader goals. At the most basic level, geographic information must be evaluated for how well it portrays the actual variation in the chosen phenomena. Thus, the simplest evaluation assess the *accuracy* of the database. However, accuracy cannot be the only measure; the next level of concern must consider the resources expended to obtain that level of accuracy—the *efficiency* of the system. Finally, evaluation should consider the *effectiveness* in serving the larger purpose. This chapter will begin with accuracy and efficiency issues. Effectiveness issues will receive greater treatment in Chapter 11.

This section will begin with a review of evaluation of the data itself—data quality assessment, followed by evaluation of the allocation of resources—computer, financial, and human. These technical concerns are important to describe the content of the information, to set limits to its possible uses, and to become a part of the other kinds of evaluation. Each component may be taken as the sole measure at a particular stage in developing a GIS. It is important to retain a comprehensive view that balances these concerns.

Data Quality Assessment

Issues of data quality have been discussed in virtually every chapter. The choice of measurement framework described in Chapter 2 and the structures used for representation in Chapter 3 form the basis for evaluation. Each of the chapters in Part 2 showed how operations could provide means to study data quality. All these tests should be collected in a coherent statement concerning the data quality, using the five categories of the SDTS data quality report (described in Chapter 3).

Measurement Frameworks and Accuracy It is important to remember the role of measurement frameworks in organizing the approach to data quality. Geographic information always involves some statement about time, space, and attributes. Because the common frameworks concentrate the measurement efforts on a single one of these, it is all too easy to focus attention of accuracy assessment on that component alone. For example, cadastral surveying is usually assessed for its positional accuracy. The procedures for surveying involve repeated geometric observations (of distances and bearings or both) that can be converted into a geometric representation such as coordinate values. The repeated measurements give a measure of the coherence of the network and an error estimate for each point. It must be remembered that such measurements are only possible for the best-defined geographic objects: the brass caps of geodetic monuments, nails driven into the pavement, or iron pipes at property corners. There are other parts of the landscape that are ignored in making this abstraction. Some of them are occasionally important to a surveyor's construction of the property boundaries. For instance, the crumbling bank of a river is an active part of the landscape where land can disappear as owners watch. Centimeter accuracy is worthwhile only in such cases when it is carefully bounded in time and linked to an estimate at another time.

As another example, many property boundaries extend to the "center of the road," but this term has multiple interpretations. The road can be located by the legal boundary of the highway **easement**, the center of the pavement as constructed by the Public Works Department, or the yellow stripe painted to control traffic. Each of

Easement: A legal agreement that grants partial rights over a portion of a property, such as a utility easement to install overhead wires or underground pipes. Also applied to rights-of-way for roads.

these centers is perfectly reasonable from a given perspective, but there is no guarantee that they fall in the same place. At the northeast corner of my property (a point described as the center of the road), these different definitions of the road center lie more than a meter apart; the physical road does not lie in the center of its legal easement. Due to constraints of a drainage ditch and a slight curve, there is 20 cm more paved area to the south (my side) than the north of the yellow stripe. And the painted stripe is 5 cm wide anyway. So where is this corner of my parcel? Three different perspectives point to different locations. While the geodetic surveyor may be horrified at such imprecision, for all practical purposes, the road serves as a much more distinct functional boundary than any hidden set of brass-capped monuments. The goal of positional accuracy must be bounded by a sense of the purpose in having the information.

As another example, a forest inventory is directed by a need to obtain attribute information. Forest managers need to know the ages of trees and their species to plan the cycle of harvest and related activities. Wildlife studies need to know the species and the structure of the canopy to predict habitat for various animal species. To respond to these needs, when a remote sensing project delivers a forest inventory, the accuracy assessment concerns the classification into various classes of tree species, age, and structure. These attribute classes are packaged inside a framework for geographic measurement and representation. There are remarkable differences between the model of a homogeneous **stand** map interpreted on an aerial photograph and the model of pixels classified based on multispectral reflectances. Often the expectations of foresters regarding the traditional stand map are carried over to image classification. Both models relate to aggregate measurements of the forest, not an inventory of each tree. The positional component of a forest stand map is treated differently and should be evaluated just as carefully as the classification. Does the forest consist of plant communities that really can be captured by 40 acre minimum mapping units or pixels of a given size? How much mixture should be expected within these geometric objects? As in all mapping, there are substantial assumptions required to obtain any measurements. An accuracy estimate is bounded by the measurement framework used to obtain the information in the first place.

Whatever the specific situation, a testing procedure can provide an external verification of an information source. Both the original data source and the testing process involve assumptions. If these do not match, the test may only tell you about the differences in the assumptions. Data quality also varies by region and landscape type. Certain rules and assumptions that may work in the large fields planted with genetically uniform hybrid seeds of the American Midwest may not apply at all in a tropical rain forest with hundreds of plant species per hectare. Each landscape has a different dynamic involving the sizes and shapes as well as the nature of the objects. Thus, testing cannot be done once and for all in some generic landscape for application worldwide.

Stand: A contiguous area of a forest considered to be homogeneous in its ability to support the intended forest crop. Used to direct forest practices such as planting, thinning, and harvest.

Strategy for Testing The concern over accuracy does not stop with data sources. All the subsequent operations and transformations can lose information in various ways as discussed throughout Part 2. Too many GIS users acquire a carefully crafted database, built around certain assumptions, and then extract some element from its context and employ the information for some purpose to which it is not suited at all. Rigorous attention to data quality testing (along with publicity for data quality statements) may help avoid these problems.

Conducting data quality assessment can be arranged in a kind of pyramid, arranged in terms of frequency. Some testing must be comprehensive. For example, tests of logical consistency check the relationships between elements of a database entirely using internal evidence. Other tests use repetition to cross-check a process. For example, a *check-plot* can be overlaid on the original graphic to test whether digitizing faithfully reproduced the original map. Moving up the pyramid, it may be possible to calibrate the accuracy and reliability of each kind of equipment and each kind of procedure. At the top of the pyramid, data quality is ultimately tested by a comparison to a source of higher accuracy. This requires detailed fieldwork or intensive resources applied to a restricted sample. Each part of a data quality program serves to support the others, giving an overall evaluation of the information.

Allocation of Resources

Taken by itself, accuracy is an unforgiving enterprise. It is always possible to obtain higher accuracy, to refine the measurement framework to reduce the discrepancy between the database and the landscape. The next level of evaluation compares the achievements in accuracy against the resources expended, thus measuring the *efficiency* of the process. This section introduces three kinds of resources (computer, financial, and human) that lead to different measures.

Computer Resources The resources expended on information can be measured on many scales. Sometimes they tend to correlate, but not always. At the most basic level, information consumes bytes of storage and operations consume processor time. Each structure for representation and each operation will have some characteristic trade-off between resources and accuracy in the result. Some of these relationships are simple to calculate. For example, the number of pixels in a raster representation goes up with the square of the refinement in the resolution. Processing time for some operations will increase in a linear manner with the number of records, but other operations (particularly the incremental operations described in Chapter 8) will be more complex. Goodchild and Rizzo (1987) found that a certain set of operations fit best to a double logarithmic model, predicting rapid increases for very large analytical problems.

In the early days of GIS even simple databases strained available computer capacities. Memory posed a substantial constraint on programming, so complex operations required sorting and other preprocessing. With batch processing systems, turnaround in one hour would seem quick. Now that the routine workstation used in GIS far exceeds the wildest dreams of supercomputers 20 years ago, constraints on processor speed and storage capacity might seem obsolete. However, the software vendors have let their systems expand to fill the memory available, often without adding much more than a user interface to software models from the older period. As computers get larger and faster, the size of problem should also expand. It will not expand as far as it could, because software that uses inefficient algorithms will not deliver the full increment in computer speed. The result is that users are placed on a permanent spiral. Current computers are always a bit underpowered for the current operating system and the expectations of response.

Financial Resources Costs have been, and will remain, an important criterion for evaluating geographic information. One way to interpret the limitations of the manual cartographic system is that costs controlled the flexibility of the products available. The amount of computing resources available is also controlled by costs, though not always allocated appropriately between the components of the computing system. An economic analysis of geographic information may be engaged at a number of levels, with potentially different results.

At one level, good management chooses the cheapest method to achieve the desired result. In this approach, the goal is to reduce costs and thus to improve efficiency. There have been a number of studies designed to prove that digital methods of cartography can reduce costs compared to previous technology (Joint Nordic Project 1987). Gillespie (1993) reviewed 30 cost reduction studies conducted for various elements of the federal government and produced a regression model that predicts the ratio between GIS-based costs and previous manual costs. Overall, GIS reduced costs, thus increasing efficiency for performing the basic tasks. The study suggests that efficiency increases for land management applications and decreases when adversarial hearings are likely or when the manual operation was costly. While small applications are less efficient, efficiency also declines as the mapped area (adjusted for scale) increases. This suggests that GIS projects are most effective at an intermediate size that balances the specific and the general. These conclusions may only relate to the particular mix of uses performed by the US federal government, not to the broad spectrum of uses.

Many difficulties arise in limiting the evaluation to costs. First, current expenses may be difficult to measure. Geographic information is an ingredient in many economic activities and is not well centralized or particularly prominent. Few studies of the costs of geographic information exist; one comprehensive example covers France in 1987 (Conseil Nationale de l'Information Géographique 1990). Still, it is clear that current costs are not trivial. The cost of computers is usually much lower than the cost of salaries for those involved in producing and consuming geographic information. If these expenses can be redirected into a more modern technology, it seems like a prudent step.

Another difficulty with a cost analysis is that it misses the other side of the balance. The modern technology may produce a different result, not an exact substitute for the old ways. Part of the advantage of digital cartography is that it changes the possibilities, making some things cheaper than others. One method of economic analysis balances costs against *benefits*. Increased accuracy is difficult to evaluate on a financial scale. The clearest measure of benefits would be *avoided costs*, the costs that the users would have spent if the information had not been available. This was the strategy taken to determine the value of the geodetic reference system in the US (Epstein and Duchesneau 1984). The study estimated the additional costs required to integrate information if there were no geodetic framework available. The joint nordic study of benefits and costs took the similar approach by asking the costs of delivering the current digital products using the old manual systems (Joint Nordic Project 1987). In principle, the results of these studies sound wonderful, but the exercise of hypothesizing mapping without geodetic reference is a bit like trying to design automobiles without the wheel.

A study of the value of geological maps (Bernkopf and others 1993) took yet a different approach, focusing on the role on information in reducing risk. Regulations and environmental controls must be imposed based on the information available. In a small study for one county, this project uses two land planning applications to measure the value of having more detailed geological information. They argue that the availability of detailed geological information would alter the nature of engineering for slope failures along new roads. It would also alter the location of a landfill. The costs of the unnecessary slope mitigation would pay for the more detailed geological survey by itself. Unfortunately, such studies are rare, and their many assumptions are difficult to test. It is not clear that highway construction would be as risk-averse as required by the model. Engineers might approximate the differences in geology from experiences of trial and error. In other words, it is hard to reconstruct a world without some form of the information available. Finally, it is unlikely that the political process would allocate savings directly to the appropriate form of information gathering.

For each of these models that balances costs, benefits, and risks, a numerical comparison requires common units of measure, just like the combinations of attributes discussed in Chapter 5. Some benefits can be converted to a cost basis more easily than others. Some simply argue that certain benefits will never be quantified (Dickinson and Calkins 1988). Economists do offer a variety of methods to reveal value indirectly (Wilcox 1990), though requiring more assumptions. It is much more complicated to evaluate the costs of contaminated water produced by a poorly located landfill than it is to project the cost of a highway retaining wall. Also, like many economic decisions, time becomes a key element. Many of the benefits of a modern information system arrive over a period of years following the initial expenditure for conversion. Altering the **discount rate** can produce different results from the same figures.

Quantitative measures of efficiency are supported as a part of a rational planning method. A single index, like the ratio of benefits over costs, is seen as a consequence of the need to choose amongst alternatives. However, it is clear that society does not

Discount rate: Economic factor that deflates a future sum to make it comparable with current expenses. Reflects expectations of interest rates and inflation.

operate all choices out of a collective benefit–cost pool. The choice of a highway retaining wall in Virginia does not really compete with hospital construction in Vermont and school lunch programs in Los Angeles, let alone clean water in Mali, refugee settlement in Cambodia, and other potential expenditures. There are other rules that become involved in making choices, such as equity and the other social concerns covered in the next chapter. From a financial perspective, it also makes much more sense to view geographic information as an investment in risk reduction. Investments involve all sorts of suppositions about future needs, not some magic single index.

Human Resources An information system cannot be mechanized to the point of measuring it entirely in terms of computers and financial costs. The quality of the information, the quality of the service provided to users depends on the intelligence of the staff. The management of an information system has to consider the role of the people in the system. In my experience, for example, a digitizing operator who understands the particular maps makes fewer errors. Milwaukee experienced such bad performance from workers hired under an unskilled jobs program that the work had to be done again (Huxold and others 1982). A major cost in operating a GIS comes from the need to train new staff, and staff turnover may be influenced by some of the choices in organizing the system. Some cartographic workplaces have been redesigned using the quality center approach that puts the workers in charge of a whole product from start to finish (Bie 1984). Many measures of productivity are not just a matter of the technology and the money expended, but also how it is organized as an institution in a larger society. This concern will be continued in the next chapter.

If there are difficulties in developing a universal calculus of efficiency, measures of effectiveness are even more controversial. To create a numerical scale, the disagreements over priorities must be adjudicated. Thus, the issues of effectiveness belong in the next chapter as a part of the social and cultural context.

IMPLEMENTING A GIS

Rather than seeking a magic bullet in a single measure of evaluation, a practical solution involves a design process that incorporates all the concerns introduced in the previous section. The design process is best studied as applied in creating a new system. In part, implementation of new systems dominated the early phase of GIS, but a new system also offers the clearest opportunity to choose between competing alternatives. Once a project selects a particular path, it becomes tied to all the consequences of the decision. For example, the initial stage might consider many alternatives for hardware and software. It becomes much more complicated to alter these decisions once the staff training, the existing hardware, and all the other factors become entrenched. Thus, it is important to make the initial selection as carefully as possible, then to revisit the decisions on a regular basis.

In some cases, the process is well considered, rational, and expensive. In others, the same items may be bought with much less formal consideration. Satisfaction and results do not necessarily follow from either approach. Many implementation projects unleash a flurry of self-congratulation from the technology-driven participants. In retrospect, some of the early claims look foolish. Careful application of evaluation at various levels might reduce inflated claims.

The formal approach to selection has its origins in the systems analysis methods used in many fields, but particularly applied to data processing systems. This approach tries to integrate all the measures of resources developed above into a comprehensive process to design and implement a new system. There are many variants of the process, often with conflicting terminology (Dangermond and Freedman 1984; Guptill 1988; Ventura 1991). The sequence given in Table 10-1 covers the major steps. This section will outline a strategy to approach implementation in three basic stages: needs assessment, requirements analysis, and construction tasks that produce the operational system.

The process actually begins before a needs assessment with *awareness* that a GIS might be of some value. This step involves education, but it often works best with

TABLE 10-1: Sequence of steps to implement a GIS

Stages	*Tasks*
Preliminaries	Create awareness
	Agreement to begin implementation process
Needs Assessment	Interviews, surveys
Data oriented	Sources, flows, transactions
Process oriented	Operations, transformations
Product oriented	Maps, graphs, tables, reports, decisions
Requirements Analysis	Match needs to current technology
Identify technical constraints	(Long-term)
System Design	Specify data models
	Create implementation plan
Generate Request for Proposals (RFP)	(Short-term)
Evaluate responses to RFP	Select vendor
Benchmark	Measures performance of system
Construction	
Pilot Project/Demonstration	Trial run of proposed design
Reevaluation	(Continual?)
Conversion	Digitize existing records; convert operations
Quality Control	Evaluate information in system
[Return to start?]	

Source: Based on Guptill (1988) and Ventura (1991).

some proponent who can demonstrate the potential utility of a GIS. Often the process of awareness begins by observing the results of another project nearby or the results of a pilot project. These elements of the context surrounding the actors will reappear as a theme in Chapter 11. Preliminary steps may not be very formal, but they create the willingness to start the first real phase, the needs assessment.

Needs Assessment

A needs assessment assembles information about the potential uses of the proposed system. It can be conducted through surveys, interviews or other methods that elicit the opinions, desires, and requirements of the relevant parties. This study should explore three aspects of needs: data, process, and products. These components are relatively difficult to disentangle. If you ask someone about their information needs, they may describe a desired output, a procedure used to make it, and the information required as input without really seeing these as distinct viewpoints. It is particularly difficult to respond in terms of long-term needs without reproducing the way things have always been done. A careful survey will try to determine why each potential user wants certain products, employs certain methods, and accesses certain sources. Any or all of these choices may need to be replaced or reworked.

Inevitably, some users are more able to articulate their needs and to phrase their needs in the terms that fit the GIS. This does not mean that these particular needs are more worthy or even the easiest ones to meet. In addition, a survey of users may focus more on the specific products provided by the previous technology, avoiding the opportunity to redesign for the different capabilities of the new technology. Still, these difficulties do not diminish the need to understand a broad range of requirements.

The needs assessment phase requires particular attention to the flow of certain key elements from one organization to another. Manual mapping systems tend to create duplicate copies of certain base maps, updated independently with slightly different content. In one study of an Ontario city, there were 17 copies of the parcel map maintained in separate offices for different purposes (Dangermond and Freedman 1984). These situations call for a redesign of the flow. However, some of the seeming duplicates may reflect subtle differences in definitions between the various participants. The user needs survey should not just assume that the word "parcel" means the same thing between the registrar of deeds, the assessor, the surveyor, the highway crew, and the conservation officer. The specific meaning will have to be recorded, so that any differences can be addressed in the later design and implementation stages.

A needs study may need to be reasonably formal when an outsider does the work. The outsider may try to elicit accurate information from the various participants, but there may be missing elements of the story. Participants may respond to the questions but leave out important information about motivations and past events. An insider may know many more pertinent details and motivations but be much less likely to be neutral or seen to be neutral. Many successful systems developed from collaborative efforts involving cooperation in defining common goals. Such a process cannot develop overnight. The next chapter details some approaches to foster cooperation and collaboration.

A needs assessment produces a list of data, processes, and products that should be accommodated in the final system. In practice, these may not be specified exactly in terms of specific records or specific output, but simply in terms of the general purposes that the information serves. For example, one study created a list of 33 "generic municipal tasks" (Table 10-2) (Dangermond and Freedman 1984). Less specific task description actually may be more helpful, since it focuses attention on the reasons for the information, not the specific map or report that has always been done a certain way. The generic functions demonstrate that municipal government works with specific transactions, like permits, inspections, and dispatching where time plays a much larger role than seen in a map-based view of the world. These tasks can be served by a GIS only through some care in design. A task, such as approving a permit, involves a flow of information. First an applicant requests the permit for a particular purpose at a particular place. The agency must assemble the material related to that place to determine whether the permit should be granted. A particular permit might involve

TABLE 10-2: Generic municipal tasks

On–Going Procedural

Acquire and dispose property
Process and issue parcel-related permits
Perform inspections
Provide legal notification
Issue licenses
Conduct street naming; create street addresses
Review site plans, subdivisions
Perform event reporting
Conduct dispatching; perform vehicle routing
...

On–Going Managerial

Create and manage mailing lists
Allocate human resources
...
Perform database management
Conduct development tracking
Disseminate public information
Respond to public inquiries

Source: Selected from Dangermond and Freedman, 1984, p.13 Table 1.

specific requirements of zoning or building regulation or some combination of environmental regulations. Each specific case triggers a different group of attributes, but the generic information handling should share many characteristics.

Once assembled, the needs assessment inventories what is available, what is wanted, and how people currently try to connect these two. The next steps attempt to satisfy these needs with a design for a revised system.

Requirements Analysis

The next step designs a system to respond to the needs through a requirements analysis, leading to a bidding process to acquire the appropriate tool kit. The core of this process should be the design process, though the bureaucratic procedures of bidding can often overwhelm everything else.

Developing on the needs assessment, the design stage creates a specification for the system, including all the components discussed so far in this book. There are choices of what to measure and how to obtain the measurements. Then there are choices of representation and the set of operations and transformations to produce the products. When designing a new system, there has to be considerable attention to the conversion process discussed in Chapter 3. Many projects are designed around the conversion process with less attention to the analytical stages than they deserve.

No magic rules govern effective system design. Design can proceed either from some simple overarching principles or from some inescapable details. These contrasting methods of attack, top-down and bottom-up, are often just a matter of perspective. Any view from the details already accepts a framework of axioms concerning measurement and the realm of the possible. Similarly, a purely top-down approach usually starts with a fairly good idea of the nature of the particular kind of system. This section presents a commonly adopted procedure centered on database design, followed by an alternative view centered on geographic measurement issues, and then the bidding process.

Database Design The system design must decide about database entities—what objects exist, relationships—how their geometry and attributes interrelate, as well as the assumptions that are accepted by all participants. In database terms, these decisions are called a **database schema**. In some cases, there is little controversy about the schema. In others it can take some detailed negotiation to determine the exact definitions of the categories that will be used and the nature of the attributes. When needs assessment has identified duplication between different organizations, the schema should decide which version should become the basis for a single, common version and how each participant will attach the appropriate attribute values for their purposes.

The schema creates a set of tables for attributes, along with their relationships.

Database schema: Logical arrangement of tables, attributes, and integrity rules to structure a database. Involves definitions of entities and their relationships.

Integrity constraints such as a single owner for each parcel become implied by the structures created. At a finer level, a schema is not enough. Each table requires a **data dictionary** to define each field (attribute), the scale for each numerical attribute, and the codes to be used to represent categories. If the meanings of these details are not decided in advance, considerable confusion can result and effort can be wasted.

Some designs are built around the specific set of tasks discovered by the needs analysis. The specific operations and transformations are specified in some detail. Such careful attention to fulfilling the original needs may miss some of the important power of the GIS approach. While the prior manual mapping era, being based on printing technology, had to produce the same product for all, the GIS can produce maps at will. The product can be tailored to the specific requirements of each user. Eventually, this means that a system design does not need to specify how to produce each potential product, as long as the schema and the tool kit are sufficiently general.

System design may seem like a technical operation, a matter of organizing the computer to produce certain products when required. Good designs, however, are rooted in the context of the organizations and the people who participate. A centralized design may seem more efficient, but often a federated database with clear rules about interconnections will make more sense in everyday operation.

One of the main constraints on a system design is in the minds of the designers. It is hard to grasp all the possibilities of a new technology. The old ways of doing work pervade the thinking for many years. Then, the specific character of a particular system can also limit the perspective of the design. Rather than trying to solve all these problems at once, system design strategies built on prototypes and iterative approximation may make more sense.

Approaches to Geographic Measurement The database design process implies an approach to measurement. Measurement of geographic phenomena can be quite costly in time, personnel, and other resources, so the design process includes consideration of measurement issues. Choosing a measurement framework often involves considering the whole database comprehensively. There are many approaches to direct and indirect measurement, lying on a continuum that characterizes the approaches to land inventory, running roughly from a *parametric* extreme to an *integrated* extreme. As GIS tools have developed, the balance between these extremes have changed without altering the fundamental dilemma.

One extreme would require a specific survey, a full measurement process, for each individual attribute. Ideally, such a survey would be tailored to each request. Unfortunately, such an approach would have serious drawbacks. Each user would have to pay for the full cost of the field data collection process. The time to perform the survey might exceed the deadlines of user needs. For example, emergency response to an oil spill cannot wait for a tidal wetland inventory to be constructed from scratch. Despite the drawbacks, there are some phenomena that must be measured by

Data dictionary: Detailed definitions of the codes employed for identifying objects and for attribute values.

special purpose surveys targeted at one element. Some of the reasons are technical or scientific—related to the methods of collection or related to the nature of the phenomenon itself.

Other reasons have less linkage to the technical problems, but arise from the human element, either institutional or societal. Certain organizations have a charter to work in a specific bounded framework. Thus, the legislature in its wisdom might commission a wetland inventory one year and an agricultural inventory the next. One will be the mission of the Conservation or Wildlife Department, while the other will be assigned to Agriculture. Although these surveys have many common elements, it is virtually assured that the two projects will remain independent for bureaucratic reasons. Two air survey companies may easily sit on the same runway, waiting for clear weather to take aerial photographs of essentially the same area, at the same altitude, and with the same type of film for two different clients.

Whether imposed by technical or institutional reasoning, a parametric approach seeks to collect each attribute in an independent data collection process, tuned to that particular set of demands. Generally, a parametric survey provides a data product that measures the phenomenon efficiently, when taken in isolation. The database consists of many layers, each tuned to its individual purpose but unlikely to be consistent.

It is easier to understand the continuum by presenting, in equally extreme form, the opposite approach. In place of the parametric approach, one could conceive of a single comprehensive survey that collected all there was to know about a given place. Such information, once collected, could be placed into a massive database, and then any particular theme could be abstracted from it.

There have been attempts to produce land inventories that combine multiple perspectives; one example comes from Australia in the 1960s. Based on the geographic traditions of earlier decades, an interdisciplinary group working in the Soils Section of the Commonwealth Scientific and Industrial Organization (CSIRO) developed a method of land evaluation that attempted to pull together the opinions of various disciplinary specialists (Christian and Stewart 1968). Rather than producing separate maps for geology, soils, vegetation, and other physical characteristics, the team of specialists collaborated on an *integrated* inventory. The resulting **land system** classes tied all the themes together into a coherent explanation of the interactions between the elements. In the integrated surveys (Figure 10-1), each land system had a specific relationship of geomorphology, soils, and vegetation within the larger regional geologic and climatic systems. The lengthy text in the legend and the mention of interactions emphasize the difference between this method and a single parametric map.

Two lines of argument support the integrated survey method (Mabbutt 1968). One has a technical basis, arguing that a single map involved less work. In particular, the process of compiling a series of different maps into a coherent analysis was quite difficult using the technology available. As Part 2 demonstrated, the modern GIS pro-

Land system: A unit of an integrated survey where common processes (involving soils, vegetation, and other factors) have created a combination of features that will support a particular group of uses.

Lee's Pinch Land System (1386 Sq. Miles)

Geology.—Triassic sandstone and minor shale.
Rainfall.—22-30 in.
Locality.—Southern mountains.
Elevation.—500-3300 ft. **Local Relief.**—Up to 2500 ft.
Wooded Area.—100%.

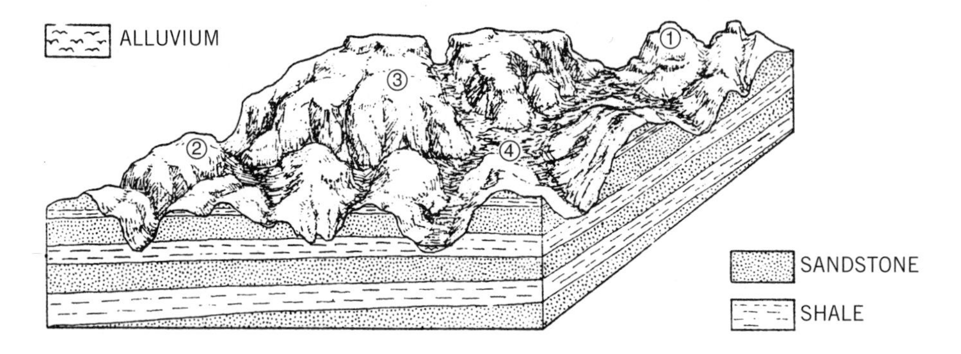

Unit	*Area (%)*	*Land Forms*	*Soils*	*Vegetation*
1	30	Rugged hills with rounded summits; irregularly benched slopes often littered with boulders and with very frequent sandstone outcrops including low cliffs up to 30 ft. high; fairly narrow flat-floored valleys 400–1000 ft deep	Mainly shallow coarse-textured skeletal soils and bare rock; in moist cool sites humic surface-soils; infrequently on interbedded shales or arkosic sandstones shallow podzolic soils (Binnie, Pokolbin); in stable sites coarse-textured earths	Shrub woodland of ironbark and gum 40–80 ft high, ironbarks common, with *E. punctata*, *E aggiomerata*, and *E. oblonga*, and with scattered or dense *Callitris endlicheri*, *Casuarina torulosa*, and *Persoonia* spp. below; shrubs usually abundant and mixed, Leguminosae common; ground cover poor, of grasses and herbs
2	30	Rugged hills margined by sandstone cliffs 50–500 ft high usually overlooking steep shaly slopes littered with boulders; cavernous weathering of the cliffs; narrow inaccessible valleys 500–2500 ft deep	Similar to unit 1; predominantly coarse-textured non-humic skeletal soils; probably more bare rock	As for unit 1, but with more herbs, shrubs, and non-eucalypt trees in ravines and at bases of cliffs
3	35	Stony, hilly plateaux with ridges and escarpments up to 200 ft high; very steep margins including cliffs up to 100 ft high; narrow gorges along the major rivers	Restricted obsevations; similar to units 1 and 2; deep yellow earth (Mulbring) in level, stable site on plateau	Shrub woodland of ironbark and gum 30 ft high, including *E. punctata*, *E. trachyphola*, and stringybarks; ground cover poor; many non-eucalypts in ravines and at bases of cliffs
4	<5	Sandy alluvium occupying valley floors in unit 1; liable to frequent flooding and deposition of sand in middle and upper reaches	Restricted observations; deep sandy stratified alluvial regosols (Rouchel); sedimentation in valley bottoms frequent and calamitous owing to low soil stability on sandstone hills	Shrub woodland or ironbark and gum with an admixture of non-eucalypt trees, sometimes cleared and under pioneer grasses

Figure 10-1: Diagram of the land units of Lee's Pinch land system from an integrated survey. Source: Story and others (1963).

vides many of the tools that were lacking. The second argument is more enduring. Whether dealing with the physical environment or human agency, many seemingly independent factors connect. Spatial distribution of these factors further entangle interpretation. Separate parametric surveys will therefore detect some of the same basic features in the landscape. When made independently, boundaries may come close, but there will be small disagreements that will generate *slivers* of marginal significance. The integrated method tries to resolve the discrepancies using fieldwork and negotiation between the expert interpreters from each discipline.

Between the parametric and the integrated approaches, the intermediary steps on this continuum are intentionally vague. The pure parametric approach is not practiced in most cases because it is usually possible to connect a number of distinct attributes to a common set of objects. Similarly, the attempt to design a single integrated survey invariably omits some phenomenon that must be handled separately.

The soil survey lies on the continuum toward the integrated end. The soils taxonomy is based on a number of criteria that include parent material, drainage, and other factors. In addition, the soil mapping unit combines the soil class with a slope class. There are compromises with mapping slope and soil using the same geometric descriptions. In Dane County, Wisconsin, about 28% of the boundaries on the soil map are simply boundaries of slope classes, not soil classes. Including slope along with soils was motivated by the same logic behind the integrated surveys. Slope was so often an element in the use of soil maps that it made sense to combine the two factors. Yet in any specific application, a large proportion of the soil boundaries will not be needed because the attribute values are judged the same.

Land inventory functions often tend toward the integrated end of the spectrum. The single survey tries to collect all the needs onto the single map. Another example of this tendency is the forest stand map. Ideally, the forest stand represents an area that is homogeneous in terms of the management goals of the forestry organization. Homogeneity involves species mixture, age composition, size, and growth potential. Of course, the real homogeneity comes from the objectives of the forest manager and management activities. The identity of stands is challenged by such events as fire and disease that may cut across the previous stand boundaries. Whether stands exist in the field or not, cultural belief in stands will eventually cause forest managers to shape the landscape into stands.

Although these issues are rarely addressed in a comprehensive manner, the balance between parametric and integrated approaches play a role in any GIS design.

Bidding Process While the system design may be developed internally in the organization or externally by a consultant, the requirements analysis eventually moves firmly into the commercial world. Using the US administrative methods, a government agency issues the design specification as a *Request for Proposal* (RFP). In other situations the procedures may differ, but there is some method for commercial vendors to compete for the contract to provide a software package and other services such as conversion. This process may be substantially constrained by the well-meaning but inflexible bidding rules for public spending, but they can also be strongly affected by the communication network and marketing efforts.

The design stage usually creates a single scheme. Bidding permits vendors to provide solutions to implement the plan. Software packages are certainly one issue for choice, but they are far from the only one. A design based on a specific measurement framework will reduce the choices to a small number of potential vendors. If the system design methodology is to be taken seriously, the competition should be organized between different strategies, not just different implementations of a single strategy.

In many respects, the RFP process places some of the most critical decisions in the hands of outside vendors who have little information about the actual application and rather little stake in the outcome. The bid produced as a response must include a complex mixture of hardware, software, maintenance for the two, training and all sorts of other factors.

Evaluating bids follows the general themes discussed in Chapter 5 when dealing with overlay of many layers. Instead of a spatial arrangement of attributes, the evaluation deals with a finite number of bids, each one characterized by the response to explicit criteria. The RFP usually specifies a set of elements that must be included, the *mandatory* elements. The RFP may include optional elements and some explicit scoring procedure to rate the responses. The scores on these various elements are added up to give a total evaluation score. According to bidding rules, the proposal with the best score should be selected. Usually there are enough elements whose measurement can be adjusted so that the evaluation team can influence this numerical procedure. Just as a weighting of environmental factors can produce some odd results, the sum of disk capacity, viewshed algorithm, macro command language, and all the other optional elements hardly obeys the rules of a ratio measure. Yet, the interface between the public agency and the commercial world requires some decision rule. It is bound to involve arbitrary elements, so that all sides want to make it explicit, even at the risk of making it devoid of any theoretical basis.

In some cases, particularly projects with a larger budget or greater uncertainty about the capabilities of the vendors, there will be a *benchmark* test. Usually this will be restricted to bidders who pass the previous evaluation process near the top of the heap. The benchmark is usually performed by the bidder's staff using a portion of the planned study area, though it would be much more rational to have unbiased benchmarks for the whole industry, to the extent that they could be devised. The purpose of the benchmark is often to establish a rating of the capacity of each software package to perform the needed tasks. The ten-step process specified by Federal Information Processing Standard 75 for benchmarking (Guptill 1988, p. 45) replicates the system design procedures of user needs, with a particular emphasis on measuring the work flow and assigning tasks to represent overall workload categories. In addition, the evaluation may focus on the role of the software in establishing the final cost of the whole project. Thus, the issue is not just machine resources, but the productivity of the staff. The estimates from the benchmark are projected forward to determine whether the particular option (hardware, software, etc.) can produce the final product within the resources allocated.

The teams who perform benchmarks operate under some of the most stressful working conditions: dirty data, short deadlines, and strong competitive pressure. With their intimate understanding of the software, they may not provide the most unbiased estimates of the potential productivity of the user's staff. One of the reasons for high costs of GIS software may come from the expectations that vendors will do these benchmarks for no cost just to get the business. Other software, such as word processing and spreadsheets, is marketed on another basis and at different prices. Consider what the response would be to an RFP for a word processor that required a benchmark test. Perhaps the need for benchmarks simply reflects a stage when the GIS tool kit is not yet well understood.

Construction

When the RFP is evaluated, a vendor is chosen, and the planning phase seems to end. The focus becomes the construction of the system. The conversion efforts of digitizing often dominate the agenda, as described in Chapter 3. The design process does not end, however. An implementation plan should integrate a host of details that address all the concerns of evaluation mentioned earlier in this chapter. Some elements of a plan are listed in Table 10-3. Many of these continue the concerns addressed in the earlier section. If the design stage specifies every detail, then the construction phase will be simple. Usually, real applications turn out to be more complex, and the construction phase may act as a prolonged cycle of design development.

TABLE 10-3: Components of an implementation plan

Managerial

- Work plans and timelines for completion of tasks
- Staff responsibilities
- Workspace arrangements
- Environmental conditions for equipment
- Security

Conversion Plan

- Priorities for automation; training
- Procedures; collection and handling, equipment, quality control
- Cost and time audits; fallback to alternatives

Application Development

- Priorities; review of performance; feedback
- Database schema; tool kit for users
- Training

Source: Selected and reworked from Ventura, 1991, p.44, Table 4.2.

After the adversarial nature of selecting a software vendor and a hardware platform, the user may think that the evaluation stage is over. The vendor certainly tries to give this impression. They run user's groups and sometimes rather lavish annual conventions of users. GIS software (at the high performance end described here) usually requires a substantial annual maintenance expenditure. This continued flow of funds depends on loyal customers. Of course, once the system is acquired, it is rather complicated to change to another vendor without substantial costs in data conversion, staff retraining, and loss of time.

Evaluation is never complete. Every choice should be revisited in the light of experience. Every agency, no matter how small will have done its own detailed benchmark after a year or two of having the system installed. By monitoring performance, it can develop a strong basis to determine if its assumptions were validated.

The suggestion of continual evaluation leads to an alternative to the one-shot acquisition contract. Instead of trying to plan the whole process and design the final system, it may be more realistic to move incrementally. The first stage can be called a pilot study. It tests a prototype design in actual production on a limited area or on a specific application that needs only part of the total content. After the pilot, the system design is expected to change, and new decisions are taken. This kind of reevaluation may occur anyway. The hardware acquired under any contract may look sorely underpowered after the pilot is finished. The vendors will have developed new options, and some may be out of business or swallowed up by another corporate structure.

Whereas the classical problem of evaluation has been how to select GIS technology, the real evaluation problem involves knowing when to dump the existing system for something better. Unless a process of continual evaluation is in place, the team will be sentenced to follow a plan developed before the full nature of the project could be known. So, paradoxically, the best stage for starting the implementation process may be at the end, not what logically seems to be the beginning. Experience gained in a limited pilot project can produce a more penetrating needs assessment and a much improved design.

Feedback on the progress of database construction comes from quality control efforts. A pilot project is one kind of quality control effort, because it tests the whole construction process while there still is time to correct the design decisions. As described in Chapter 3, the conversion process requires substantial attention to tests of various elements of data quality. Quality control can also indicate when the database design must be modified.

SERVING LARGER GOALS

This chapter has discussed mostly the processes of evaluation and implementation, not the goals. Most of the procedures have been developed inside a rather narrow view of goals, focusing on the efficiency of a system. The measures of computer use, as well as the economic analysis of benefits and costs, seek to make a system more efficient in using scarce resources. The selection process also tends to focus on the potential productivity of the system. There are other levels of evaluation that involve more complex relationships. It is common to refer to the *bottom line* as the ultimate measure, implying the financial balance sheet as the capitalist measure of good and evil. The process of implementation requires many dimensions of evaluation beyond simple accounting. The budget may actually serve more as a constraint than as the sole objective function. Information systems are judged by their effectiveness in serving larger goals, not just by their efficiency.

The original purpose of many geographic information collection processes may not involve an economic market. The property map for a county may serve to ensure that taxes are collected fairly. The wetland inventory may be motivated by a desire to protect endangered species and to preserve clean water. Some may try to quantify all these concerns in terms of some future economic benefit, but such reductionism misses the diversity of these divergent viewpoints. These goals require just as much attention as the economic efficiency with which we pursue them. Perhaps there is a software package so cheap that it underbids all competitors, but it may require a measurement framework that will not support the ultimate goal. It may not permit some critical transformations. In that case, the economic analysis should not govern the choice. The criteria of effectiveness overrule the findings of efficiency. Technology has limited the availability of geographic information in many respects, but now that the limits are less restrictive, systems can be designed around the ultimate goals. These ultimate goals form the topic of the final chapter.

CHAPTER 11

SOCIAL AND INSTITUTIONAL CONTEXT

CHAPTER OVERVIEW

- Place geographic information in historical and cultural context.
- Sketch the geography of geographic information worldwide.
- Consider institutional, social, and cultural consequences of GIS.

Throughout this book, there have been hints that the technical elements of geographic information must be interpreted within a broader context. For example, the simplicity of the wetland zoning for Wisconsin (Chapter 9) may not be scientifically perfect, but simplification is required for administration under the existing legal framework. Similarly, the Pennsylvania LLRW disposal siting process (Chapter 5) is heavily constrained by the forces at play. In both cases, the technical choices were chosen to suit the purposes of the particular institutions. Such connections are just the starting point to understand the two-way connection between information and its context.

The modern world has been termed an *information society*, replacing an earlier era of industries that produced tangible goods. This development has strengthened the belief that information is objective. Proponents of GIS as a technology see objectivity as a major virtue of the GIS revolution. These proponents, consciously or unconsciously, assert that "better information will make better decisions." The connection may turn out to be less direct and less automatic. Every piece of geographic information invokes some set of assumptions, a framework of axioms. Some axioms may be more widely accepted than others, but they remain open to reexamination. It makes more sense to accept that geographic information is *constructed* inside a framework of institutional, social, and cultural relationships. Thus, it can be taken as objective only by those who accept a common set of assumptions. There is nothing magic about an automated system that will resolve deep social, political, and cultural differences. Yet, some of the positive intentions of the proponents make sense. Many well-accepted social goals have been thwarted by the lack of appropriate information. Rather than the blanket assertion that more information automatically improves the situation, members of society must take some responsibility for their ignorance and the ethical choices between different activities.

Some proponents have also presented GIS in the time-worn model of inexorable progress. In particular, they are likely to brand opposition as obstruction of the inevitable. Again, this book should demonstrate that there is nothing inevitable about the particular outcome of a geographic information system. There are many choices, and they must be justified on economic, political, social, cultural, and ethical grounds.

This chapter cannot exhaust the intricacies of the context surrounding geographic information. Nevertheless, a review of the technological developments will situate the current circumstances within historical sequence. The next section of this chapter presents a "geography of geographic information" that examines variations in mapping around the world. These empirical differences then connect to a number of institutional, social, and cultural dimensions.

Throughout these topics, the common theme returns to the starting point of the book. All information involves more than entity and attribute; object and relationship. The facts are embedded in a system of axioms, assumptions required to engage in a measurement. While the technology provides new ways to manipulate measurements in ways that integrate many viewpoints, these can be effective only inside an understanding of the assumptions.

TECHNOLOGICAL CHANGE IN HISTORICAL CONTEXT

This book began (in Chapter 1) by connecting the current use of geographic information to its deep historic roots. Maps have a long history that is strongly influenced by development of various media of dissemination, measurement techniques, and societal demands. In addition, society has been influenced by the geographic information available at a given time. Few societies have ever had enough or chosen to stay with the level that they inherited from previous generations. This tendency does not imply inexorable progress on any particular trajectory, but that geographic information is never good enough.

For a number of centuries, the various trends of mapmaking technology had a common theme. Each development fostered a larger market for a more uniform product. At one time, the parchment manuscripts of portolan charts were carefully guarded family secrets, slowly refined by the accumulated experience from each voyage. Each chart had its distinct value because it was different from the rest. The tailored, private map was run out of existence by a mass market of printed replicates and the organized collection of information by the state. By the beginning of the twentieth century, the printing press could reproduce thousands of identical charts and maps.

The marginal cost for each additional unit was tiny, but they all had to be the same to obtain this economy. Mapping agencies became a routine element of scientific administration. The twentieth-century developments of aerial photography, photogrammetry, and satellite remote sensing continue the push in the direction of greater centralization and standardized products. Yet, the technological imperative is not always toward greater uniformity. As Sherman and Tobler's (1957) multipurpose cartography reveals, photographic methods and offset printing of mid-century offered the ability to tailor products to different requirements, within certain constraints. These tendencies were not strong enough in mid-century to overcome the power of mass distribution of identical products.

Many of the mapmaking developments of the last century made the information handling process more and more remote. The plane table was replaced by aerial photograph and satellite image. In this process, the cartographic production process developed more specializations and lost touch with fieldwork. Divisions of labor appeared in what had been an integrated field procedure. Whereas each field surveyor used to be recognized by name on the topographic quadrangle, photogrammetric work became anonymous, a matter of replaceable parts.

Recently, the technology of GPS satellite surveying may reverse the trend toward remoteness. Accurate positional measurements will again be attainable in the field, where the attribute accuracy may be easier to obtain. This new technology may reverse some of the distinctions in the mapping labor force, but it may encounter resistance from the surveyor's guild. Implications for the social and institutional structure of the industry are not yet clear.

The modern geographic information system can also develop in different directions. There are certain tendencies to centralize through the economic incentive of using a common database. There are also the capabilities to produce a different product for each consumer, the characteristic of **flexible production** that is replacing the standardized product mentality in many industries worldwide (Scott and Storper 1986). Which influence eventually overcomes the other cannot be decided from our current perspective. The results may not be inherent in the technologies at all but played out in the complexities of each context.

GEOGRAPHY OF GEOGRAPHIC INFORMATION

Just as the implementations of many technologies differ around the world, the production and consumption of geographic information is far from uniform (Brandenburger and Ghosh 1990). The first issue is the supply of information. The cartographic enterprise, even when restricted to national topographic surveys, produces remarkably different amounts of map products (Figure 11-1). Europe, in general, is mapped

Flexible production: A system of economic organization characterized by rapid changes in production plans, "just in time" delivery, and other techniques that break from the economy-of-scale, mass-production approach characterized by the assembly line.

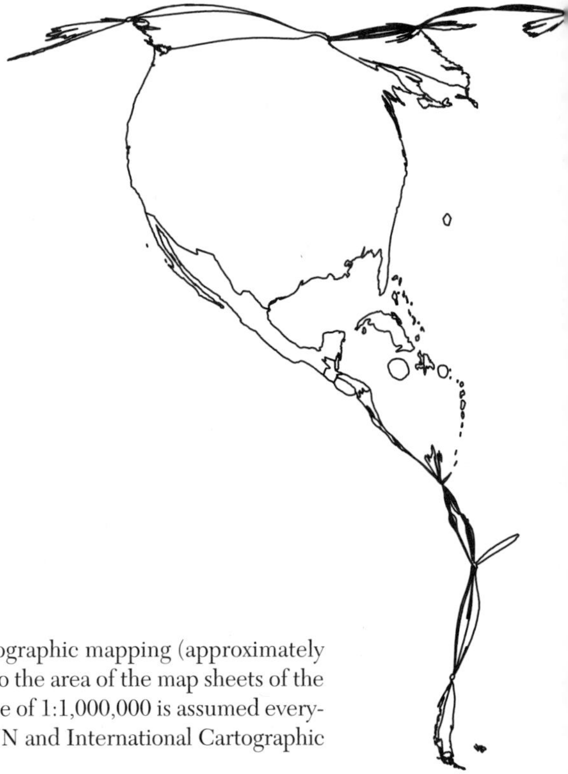

Figure 11-1: Cartogram of the worldwide coverage of topographic mapping (approximately mid 1990s). Each country has been distorted in proportion to the area of the map sheets of the most detailed coverage for the whole country. The base value of 1:1,000,000 is assumed everywhere. Most detailed complete coverage assembled from UN and International Cartographic Association reports. (International boundaries of 1987).

at a much more detailed scale than any other region. Within Europe, there are large differences in other aspects of the mapping enterprise. For example, the Swiss have a seven-year revision cycle for all topographic maps, while the adjacent Italian agency currently revises its maps on a 1,000-year cycle. The rate of topographic map revision in only a few countries would be adequate to support urban planning and similar applications (Figure 11-2). The differences go beyond the mapping process. Maps continue to be controlled as state secrets in many countries. In some they are provided at quite low cost, while in others they are bid up to some presumed market value. All these differences contribute to large variations in the supply of information.

Even under a regime of public access to the information of the US federal government, finding geographic resources can be quite complex. The Federal Geographic Data Committee organized a Federal Geographic Data Clearinghouse in 1995. While this is certainly an advance, it is interesting that it was not done earlier. This

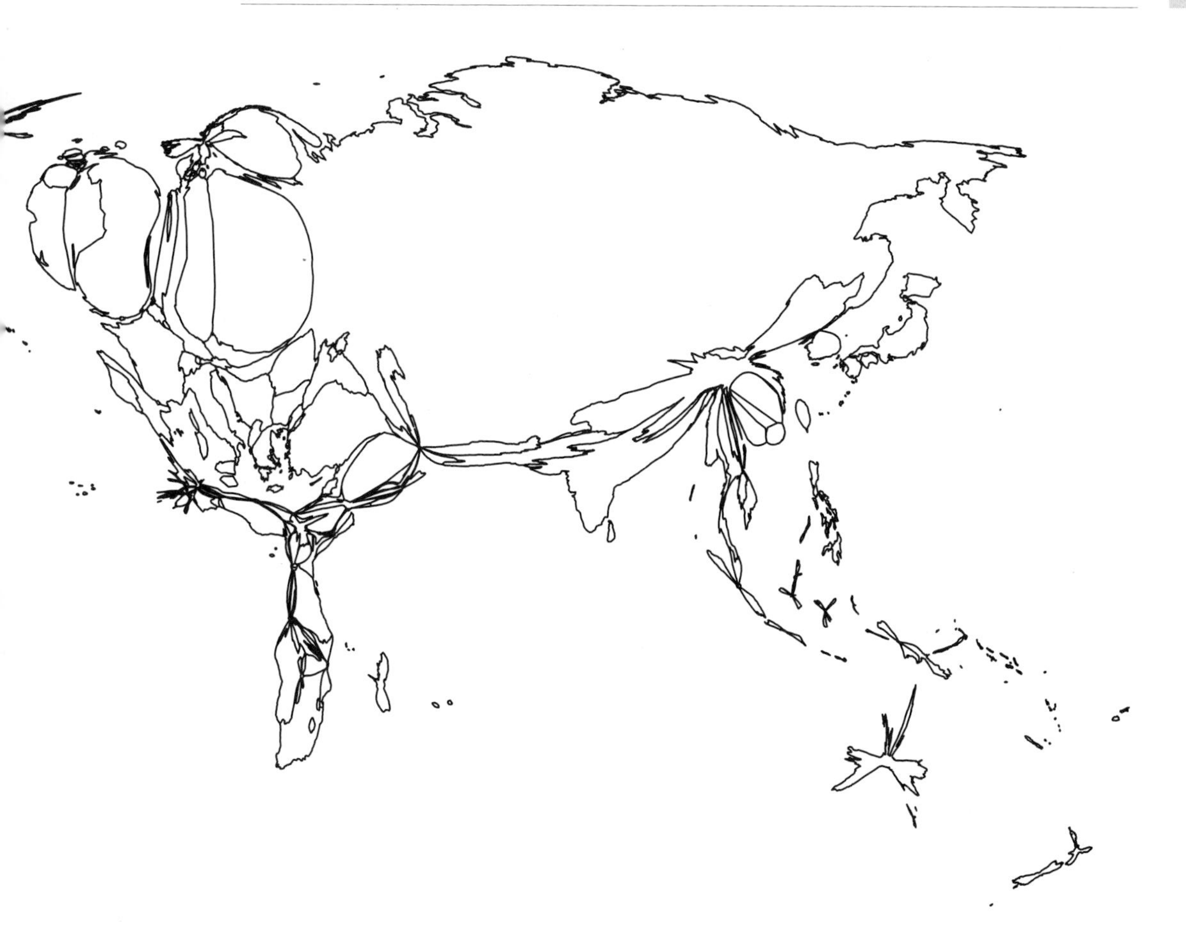

clearinghouse will cover only a portion of the resources produced in the United States, since so much of the GIS activity falls into the responsibility of the states, counties, and municipalities. These issues of the bureaucratic dispersion of geographic information form another component of the geography of its diversity. The federal structure of the United States leads to decentralized responsibilities among 60,000 local jurisdictions. Yet, the equally federal organization of Switzerland has produced a detailed set of national standards. Some national systems are under the control of a single national survey, while others divide responsibility between various ministries on thematic, rather than spatial, grounds.

There is no comprehensive study of the comparative economics of geographic information in various countries. A few detailed surveys establish the size of current expenditure on geographic information. For example, France spent 5.7 billion francs in 1987 (about US $1 billion) and employed 24,000 in related jobs (Conseil Nationale

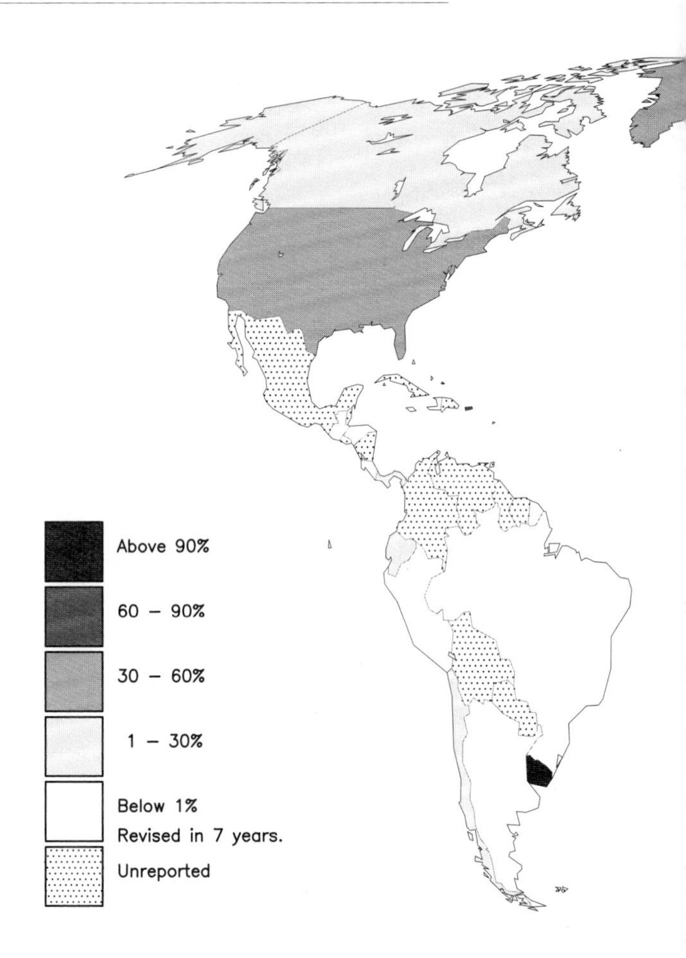

Figure 11-2: Percent of mapped area revised 1980–1987. Source: Brandenburger and Ghosh (1990).

de l'Information Géographique 1990). From the information available, the labor force working in the field of mapping varies rather dramatically, at least when crudely measured by the density of mapping employment over a whole country (Brandenburger 1993). The regions of the Third World with incomplete coverage of maps show low densities of employment, with a few bureaucratic exceptions such as Sri Lanka. Much of Africa has a density below 1 employee/1000 km^2. The industrialized countries generally show the highest densities. The high values for Italy (338 / 1000 km^2) and Japan (940/1000 km^2) show that a privately funded property records system plays a large role in some industrialized societies.

To understand the differences between geographic information in various places, remember that a number of factors comprise the context for the information system. The next sections will consider some of the most prominent elements, beginning with the inner workings of bureaucratic organizations outward through social and cultural context.

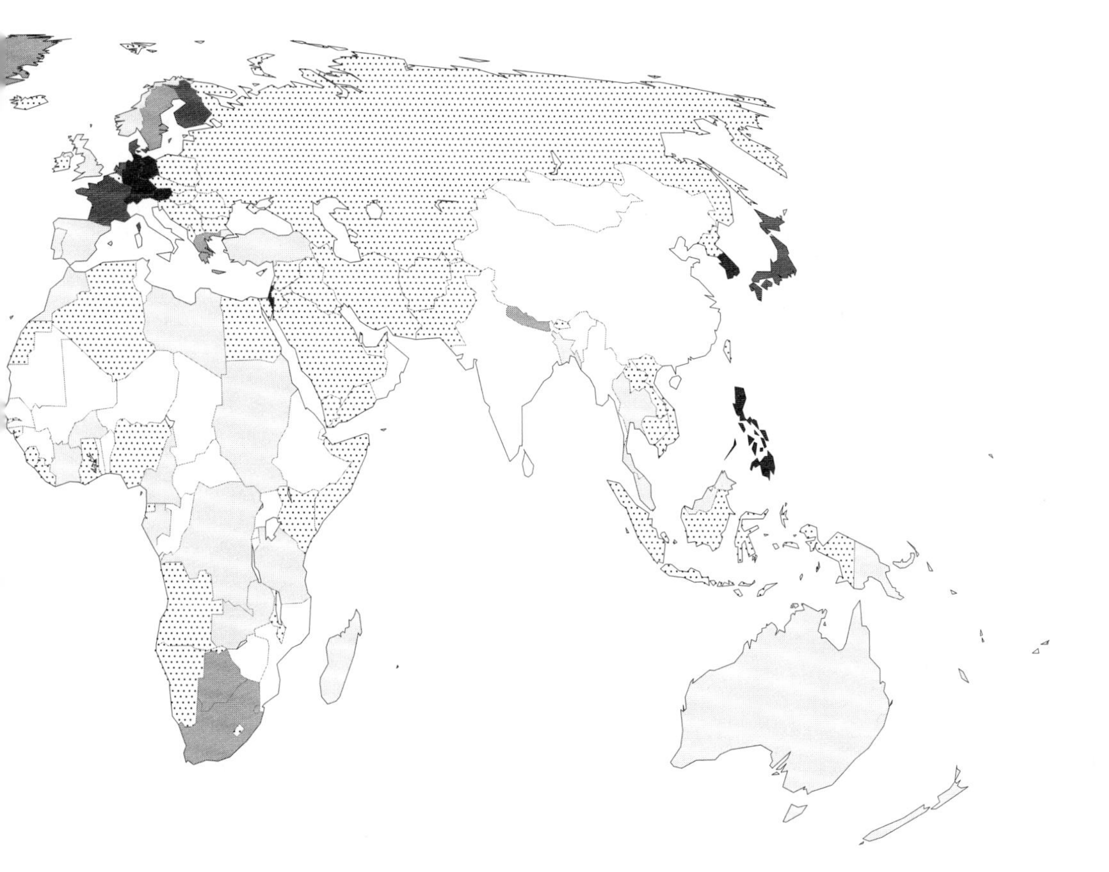

GEOGRAPHIC INFORMATION IN THE BUREAUCRACY

At the practical level, it is important to understand geographic information in the context of institutions. Very little geographic information is collected by individuals working just for their personal reasons. In part, an organization is required to coordinate and to cover a large area consistently. More importantly, the needs for geographic information arise from larger units in the society. National surveys arose alongside the nation-state (Kain and Baigent 1992). Topographic maps form a part of the methodology of scientific administration (tracing back to Napoleon in Europe and the utilitarian colonial regime in British India), as much as they relate to artillery warfare. But societies have become more complex than the nineteenth-century solution of a single map for all purposes. The single monolithic national survey agency is mostly confined to the least developed countries of the world (Figure 11-3). For most countries, institutional relationships involve multiple groups.

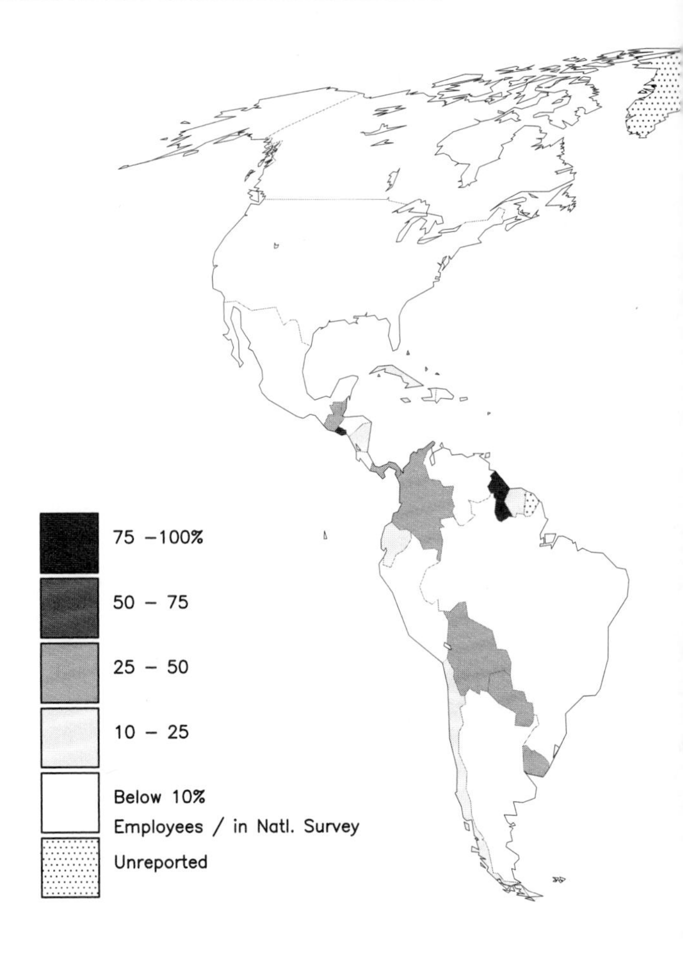

Figure 11-3: Percentage of mapping employment in the national topographic survey. Source: Brandenburger (1993).

Institutional Continuity Institutions have their own dynamics. Whether they are in the private or the public sector, there is a strong tendency to ensure their own long-term preservation. Despite attempts by commercial or government managers to curtail a bureaucracy, certain patterns tend to persist. Once a department is created to serve a particular need, it can often motivate its staff to pursue that goal single-mindedly. This is particularly true if the head of the organization is an independently elected official. In most counties of the United States, the property records are recorded by an elected official titled "Registrar of Deeds" or variations on the theme. Frequently, this office also registers births and deaths, but may not be responsible for the maps recording the property parcels described in the deeds. Usually, this agency will not feel any particular need to join a project designed to modernize the whole collection of land records (Sullivan and others 1984). Ironically, the decentralized nature of the United States may make it easier for these property registration offices to collaborate with other agencies, if the appeal is based on local pride and tax rates.

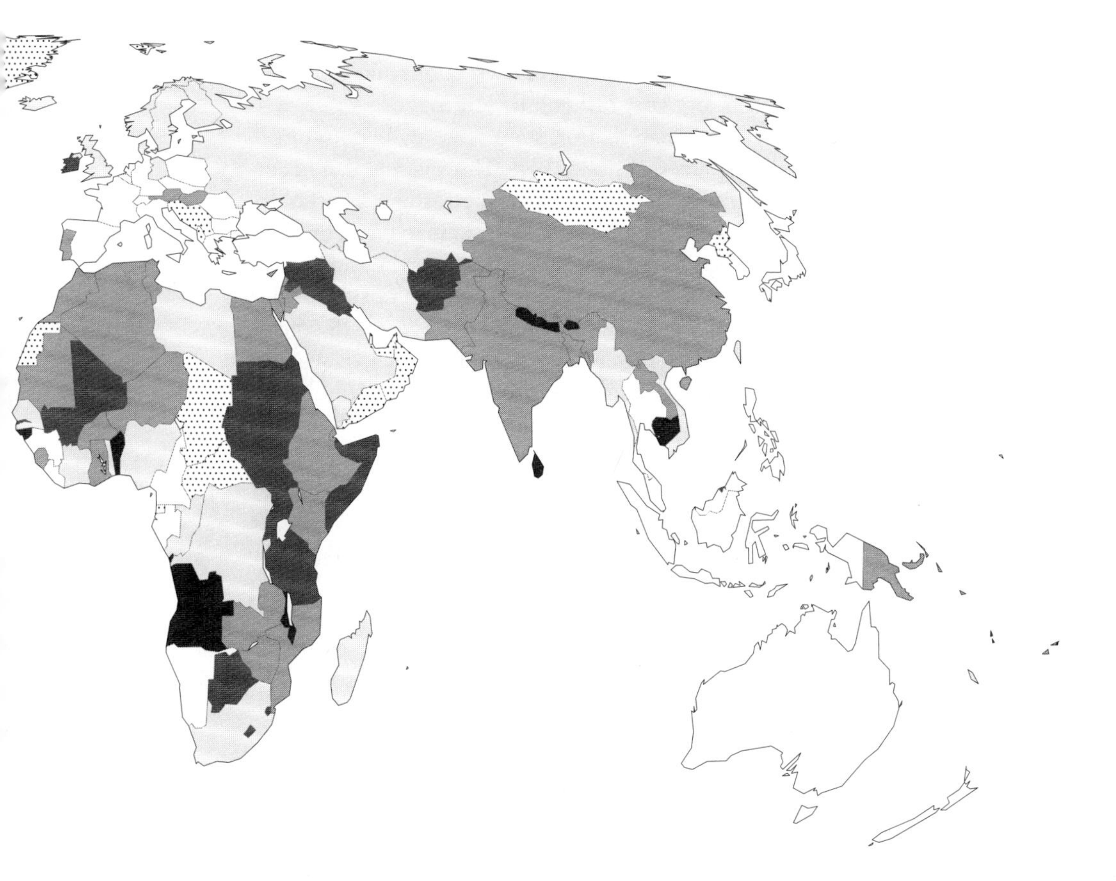

In countries with national registration offices, an automation project may serve the single-purpose function of the office nationwide (Andersson 1987), often without any coordination with other institutions.

Existing agencies developed their internal structure and their external relationships in the prior era of technology. The number of staff and the skills required depend on the form of technology employed. For a number of years, it has been relatively easy to copy maps. Once the first agency makes the first coverage, each group then develops its own additions and corrections on its copy. Information spreads outward, but it is much more complicated to collect and integrate. Eventually, each organization ends up placing corrections on its own map set. This process led to the creation of those 17 copies of the parcel maps for the Ontario city mentioned in the previous chapter. While this may seem like a gigantic waste of resources, each decision may have been perfectly rational within the bounded perspective of each office. Copying the parcel map offered a cheap way to start. Once the staff members learned the procedures, they would train their successors in this particular way of operation without seeking a global redesign. The duplicated effort of corrections and the inconsistencies between the different copies were very difficult to control.

In developing the new technology, duplicated effort seems like an obvious target. The goal of technical efficiency is often placed above all else. However, a plan to centralize the 17 parcel maps in a single office is bound to generate resistance. Part of the resistance comes from the loss of resources, since no manager wants to lose a group of subordinates. Such fears of change should not be underestimated, but they are not the only risk. Each one of those 17 versions of the parcel map had been customized to some extent. Each version may appear to the outsider (the professional GIS consultant) as variations on the parcel map, but there may be very subtle reasons why the newly centralized version will not really satisfy all the requirements. Parcel maps around the world involve various definitions that are never as consistent as they look. The idea of a single owner is frequently modified by easements. The treatment of private roads is thus a matter of interpretation where different perspectives lead to different solutions. Although the parcel map may seem to be an exhaustive coverage, each agency will define its categories differently. Thus, the French **cadastral** authorities do not provide parcel identifiers for certain state-owned parcels like the Louvre that have never been bought or sold (Rouet 1993). American assessment databases often exclude various kinds of tax-exempt parcels like roads, schools, and churches. The universe for these organizations is filtered by their view of their purpose.

Institutional Definitions of Time and History Time presents another difficulty. For example, in the application of cadastral databases, the subdivision of old parcels is rather common, and aggregation occurs as well. Thus, no identifier remains permanent. Differences in treatment will depend on the agency's need to refer to the old objects. Another kind of time involves differences between the administration of the current world and planning for the future. Administrative entities, like the tax assessor, require a database focused on the current situation, while planners must frequently deal with hypothetical future possibilities. In King County, Washington, the (elected) tax assessor operates the main attribute database of parcels, organized hierarchically by municipality because the tax rates are specific to these units. The planners are required to determine the potential impact on taxes if certain areas became incorporated as new cities. The assessor's office could not let these hypothetical codes contaminate the database, so the calculation was done on a copy. During this period, the assessor's office continued to maintain its database with routine changes in ownership and everything else. When an incorporation vote passed, the assessor could not switch over to the planner's version, and the city attribute had to be reentered (Horning 1990). These are concrete examples of the complexities in perspective that keep institutions apart. Perhaps the development of more flexible software models can resolve some of these issues at a technical level, but no user

Cadastral: Related to the records of landownership.

needs study will reveal the intricacies of all the potential incompatibilities. Each group may use common words, like "parcel," but not mean the same thing.

Cooperation and Coordination Many institutional problems boil down to a lack of cooperation. Each unit pursues its limited goals without regard to the common good. To some extent, this tendency is inevitable. There are many possible strategies to ensure cooperation in the common project of a multipurpose information system. In the early days, when computers were rare and difficult to maintain, the system was usually centralized in a single site. The technical requirements reenforced an organizational plan with a single lead agency. In some cases, a single site developed into a service bureau for general use. This model has been successful in Minnesota's Land Management Information Center (Robinette 1984), but largely because the group made the transition to a multipurpose era with a strong emphasis on service to clients. In the 1970's, 32 other states established state-wide natural resource information centers (Mead 1981), but only the Minnesota organization seems to have made the transition to the era of decentralized computing. Many of the other 31 programs were canceled, suspended, or completely overhauled as goals shifted, funding vanished, staff changed, and technology developed.

Technical design should not be the only guide, since each organization must feel a part of the project. Smooth cooperation is not just an accidental by-product, it must be constructed consciously. One possible solution involves an analysis of the **mandates** for each organization. In the case of public agencies, these will be codified in statute or administrative rule, as demonstrated in earlier chapters. In the case of the private sector there may be equally clear guidelines for the mission of each element within a corporation. From the mandates, some of the conflict over different views of the world may become more apparent. Each participant in the system should be declared **custodian** (Chrisman 1987) for those elements of the database that they generate. The others should agree to use the custodian's version of that element in their own operations. This network of reciprocal agreements ties everyone together with mutual responsibility, not through top-down control. If each custodian does its job, the whole system will operate as a kind of partnership. A failure in one component will influence the others, so that the community of users will have a stake in the success and productivity of the whole network. The connection between mandates and custodians was explicitly used in the study leading to the Wisconsin Land Records Program (Wisconsin Land Records Committee 1987). The organization of the Federal Geographic Data Committee, with its thematic division of responsibilities, follows the same concept with somewhat different terminology (Figure 11-4). The formal organization is the first step. It remains to be seen if a cooperative network actually develops from these attempts.

Mandates: Organizing principles of purpose that drive an organization; in a bureaucracy, the laws, administrative rules, and regulations that define the purpose and content of actions.

Custodian: An organization that takes responsibility to generate a particular kind of information for a defined geographic region and agrees to make it available to others.

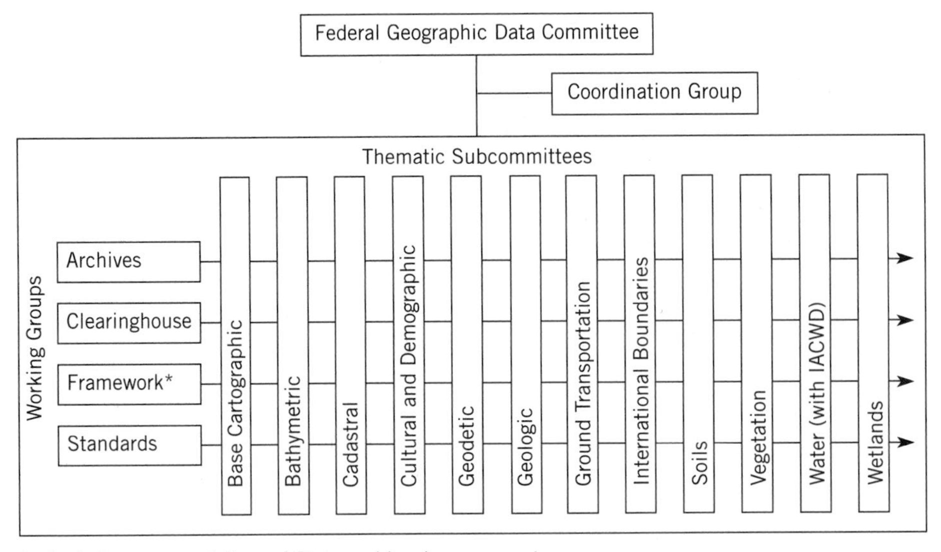

* - Includes representatives of State and local government.

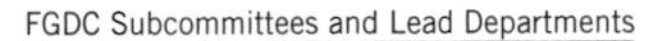

Base Cartographic - Interior
Bathymetric - Commerce
Cadastral - Interior
Cultural and Demographic - Commerce
Geodetic - Commerce
Geologic - Interior
Ground Transportation - Transportation
International Boundaries - State
Soils - Agriculture
Vegetation - Agriculture
Water - cosponsored with the Interagency Advisory Committee on Water Data
Wetlands - Interior

Figure 11-4: Diagram of activities of Federal Geographic Data Committee.

Coordination of disparate institutions is not easy. Each unit will go through changes in staff, in direction, and in budget. The focus on common goals can be sorely challenged. Some recipes call for a champion, a leader to instill with charisma what cannot be done by rational persuasion. In many cases, such a person seems to be effective, but often the champion is particularly careful to include each member in the collective operation. Leadership requires more than a forceful visionary. In many organizations, the proponents of the new technology have banded together across hierarchical institutional boundaries in a structure resembling a subversive cell plotting to overthrow the existing order. Revolution from the bottom may be more likely to succeed than a top-down decree.

INFORMATION IN ITS SOCIAL CONTEXT

While institutional arrangements are quite crucial, it is somewhat naive to consider them in isolation. Organizations may have great inertia, but their overall paths respond to larger social pressures (which are, in turn, the result of the actions of people). The mandates that guide bureaucracies are established by political and social processes. These processes vary substantially around the world, but there are some overall regularities. This section will review a few that relate to geographic information.

Equity

In Chapter 10, the discussion of the process of evaluating geographic information concentrated mostly on issues of efficiency. Of course efficiency is important, but it does not establish the purpose for the activity. One of the prime reasons to collect comprehensive geographic coverage is to ensure equity in various public affairs. Siting procedures, like the Pennsylvania waste depository, and administrative procedures, like tax assessment, must establish clear track records for procedural fairness. The Pennsylvania statute begins by saying that the whole state is eligible unless excluded for one of the criteria explicitly articulated. To some extent, all the mechanism of a GIS analysis for the whole state serves as massive overkill to find one 500-acre site that satisfies the criteria. However, the analysis demonstrates that all regions of the state were duly considered. It is a form of political insurance, though it is not yet clear that it will be sufficient to allay opposition.

In the implementation of the Dane County Soil Erosion Plan (Ventura 1988), the officials appointed to the Land Conservation Committee (many of them farmers) were particularly impressed by the comprehensive coverage of the whole county. Because the plan selected certain farms as being out of compliance, the officials felt much more secure that the selection would be defensible. Perhaps the glamour of the computer contributed, but the comprehensiveness of the analysis did play a role. This role cannot be confused with (or measured on the same scale as) financial costs and benefits.

Procedural equity is only the simplest of these goals. A deeper consideration is "substantive" equity that deals with the fairness of the results. Some of the most fundamental reasons to collect geographic information derive from these goals. Tax assessment is guided by a political desire to extract the tax in some proportion to the ability to pay.

Property assessment, as demonstrated by the wasteland assessment tests described in Chapter 9, is often far from an exact science. Many errors creep into the system, and many assumptions are made to fill in for lack of information. The flaws in the system do not invalidate the social purpose. One of the main reasons for collecting detailed local information is to ensure that the results of taxation and regulation are as fair as possible. Each society must judge how much equity it is willing to fund. These choices are not uniform around the world.

Access to Information

While the issues of equity involve fairness in process or result, there is also social and political pressure to limit the access to information. In the issues of secrecy, privacy, and access, there are very large differences around the world. For example, the United States requires that most public agencies operate according to various sunshine laws that ensure public access to meetings, reports, and records. These ensure that the government (at the local level) has very little secrecy. By contrast, private citizens in the US expect certain protections of their individual privacy. For example, census returns are locked up for 72 years, and spatial aggregates are *suppressed* if they will reveal individual information. Interestingly, the local property tax records that are made public due to the sunshine laws may reveal some of the same details that the census reports suppress. Unfortunately, most of the personal data restricted in government hands is available from commercial sources such as credit agencies. These arrangements vary dramatically in different countries. The UK Official Secrets Act makes it quite difficult to obtain records that would be public in the US. In the Netherlands and Scandinavia, the address of each person is recorded publicly, making migration studies much more precise than they can be under US privacy controls. However, in these societies there is strong regulation of personal information in commercial databanks.

Regimes of public access to geographic information very dramatically. The US federal government places most of its data in the *public domain,* meaning that there is no copyright restriction on its use. This is hardly universal. US remote sensing satellites were partially privatized according to the political will of the Reagan administration, and there are substantial fees and restrictions on the Landsat Thematic Mapper data. Some local governments in the US have the public domain policy, but many are trying to recover costs and to exempt these databases from open records laws.

The US is one of very few countries with a public domain approach. Mapping agencies around the world are much more likely to have copyright restrictions and cost recovery programs of some form (Rhind 1992). The nature of these programs is a matter of national political pressures and the influence of various pressure groups as well as the tendency to continue earlier policies. New Zealand has conducted one kind of experiment by instituting a major increase in the cost recovery required for the Department of Survey and Land Information. The elasticity of demand seems to vary among their products. Total revenue did increase slowly for topographic maps, while numbers of maps sold decreased dramatically. For their street map product, decline in sales was so large that revenue declined as well. Presumably, there are more commercial competitors for street maps than there are for topographic maps (Rhind 1992).

A cost recovery program is often motivated by a market capitalist view of information as a commodity. On the other side, the public domain approach sees information as an element of the infrastructure required to operate a modern economy. The market-driven policies in New Zealand intend to shift the funding from the general population to those who use the service directly. Another political philosophy would lament the loss of a broad spectrum of map users. The market approach fits poorly when the residual users are other government agencies that must use these maps and pass on the costs in increased taxes. Neither approach can be vindicated on grounds of efficiency, because they adopt such different political values. It is certain that the eventual character of the geographic information systems in each country will be strongly influenced by the political choices made in structuring the basic institutions.

The issue of access to information must also be addressed in social terms. Who will have access to these information resources and what difference will it make? If information access is distributed in proportion to economic power and social status, then it may simply become a means for the rich to get richer. If complex equipment and skills are required, it may also filter out certain elements in society. The conversion from map to database certainly increases the level of capital investment required to access the information. There are currently some projects to provide access to geographic information in schools, public libraries, and other public places (for example, the Alexandria Project conducted by University of California–Santa Barbara, other campuses, and various vendors). There is still a risk that these access points will provide the capability only to view the works of others, not to engage in the operations and transformations required to forge a new product. Libraries serve an important role in providing access, but they depend on an educated public that can read and interpret the contents. In the case of geographic information, the skills to manipulate complex representations may not be sufficiently widespread in the population.

Balancing Competing Concerns

In the simplified view of many early GIS proponents, the new technology would support objective planning. Decisions could now be based on facts, not political opinion or emotions. This view oversimplifies the process. Each database incorporates a number of assumptions that may lie at the heart of the disagreement. In many complex environmental issues, the advent of GIS did not calm the debate. In the controversy over ancient forests and spotted owl habitat in the Pacific Northwest, the US Forest Service built a large database, but so did the Wilderness Society. The timber companies have their own, and so does Washington Department of Natural Resources. Each stakeholder uses its GIS to try to argue for its particular definition of old growth. Each produces color maps to illustrate its version of a protection plan. Much of the debate turns on minutiae regarding definitions and measurements. This is certainly not the outcome predicted back in the days of centralized databases and scarce computing resources. Perhaps the whole process of argument must be reinvented, so that the adversarial legalisms of the Environmental Impact Statement can be replaced with a scientific dialogue and so that persuasion invokes dynamic models, data transformations, and interactive displays. Such a development is unlikely, considering the inertia of the legal system and the entrenched interests of the stakeholders.

Rather than producing better decisions, the information society has a tendency to postpone decisions and to grind them out with more and more deliberation over the preliminaries. The Not In My Back Yard syndrome (NIMBY for short) evokes a kind of knee-jerk resistance by any affected party who can afford to hire a lawyer. NIMBY is deprecated by professional planners, but the reaction is logical, given the nature of the debate. If the goal is to allow local participation, residents must be given some way to influence the decision. A public hearing provides no substitute for understanding of the whole slow process leading from measurement through representation and operations to a final product. Currently, the only tactic is to demolish a proposal with interminable proceedings.

As a summary, the social purposes of an information system include many important concerns beyond the issues of evaluation. This section has reviewed equity, access to information, and competing concerns. In another society or another era, other issues may be more important. All the technical steps should serve these goals.

INFORMATION AND CULTURE

The outermost ring in the scheme for this book involves culture. Culture can be broadly defined as the integrated web of ideals, beliefs, and values that give meaning, purpose, and direction to individuals in a society. While social issues in the previous section involve formally recognized purposes, culture provides the framework of meanings underlying the social system. Culture also operates at various scales. The system of rules governing how a task is done may be transmitted from person to person inside a particular office. This process has to be treated as cultural as much as the broad goals that operate at a national scale.

Culture: Continuity and Change

Culture has a certain coherence over time. People transmit their system of meanings to the next generation, but in a changed form. Culture should not be viewed as a superorganic structure that denies human expression and creativity. Change in cultures can be quite rapid. For example, a common system of meanings was applied to regulate the landscape and geographic abstractions (maps) in the British empire of 1775. The colonists in North America were as much a part of that British system as the English, Scots, and Welsh. With the creation of a separate political structure, however, the two systems began to diverge. Over the nineteenth century, the United Kingdom developed into a rather centralized nation-state, replacing its local peculiarities with a single administrative system. One mapping agency developed from a military origin. Maps were placed under Crown copyright. In the United States, many of these decisions went in the opposite direction. The United States kept certain elements of British common law that have been lost in the United Kingdom. The government is divided into many levels, each with responsibility to define its own system. The military mapping agencies remained distinct from the civilian. Maps are placed in the public domain, free of copyright. It is hard to find two countries that share so much in some aspects of their culture, but whose mapping systems are so entirely different. The lesson is that minor differences in institutions can become amplified, even when the popular culture seems quite similar.

Culture may seem to have little relevance for the practical study of geographic information. Any difference could simply be written off as a cultural difference, without any sort of analysis. Cultural differences are a lame excuse for a limited understanding of motivations and purposes. Perhaps the most important application of cultural concepts is not at the national level, but in explaining the differences of practice between disciplines and organizations.

The Practice of GIS

An organization, such as an agency, can operate as an independent subculture in many respects. Many organizations would rather construct new ideas internally than have them imposed from outside. New employees are trained to accept the common rules and then to impart them to the next generation. These rules include measurement frameworks for geographic information, operations, transformations, and data quality standards. "Standard Operating Procedures" maintain consistency, often at the expense of innovation.

One kind of organization is the profit-making corporation. Most geographic information is in the hands of the state, but certain roles are provided by the commercial sector. Surveyors have traditionally been individual practitioners. Software innovation, which used to come from research institutions, now centers on private companies. These vendors play an extremely critical role. Software builds upon systems of representation, implements operations and transformations, and enforces views of data quality. Without a strong sense of ethics, these companies could get the wrong message. Corporations are forced to adopt the "Not Invented Here" strategy in order to protect the private control of their software technology. The balance of economic power makes it difficult for the research community to develop new techniques under the conditions where the tools are all considered protected commercial secrets. It is not clear that the current arrangement is designed in order to ensure the most efficient development of new tools. Yet, there is little that a lone research scholar can do to change the current situation.

Some of the deepest rifts concerning geographic information split along disciplinary boundaries. A discipline organizes the world according to its purposes. Accountants organize the world into assets and expenses, using double-entry bookkeeping. Surveyors see their role in making measurements, and the process of adjustments is part of a common heritage. Cartographers are trained to communicate certain messages using a set of conventional tools. The list could be expanded infinitely in this world of continually sharpened specialization. The environment as a connected system is viewed in so many formalized ways that the relationships tend to be forgotten. Assumptions become ingrained and are rarely challenged. Each discipline maintains its boundaries in different ways. Some have mobilized their legal and political power to demarcate their work as distinct. One cannot practice medicine, law, or surveying without the appropriate certification. These groups can often act not just as a community sharing a certain set of principles, but as a guild defending their economic status from all challengers. Other disciplines are much more open, with no particular political power to protect. The term "cartographic license" implies the opposite of a strict set of professional principles; it evokes the artistic element, bending the facts to communicate the underlying message.

It is impossible to consider the development of geographic information without giving disciplines their role. These organizations are long-lived, with the capacity to transmit their viewpoints to the next generation. The educational system is a strong element in the process of acculturation and certification. Yet, in some manner, the educational system also acts as critic. Research can, on occasion, challenge the beliefs that are held without much proof. Interdisciplinary collaboration can, with the usual difficulties, cross-fertilize to the benefit of all.

SUMMARY

The systems used to handle geographic information are designed for more complex goals than simple efficiency. The people who create a GIS do so inside complex social and cultural institutions, where the past may influence current thinking. Yet, the context surrounding GIS is changing rapidly, as people and institutions adopt new ways of thinking.

The central message of this book involves integration of diverse forms of geographic information. At the root, the diversity comes from different goals of different people as they measure the environment and try to comprehend its interactions. Each step in this book has tried to incorporate the divergence of different techniques and points of view. The sequence moved from technical issues of measurement outward to the differences in culture. Yet, each discussion highlighted different aspects of the same choices. The system of meaning, transmitted through a discipline or an agency, determines a specific combination of measurement framework, transformations, and operations. Together these components carry out a particular purpose within the constraints of resources and knowledge. Each component can be improved by careful examination and reconsideration of the underlying assumptions.

Continuing the Exploration

The exploration begun by this book can continue using the World Wide Web. There are many examples and resources of a transitory nature that should not be printed in a book, lest they become obsolete before the book is printed. The web site for this book `http://www.wiley.com/college/chrisman` will provide current resources for continued exploration.

BIBLIOGRAPHY

Adams, J. B., Smith, M. O., and Johnson, P. E. 1986. Spectral mixture modeling: A new analysis of rock and soil types at the Viking Lander I site. *Journal of Geophysical Research* 91: 8098–8112.

Alexander, C., and Manheim, M. L. 1962. The use of diagrams in highway route location: An experiment. *Civil Engineering Systems Laboratory publication 161.* Cambridge, MA: Department of Civil Engineering, MIT.

American Society for Photogrammetry and Remote Sensing. 1989. Interim Standards for large scale line maps. *Photogrammetric Engineering and Remote Sensing* 55: 1038–1040.

Anderson, J. R., Hardy, E. E., Roach, J. T., and Witmer, R. E. 1976. A land use and land cover classification for use with remote sensor data. *Geological Survey Professional Paper 964.* Washington, DC: US Geological Survey.

Andersson, S. 1987. The Swedish Land Data Bank. *International Journal of Geographical Information Systems* 1(3): 253–263.

Arbia, G. 1989. *Spatial data configuration in statistical analysis of regional economic and related problems*. Dordrecht: Kluwer Academic.

Band, L. E. 1986. Topographic partition of watersheds with digital elevation models. *Water Resources Research* 22: 15–24.

Beard, M. K. 1987. How to survive on a single detailed data base. *Proceedings, AUTO–CARTO* 8:211-220.

Beard, M. K., and Chrisman, N. R. 1988. Zipper: A localized approach to edgematching. *The American Cartographer* 15: 163–172.

Bello-García, A., González-Nicieza, C., Ordieres-Meré, J. B., and Menéndez-Díaz, A. 1992. A contour line based triangulating algorithm. *Proceedings of the 5th International Symposium on Spatial Data Handling* 2: 411–423.

Bernkopf, R. L., Brookshire, D. S., Soller, D. R., McKee, M. J., Sutter, J. F., Matti, J. C., and Campbell, R. H. 1993. Societal value of geologic maps. *Circular 1111.* Washington, DC: US Geological Survey.

Berry, B.J.L. 1964. Approaches to regional analysis: A synthesis. *Annals of the Association of American Geographers* 54: 2–11.

Berry, B.J.L. 1973. Paradigms for modern geography. In *Directions in geography*, ed. R. J. Chorley, pp. 3–22. London: Methuen.

Berry, B.J.L., and Baker, A. M. 1968. Geographical sampling. In *Spatial analysis*, eds. B.J.L. Berry and D. F. Marble, pp. 91–100. Englewood Cliffs, NJ: Prentice-Hall.

Bie, S. 1984. Organizational needs for technological advancement. *Cartographica* 21(3): 44–50.

Blakemore, M.J. 1984. Generalisation and error in spatial data bases. *Cartographica* 21(3 and 4): 131–139.

Blakemore, M. J., and Harley, J. B. 1981: Concepts in the history of cartography. *Cartographica* 17(4): 1–120.

Boyer, C. B., and Merzbach, U. C. 1991. *A History of mathematics,* 2nd ed. New York: John Wiley & Sons.

Brandenburger, A. J. 1993. Study of the world's surveying and mapping human power and training facilities. *World Cartography* 22: 72–138.

Brandenburger, A. J., and Ghosh, S. K. 1990. Status of the world's topographic mapping, geodetic networks and national mapping activities. *World Cartography* 20: 1–116.

Brassel, K. E., and Weibel, R. 1988. A review and conceptual framework of automated map generalization. *International Journal of Geographical Information Systems* 2(3): 229–244.

Bunge, W. 1962. *Theoretical geography*. Lund: Gleerup.

Burrough, P. A. 1986. *Principles of geographical information systems for land resource assessment*. Oxford: Clarendon Press.

Burrough, P. A. 1989. Fuzzy mathematical methods for soil survey and land evaluation. *Journal of Soil Science* 40: 477–492.

Burrough, P. A. 1992. Development of intelligent geographical information systems. *International Journal of Geographical Information Systems* 6(1): 1–11.

Carver, S. J. 1991. Integrating multi-criteria evaluation with geographic information systems. *International Journal of Geographical Information Systems* 5(3): 321–339.

Castelle, A. J., Conolly, C., Emers, M., Metz, E. D., Meyer, S., Witter, M., Mauermann, S., Erickson, T., and Cooke, S. S. 1992. Wetland buffers: Use and effectiveness. *Publication 92-10.* Olympia: Shorelines and Coastal Zone Management Program, Washington Department of Ecology.

Cayley, A. 1859. On contour and slope lines. *Philosophical Magazine* 18: 264–268.

Chem-Nuclear Systems, Inc. 1992. Pennsylvania low-level radioactive waste disposal facility phase I (screening) plan. Harrisburg: Pennsylvania Department of Environmental Resources.

Chem-Nuclear Systems, Inc. 1994. Pennsylvania low-level radioactive waste disposal facility site screening interim report: stage three–Local disqualification. Harrisburg: Pennsylvania Department of Environmental Resources.

Chrisman, N. R. 1974. The impact of data structure on geographic information processing. *Proceedings AUTO–CARTO I,* pp. 165–177.

Chrisman, N. R. 1975. Topological data structures for geographic representation. *Proceedings AUTO–CARTO II,* 1: 346–351.

Chrisman, N. R. 1982a. Methods of spatial analysis based on error in categorical maps. Ph.D. dissertation, University of Bristol.

Chrisman, N. R. 1982b. A theory of cartographic error and its measurement in digital data bases. *Proceedings AUTO–CARTO 5,* pp. 159–168.

Chrisman, N. R. 1984a. On storage of coordinates in geographic information systems. *Geoprocessing* 2: 259–270.

Chrisman, N. R. 1984b. The role of quality information in the long-term functioning of a geographic information system. *Cartographica* 21 (3 and 4): 79–87.

Chrisman, N. R. 1987. Design of information systems based on social and cultural goals. *Photogrammetric Engineering and Remote Sensing* 53: 1367–1370.

Chrisman, N. R., and Lester, M. K. 1991. A Diagnostic test for error in categorical maps. *Proceedings AUTO-CARTO 10,* pp. 330–348.

Christian, C. S., and Stewart, G. A. 1968. Methodology of integrated surveys. In *Aerial surveys and integrated studies*, pp. 233–280. Toulouse, France: UNESCO.

Clarke, K. C. 1995. *Analytical and computer*

cartography, 2nd ed. Englewood Cliffs, NJ: Prentice Hall.

Cliff, A. D., and Ord, J. K. 1981. *Spatial processes: Models and applications*. London: Pion.

Codd, E. F. 1981. Data models in database management. *SIGMOD Record* 11(2): 112–114.

Coleman, A. 1961. The second land use survey: Progress and prospect. *Geographical Journal* 127: 168–186.

Congalton, R. G. 1991. A review of assessing the accuracy of classifications of remotely sensed data. *Remote Sensing of Environment* 37: 35–46.

Conseil Nationale de l'Information Géographique. 1990. Evaluation des moyens consacrés en France a l'information géographique pour 1987. Paris: CNIG.

Coppock, J. T., and Rhind, D. W. 1991: The history of GIS. In *Geographical information systems: Overview, principles and applications*, eds. D. J. Maguire, M. F. Goodchild, and D. W. Rhind, pp. 21–43. Harlow, Essex: Longmans.

Cornwell, B., and Rohardt, S. 1983. An extension of map overlay concepts using an image processing system. *Proceedings AUTO-CARTO 6,* 1: 335–344.

Csillag, F., and Kummert, A. 1990. Spatial complexity and storage requirements of maps represented by region quadtrees. *Proceedings of the 4th International Symposium on Spatial Data Handling* 2: 928–937.

Dames and Moore. 1975. Data management for power plant siting: Delmarva interface study. New York: Dames and Moore.

Dane County (Wisconsin) Regional Planning Commission. 1980. Dane County Solid Waste Plan. Madison, WI: Dane County.

Dangermond, J. 1979. A case study of the Zulia Regional Planning Study, describing work completed. In *Harvard Library of Computer Graphics*, vol. 3, ed. P. J. Moore, pp. 35–62. Cambridge, MA: Harvard Laboratory for Computer Graphics.

Dangermond, J., and Freedman, C. 1984. Findings regarding a conceptual model of a municipal data base and implications for software design. In *Seminar on the multipurpose cadastre: Modernizing land information systems in North America*, ed. B. J. Niemann Jr., pp. 12–49. Madison: University of Wisconsin–Madison, Institute for Environmental Studies.

Davis, J. C. 1973. *Statistics and data analysis in geology*. New York: John Wiley & Sons.

Dickinson, H. J., and Calkins, H. W. 1988. The economic evaluation of implementing a GIS. *International Journal of Geographical Information Systems* 2: 307–327.

Dijkstra, E. W. 1959. A note on two problems in connection with graphs. *Numerische Mathematik* 6: 37–53.

Donley, M., Allan, S., Caro, P., and Patton, C. 1979. Atlas of California. Portland, OR: Academic Book Center.

Dougenik, J. A. 1980. WHIRLPOOL: A geometric processor for polygon coverage data. *Proceedings, AUTO-CARTO IV*, pp. 304–311.

Douglas, D. H. 1993. Least cost path in GIS. *Research Note 61.* Department of Geography, University of Ottawa.

Douglas, D. H., and Peucker, T. K. 1973. Algorithms for the reduction of the number of points required to represent a digitized line or its charicature. *The Canadian Cartographer* 10(2): 110–122.

Dueker, K. J., and Kjerne, D. 1989. *Multi-*

purpose cadastre: Terms and definitions. Falls Church, VA: ASPRS and ACSM.

Dueker, K. J., and Vrana, R. 1992. Dynamic segmentation revisited: A milepoint linear data model. *Urban and Regional Information Systems Association Journal* 4(2): 94–105.

Eastman, J. R., Toledano, J., Jin, W., and Kyem, P.A.K. 1993. Participatory multiobjective decision-making in GIS. *Proceedings, AUTO-CARTO 11,* pp. 33–42.

Egenhofer, M. J., and Herring, J. R. 1991. High level spatial data structures for GIS. In *Geographical Information Systems: Overview, Principles and Applications*, eds. D. J. Maguire, M. F. Goodchild, and D. W. Rhind, pp. 227–237. Harlow, Essex: Longmans.

Ellis, B. 1966. *Basic concepts of measurement*. Cambridge: Cambridge University Press.

EPA Environmental Research Laboratory. 1994. *Image processing workbench.* Corvallis, OR: ftp://morpheus.cor.epa.gov/pub/misc/ipw.

Epstein, E. F., and Duchesneau, T. D. 1984. The use and value of a geodetic reference system. Rockville, MD: National Geodetic Survey, NOAA.

Fabos, J., and Caswell, S. J. 1977. METLAND landscape planning research. *Research Bulletin* 637. Amherst: Massachusetts Experiment Station.

FAO. 1976. *A Framework for land evaluation*. Rome: Food and Agriculture Organization.

Fisher, P. F. 1991. Modelling soil map-unit inclusions by Monte Carlo simulation. *International Journal of Geographical Information Systems* 5(2): 193–208.

Fisher, P. F. 1993. Algorithm and implementation uncertainty in viewshed analysis. *International Journal of Geographical Information Systems* 7(4): 331–347.

Fitzpatrick-Lins, K. 1981. Comparison of sampling procedures and data analysis for a land use and land cover map. *Photogrammetric Engineering and Remote Sensing* 47: 343–351.

Flowerdew, R., and Green, M. 1992. Developments in areal interpolation methods and GIS. *Annals of Regional Science* 26: 67–78.

Frank, A. U., and Kuhn, W. 1986. Cell graphs: A provable correct method for the storage of geometry. *Proceedings of the Second International Symposium on Spatial Data Handling,* pp. 411–436.

Gillespie, S. R. 1993. The value of GIS to the federal government. Washington, DC: US Geological Survey.

Gold, C. M. 1988. PAN graphs. An aid to GIS analysis. *International Journal of Geographical Information Systems* 2(1): 29–41.

Gold, C. M. 1992. An object-based dynamic spatial model, and its application in the development of a user-friendly digitizing system. *Proceedings of the 5th International Symposium on Spatial Data Handling* 2: 495–504.

Goodchild, M. F. 1978. Statistical aspects of the polygon overlay problem. In *Harvard papers on geographic information systems*, vol. 6 , ed. G. Dutton. Reading, MA: Addison-Wesley.

Goodchild, M. F. 1987. A spatial analytical perspective on geographical information systems. *International Journal of Geographical Information Systems* 1(4): 327–334.

Goodchild, M. F., and Gopal, S., eds. 1989. *Accuracy of spatial databases.* London: Taylor & Francis.

Goodchild, M. F., Sun, G., and Yang, S. 1992. Development and test of an error model for categorical data. *International Journal of Geographical Information Systems* 6(2): 87–104.

Goodchild, M. F., and Rizzo, B. R. 1987. Performance evaluation and work-load estimation for geographic information systems. *International Journal of Geographical Information Systems* 1: 67–76.

Guptill, S. C. 1988. A process for evaluating geographic information systems. *US Geological Survey Open File Report 88-105.* Technology Exchange Working Group, Federal Interagency Coordinating Committee on Digital Cartography.

Hägerstrand, T. 1970. What about people in regional science? *Papers, Regional Science Association* 24: 1–21.

Haggett, P. 1965. *Locational analysis in human geography*. London: Edward Arnold.

Hammond, E. H. 1964. Classes of land-surface form. In *National atlas of the United States of America*, pp. 62–63. Washington, DC: US Geological Survey.

Harley, B., and Woodward, D., eds. 1989. *History of cartography.* Chicago: University of Chicago Press.

Harvey, F. 1994. Defining unmovable nodes/segments as a part of vector overlay: The alignment overlay. *Spatial Data Handling 94* 1: 159–176.

Herring, J. 1987. TIGRIS: Topologically Integrated Geographic Information System. *Proceedings, AUTO-CARTO 8*, pp. 282–291.

Hobbs, B. F. 1985. Choosing how to choose: Comparing amalgamation methods for environmental impact assessment. *Environmental Impact Assessment Review* 5: 301–319.

Hopkins, L. D. 1977. Methods of generating land suitability maps: A comparative evaluation. *American Institute of Planners Journal* 43: 386–400.

Horn, B.K.P. 1981. Hill shading and the reflectance map. *Proceedings IEEE* 69(1): 14–47.

Horning, G. W. 1990. Information integration for geographic information systems in a local government context. M.A. thesis, University of Washington.

Huxold, W. E. 1991. *An Introduction to urban geographic information systems*. Oxford: Oxford University Press.

Huxold, W. E., Allen, R. K., and Gschwind, R. A. 1982. An evaluation of the city of Milwaukee automated geographic information and cartographic system in retrospect. *Proceedings Harvard Computer Graphics Week* 1–25.

Jackson, M. J., and Woodsford, P. A. 1991. GIS data capture hardware and software. In *Geographical information systems: Overview, principles and applications*, eds. D. J. Maguire, M. F. Goodchild, and D. W. Rhind, pp. 239–249. Harlow, Essex: Longmans.

Joint Nordic Project. 1987. Digital map data bases, economics and user experiences in North America. Helsinki, Finland: National Board of Survey.

Kain, R.J.P., and Baigent, E. 1992. *The cadastral map in the service of the state*. Chicago: University of Chicago Press.

Keeney, R. L., and Raifa, H. 1976. *Decisions with multiple objectives: Preferences and value tradeoffs*. New York: John Wiley & Sons.

Kiefer, R. W. 1967. Terrain analysis for metropolitian fringe area planning. *Journal of the Urban Planning and Development Division, American Society of Civil Engineers* 93(UP4): 119–139.

Kjerne, D., and Dueker, K. J. 1986. Modeling cadastral spatial relationships using an object-oriented language. *Proceed-*

ings of the Second International Symposium on Spatial Data Handling, pp. 142–157.

Langran, G. E. 1991. *Time in geographic information systems*. London: Taylor & Francis.

Leopold, A. 1949. *Sand country almanac and sketches here and there*. New York: Oxford University Press.

Lester, M. K., and Chrisman, N. R. 1991. Not all slivers are skinny: A comparison of two methods for detecting positional errors in categorical maps. *Proceedings GIS/LIS 91,* 2: 648–658.

Leung, Y., Goodchild, M. F., and Lin, C.-C. 1992. Visualization of fuzzy scenes and probability fields. *Proceedings of the 5th International Symposium on Spatial Data Handling* 2: 480–490.

Lewis, P. 1963. *Recreation in Wisconsin.* Madison: State of Wisconsin, Department of Resource Development.

Lillesand, T. M., and Kiefer, R. W. 1994. *Remote sensing and image interpretation*, 3rd ed. New York: John Wiley & Sons.

Linse, A.R. 1993 Geoarchaeological scale and archaeological interpretation. In *Special Paper 283*, eds. J. K. Stein and A. R. Linse, pp. 11–27. Boulder, CO: Geological Society of America.

Loveland, T. R., Merchant, J. W., Ohlen, D. O., and Brown, J. F. 1991. Development of a land-cover characteristics database for the conterminous United States. *Photogrammetric Engineering and Remote Sensing* 57: 1453–1463.

Mabbutt, J. A. 1968. Review of concepts of land classification. In *Land Evaluation*, ed. G. A. Stewart, pp. 11–28. Melbourne: Macmillan.

Maguire, D. J. 1991. An overview and definition of GIS. In *Geographical information systems: Overview, principles and applications*, vol. 1 , eds. D. J. Maguire, M. F. Goodchild, and D. W. Rhind, pp. 9–20. Harlow, Essex: Longmans.

Males, R. M., and Gates, W. E. 1978. ADAPT—A spatial data structure for use with planning and design models. In *Harvard Papers on Geographic Information Systems*, vol. 3 , ed. G. Dutton. Reading, MA: Addison-Wesley.

Maling, D. H. 1973. *Coordinate systems and map projections*. London: Philip.

Manning, W. 1913. The Billerica town plan. *Landscape Arcitecture* 3: 108–118.

Mark, D. M. 1984. Automated detection of drainage networks from digital elevation models. *Cartographica* 21(3): 168–178.

Mark, D. M., and Csillag, F. 1989. The nature of boundaries on "area-class" maps. *Cartographica* 26: 65–78.

Marks, D., Dozier, J., and Frew, J. 1984. Automated basin delineation from digital elevation data. *GeoProcessing* 2: 299–311.

Matthews, E. 1993. Global geographic databases for modelling trace gas fluxes. *International Journal of GIS* 7: 125–142.

McCartney, J. W., and Thrall, G. I. 1991. Real estate acquisition decisions with GIS: Ranking property for purchase with an example from Florida's St. John's Water Management District. *Proceedings GIS/LIS 91,* 1: 90–99.

McHarg, I. L. 1969. *Design with Nature*. Garden City, NY: Natural History Press.

McMaster, R. B., and Shea, K. S. 1992. *Generalization in digital cartography*. Washington, DC: Association of American Geographers.

Mead, D. A. 1981. Statewide natural resource information systems — A status report. *Journal of Forestry* 79: 369–372.

Michell, J. 1993. The origins of the representational theory of measurement: Helmholtz, Hölder and Russell. *Studies in the History and Philosophy of Science* 24: 185–206.

Morrison, J. (editor). 1987: A draft proposed standard for digital cartographic data transfer. *The American Cartographer* 15(1): 1–140.

Murray, T., Rogers, P., Sinton, D., Steinitz, C., Toth, R., and Way, D. 1971. Honey Hill: A systems analysis for planning the multiple use of controlled water areas. *Institute of Water Resources 71-9; NTIS AD 736 343 and 344.* Washington, DC: Army Corps of Engineers.

Nyerges, T. L. 1990. Locational referencing and highway segmentation in a geographic information system. *Institute of Transportation Engineers Journal* 60 (3): 27–31.

Nyerges, T. L. 1991. Analytical map use. *Cartography and Geographic Information Systems* 18: 11–22.

Oliver, M. A., and Webster, R. 1990. Kriging: A method of interpolation for geographical information systems. *International Journal of Geographical Information Systems* 4(3): 313–332.

Parkes, D., and Thrift, N.J. 1980. Times, spaces and places. New York: John Wiley & Sons.

Perkal, J. 1956. On epsilon length. *Bulletin de l'Academie Polonaise des Sciences* 4: 399–403.

Peucker, T. K., and Chrisman, N. R. 1975. Cartographic data structures. *The American Cartographer* 2: 55–69.

Preparata, F. P., and Shamos, M. I. 1985. *Computational geometry: An introduction*. New York: Springer Verlag.

Pullar, D. V. 1991. Spatial overlay with inexact numerical data. *Proceedings, AUTO–CARTO 10,* 313–329.

Pullar, D. V. 1993. Consequences of using a tolerance paradigm in spatial overlay. *Proceedings, AUTO–CARTO 11, pp.* 288–296.

Raisz, E. 1948. *General cartography*. New York: McGraw-Hill.

Rhind, D. 1992. Data access, charging and copyright and their implications for geographical information systems. *International Journal of Geographical Information Systems* 6(1): 13–30.

Riggle, M. A., and Schmidt, R. R. 1991. The Wisconsin groundwater contamination susceptibility map. *Urban and Regional Information Systems Association Journal* 3: 85–88.

Robinette, A. 1984. Institutional innovations: Moving from academia to serving government. In *Seminar on the multipurpose cadastre: Modernizing land information systems in North America*, ed. B. J. Niemann, Jr., pp. 155–166. Madison: University of Wisconsin–Madison, Institute for Environmental Studies.

Robinson, A. H. 1953. *Elements of cartography*. New York: John Wiley & Sons.

Robinson, A. H., Morrison, J. L., Muehrcke, P. C., Guptill, S. C., and Kimerling, J. 1995. *Elements of Cartography,* 6 ed. New York: John Wiley & Sons.

Rosenfeld, A., and Kak, A. 1976. *Digital picture processing*. New York: Academic Press.

Rosenfield, G. H., and Fitzpatrick-Lins, K. 1986. A coefficient of agreement as a measure of thematic classification accuracy. *Photogrammetric Engineering and Remote Sensing* 52: 223–227.

Rosing, K., Hillsman, E., and Rosing-Vogelaar, H. 1979. A note on comparing optimal and heuristic solutions to the P-Median problem. *Geographical Analysis* 11: 86–89.

Rouet, P. 1993: Réflexions sur un modèle de données spatiales de référence pour un cadre urbain. *Revue de Géomatique* 3: 363–404.

Rushton, G. 1979. *Optimal location of facilities*. Wentworth, NH: COMPress.

Samet, H. 1990. *The design and analysis of spatial data structures*. Reading, MA: Addison-Wesley.

Scott, A. J., and Storper, M., eds. 1986. *Production, work, territory: The geographical anatomy of industrial capitalism.* London: Allen & Unwin.

Shapiro, C. 1995. Coordination and integration of wetland data for status and trends and inventory estimates: Progress Report. *Techncial Report 2.* Reston, VA: Federal Geographic Data Committee, Wetlands Subcommittee.

Sherman, J., and Tobler, W. 1957. The multiple use concept in cartography. *Professional Geographer* 9(5): 5–7.

Shyue, S. W. 1989. High breakdown point robust estimation for outlier detection in photogrammetry. Ph.D. dissertation, University of Washington.

Silfer, A. T., Kinn, G. J., and Hassett, J. M. 1987. A geographic information system utilizing the Triangular Irregular Network as a basis for hydrologic modeling. *Proceedings AUTO–CARTO 8*, 129–136.

Sinton, D. F. 1978. The inherent structure of information as a constraint to analysis: Mapped thematic data as a case study. In *Harvard Papers on Geographic Information Systems*, vol. 6, ed. G. Dutton, pp. 1–17. Reading, MA: Addison-Wesley.

Snyder, J. P. 1987. Map projections — A working manual. *Professional Paper 1395.* Washington, DC: US Geological Survey.

Steinitz, C., Parker, P., and Jordan, L. 1976. Hand-drawn overlays: Their history and prospective uses. *Landscape Architecture* 66: 444–455.

Stevens, S. S. 1946. On the theory of scales of measurement. *Science* 103: 677–680.

Storie, T. F. 1933. An index for rating agricultural value of soils. *Bulletin 556.* California Experiment Station.

Story, R., Galloway, R. W., van de Graaf, R.H.M., and Tweedie, A. D. 1963. General report on the lands of the Hunter Valley. *Australian Land Research Series 8.* Melbourne: CSIRO.

Sullivan, J. G., Chrisman, N. R., and Niemann, B. J., Jr. 1985. Wastelands versus wetlands in Westport Township, Wisconsin. *Proceedings URISA* 1: 73–85.

Sullivan, J. G., Niemann, B. J., Jr., Chrisman, N. R., Moyer, D. D., Vonderohe, A. P., and Mezera, D. F. 1984. Institutional reform before automation: The foundation for modernizing land records systems. *The Decision maker and Land Information Systems, Fédération Internationale des Géometres Symposium,* 383–391.

Swiss Federal Bureau of Statistics. 1992. Die Bodennutzung der Schweiz: Arealstatistik 1979/85 Kategorienkatalog. Berne: Bundesamt für Statistik.

Tobler, W. 1979a. Cellular geography. In *Philosophy in geography*, eds. S. Gale and G. Olsson, pp. 379–386. Dordrecht: Reidel.

Tobler, W. 1979b. A transformational view of cartography. *The American Cartographer* 6: 101–106.

Tomlin, C. D. 1990. *Geographic information systems and cartographic modeling.* Englewood Cliffs, NJ: Prentice Hall.

Tomlinson, R. F. 1967. An introduction to the geographic information system of the Canada Land Inventory. Ottawa: Department of Forestry and Rural Development.

Travis, M. R., Elsner, G. H., Iverson, W. D., and Johnson, C. G. 1975. VIEWIT: Computation of seen areas, slope and aspect for land-use planning. *Technical Report PSW-11/1975.* Washington, DC: USDA Forest Service.

Tyrwhitt, J. 1950. Surveys for Planning. In *Town and country planning textbook*, ed. Association for Planning and Regional Reconstruction. London: Architectural Press.

Ullman, E. L. 1954. Geography as spatial interaction. *Interregional Linkages, Proceedings of the Western Committee on Regional Economic Analysis*, pp. 63-71.

Ullman, J. 1982. *Principles of database systems*. Rockville, MD: Computer Science Press.

Unwin, D. 1981. *Introductory spatial analysis*. London: Methuen.

United States Bureau of the Budget. 1947. *National map accuracy standards*. Washington, DC: Government Printing Office.

United States Federal Geographic Data Committee. 1994. *Content standards for digital geospatial metadata.* Washington, DC: Federal Geographic Data Committee.

United States Geological Survey (USGS). 1970. *National atlas of the United States of America*. Washington, DC: Government Printing Office.

Ventura, S. J. 1988. *Dane County soil erosion plan.* Madison, WI: Dane County Land Conservation Committee.

Ventura, S. J. 1991. *Implementation of land information systems in local government—Steps toward land records modernization in Wisconsin.* Madison: Wisconsin State Cartographer's Office.

Voogd, H. 1983. *Multicriteria evaluation for urban and regional planning*. London: Pion.

Wang, F., Hall, G. B., and Subaryono. 1990. Fuzzy information representation and processing in conventional GIS software: Database design and application. *International Journal of Geographical Information Systems* 4(3): 261–283.

Warntz, W. 1965. A note on surfaces and paths: Applications to geographical problems. *Discussion Paper X.* Ann Arbor: Michigan Inter-University Community of Mathematical Geographers.

Warntz, W. 1966. The topology of socioeconomic terrain and spatial flows. *Papers, Regional Science Association* 17: 47–61.

Washington State. 1987. A better future in our woods and streams: Final Report. *Timber Fish Wildlife Agreement.*

Washington State. 1988. *Forest practice regulations: Riparian management zones.* Washington Administrative Code 222-30 paragraph 020(4).

Washington State. 1990. *Minimum guidelines to classify agriculture, forest, mineral lands and critical areas.* Washington Administrative Code 365-190 paragraph 080.

Wilcox, D. L. 1990. Concerning "The economic evaluation of implementing a GIS." *International Journal of Geographical Information Systems* 4: 203–210.

Williams, T.H.L. 1985. Implementing LESA on a geographic information system—A case study. *Photogrammetric Engineering and Remote Sensing* 51: 1923–1932.

Wischmeier, W. H., and Smith, D. D. 1978. Predicting rainfall losses—A guide to conservation planning. *Agricultural Handbook 537.* Washington, DC: US Department of Agriculture.

Wisconsin Land Records Committee. 1987. *Final report—Modernizing Wisconsin's land records.* Madison: University of Wisconsin–Madison, Institute for Environmental Studies.

Wright, J. K. 1936. A method of mapping densities of population with Cape Cod as an example. *Geographical Review* 26: 103-110.

Young, A., and Goldsmith, P. F., 1977. Soil survey and land evaluation in developing countries. *Geographical Journal* 143: 407–431.

Zaslavsky, I. 1995. Logical inference about categorical coverages in multi-layer GIS. Ph.D. dissertation, University of Washington.

Zhu, Z. 1994. Forest density mapping in the lower 48 states: a regression procedure. Research Paper SO–280. New Orleans: US Forest Service.

Zhu, Z., and Evans, D. L. 1994. US forest types and predicted percent forest cover from AVHRR data. *Photogrammetric Engineering and Remote Sensing* 60: 525-531.

SOURCES AND CREDITS

Figures and tables in this book were produced by John Wiley & Sons or the author, unless noted below. The author thanks all these sources for the permission to use the material.

Chapter 1

Figure 1-1: Redrawn from Parkes and Thrift 1980. *Times, Spaces and Places*. By permission of John Wiley & Sons.

Figure 1-6: Reproduced from page 46, Donley and others 1979. *Atlas of California*. Portland, OR: Academic Book Center. Reproduced by permission of authors.

Figure 1-7: Reproduced from page 46, Donley and others 1979. *Atlas of California*. Portland, OR: Academic Book Center. Reproduced by permission of authors.

Figure 1-11: Illustration produced by author and Jerome Sullivan, University of Wisconsin Land Information and Computer Graphics Facility 1984.

Chapter 2

Figure 2-3: Reproduced from page 304, USGS 1970. *National Atlas*.

Figure 2-4: Reproduced from page 509, Robinson and others 1995. *Elements of Cartography*. By permission of John Wiley & Sons.

Figure 2-6: Redrawn from Defense Mapping and USGS sources.

Figure 2-12: Map base reduced from USGS 1:100,000 topographic map 41094-C4-CF-100.

Figure 2-13: Reduced from Defense Mapping Agency Joint Operations Graphic.

Figure 2-15: Reproduced from page 83, Ullman 1980. *Geography as Spatial Interaction*. Used with the permission of the University of Washington Press.

Figure 2-18: Reproduced from page 241, USGS 1970. *National Atlas*.

Chapter 4

Figure 4-6: GIRAS data from USGS; Detailed land use from Beard (1983) Figure 10. Reproduced by permission of author.

Figure 4-8: Aerial photograph taken by Washington Department of Natural Resources. Reproduced by permission.

Chapter 5

Table 5-4: Selected criteria from Table 4-1, CNSI (1994). Provided by Chem-Nuclear Systems.

Figure 5-10: Reproduced from Plate 2, CNSI (1994). Provided by Chem-Nuclear Systems Inc.

Table 5-5: From Table 4-1, CNSI (1992). Provided by Chem-Nuclear Systems.

Figure 5-11: Redrawn from Figure V-3, *Dane County Solid Waste Plan* (1980). By permission of Dane County Regional Planning Commission, Madison, Wisconsin.

Table 5-7: Reproduced with permission, The American Society of Photogrammetry and Remote Sensing. Williams THL. Implementing LESA in a GIS: A case study. Table 1. *Photogrammetric Engineering and Remote Sensing* 1985 51 (12) pp. 1928.

Table 5-8: From Riggle and Schmidt (1991). Reprinted by permission of The University of Wisconsin Press.

Chapter 6

Figure 6-2: Produced by King County Metro, Seattle, Washington, used by permission.

Figure 6-6: Data reported in Linse (1993), used by permission of author.

Figure 6-8: Illustration produced by Professor Gold, Université de Laval. Used by permission of author.

Chapter 8

Figure 8-3: Reproduced from page 68, *ESRI Maps* 1991. Used with permission of Bureau of Reclamation and ESRI. Produced with ESRI software.

Figure 8-9, 8-11: Data developed in class project by Mark Wilbert and Robert Gray. Used with permission.

Chapter 9

Figure 9-9: Reproduced from Figures 1, 2, and 3, page 105. Wright (1936) A method of mapping densities of population with Cape Cod as an example, vol 26 *Geographical Review*. By permission of the American Geographical Society.

Figure 9-10, 9-11, 9-12: Reproduced from Sullivan and others (1985). By permission of Urban and Regional Information Systems Association, (202) 289-1685.

Chapter 10

Figure 10-1: Reproduced from page 38. Story and others (1963) *Australian Land Research Series #8*. With permission of CSIRO, Australia.

Table 10-2: Extracted from page 13, Dangermond and Freedman (1984) in *Wisconsin Land Report #1*. Used with permission of University of Wisconsin Institute of Environmental Studies.

Table 10-3: Extracted from Table 4.2, page 44, Ventura (1991). With permission of author.

INDEX

H

I

J

K